ABSTRACT ALGEBRA

ABSTRACT ALGEBRA

Shaoqiang Deng
Nankai University, China

Fuhai Zhu
Nanjing University, China

SCIENCE PRESS

World Scientific

Published by

World Scientific Publishing Co. Pte. Ltd.

5 Toh Tuck Link, Singapore 596224

USA office: 27 Warren Street, Suite 401-402, Hackensack, NJ 07601

UK office: 57 Shelton Street, Covent Garden, London WC2H 9HE

Library of Congress Cataloging-in-Publication Data
Names: Deng, Shaoqiang, author. | Zhu, Fuhai, author.
Title: Abstract algebra / authors, Shaoqiang Deng, Nankai University, China,
    Fuhai Zhu, Nanjing University, China.
Description: New Jersey : Word Scientific, [2024] | Includes bibliographical references and index.
Identifiers: LCCN 2023039692 | ISBN 9789811277665 (hardcover) |
    ISBN 9789811278709 (paperback) | ISBN 9789811277672 (ebook for institution) |
    ISBN 9789811277689 (ebook for individuals)
Subjects: LCSH: Algebra, Abstract--Textbooks.
Classification: LCC QA162 .D46 2024 | DDC 512/.02--dc23/eng/20230907
LC record available at https://lccn.loc.gov/2023039692

British Library Cataloguing-in-Publication Data
A catalogue record for this book is available from the British Library.

抽象代数
Originally published in Chinese by China Science Publishing & Media Ltd.
Copyright © China Science Publishing & Media Ltd., 2017

For any available supplementary material, please visit
https://www.worldscientific.com/worldscibooks/10.1142/13449#t=suppl

Desk Editors: Soundararajan Raghuraman/Nijia Liu

Typeset by Stallion Press
Email: enquiries@stallionpress.com

Preface

This is the second one of the overall designed series of textbooks for algebraic courses of Nankai University, which can be used as a textbook for the course *Abstract Algebra* for students specialized in Pure or Applied Mathematics. The main topics of the book include the theory of groups, rings, modules and fields. It is supposed that readers be familiar with the fundamental contents of the course *Advanced Algebra* for freshmen in the university. However, almost no contents of this book really needs the knowledge of *Advanced Algebra*, except for some examples which are related to matrices or vector spaces. The main goal of this book is to provide readers with necessary prerequisite for further algebraic courses in the university and to set up a solid algebraic foundation for the study of algebra theory.

The book consists of four chapters. In the first chapter, we present the fundamental theory of groups, including the topics of groups, subgroups, quotient groups, homomorphisms and isomorphisms of groups, transformation groups and permutation groups, extensions of groups, group actions on sets and the Sylow's Theorems; the second chapter is devoted to the theory of rings, including the subjects of rings, subrings, ideals, quotient rings, homomorphisms of rings, quaternions, factorizations in a domain, prime ideals and maximal ideals, principal ideal domains and Euclidean domains, and the theory of polynomials over a ring; the third chapter is concerned with the module theory. Besides the fundamental theory of modules, we mainly introduce the structure theory of finitely generated modules over a principal ideal domain, and the applications of module theory to the classification of finitely generated abelian groups, as well as to the canonical form theory of linear transformations on a finite dimensional vector space. Finally, in the fourth chapter, we deal with Galois theory, including the extensions of fields, the fundamental theorem of the Galois theory, and the necessary and sufficient condition for the existence of radical solutions of an algebraic equation.

In the composition of this book, we always pay special attention to the clarification of the originality and the mathematical thoughts involved in the definitions and theorems of the context. Moreover, we will always emphasize the ideas of the proofs of propositions and theorems. To this end, we usually put forward some indicative explanation before introducing a new notion or result, even if sometimes such explanations seem to be a little bit trifling or wordy to professional mathematical researchers. In our point of view, the current mathematical education pays too much attention to techniques and logic, but there is a tendency to neglect the dissemination of mathematical thoughts. Of course, it remains to be tested by practice to see whether our attempt is appropriate or not. We are very eager to receive suggestions or even criticisms from readers.

This book can be used as a textbook for a two-term course *Abstract Algebra* in universities in China, especially suitable for high-level universities. We stress that the content of this book is somehow difficult, so it is not necessary to teach all the content for the course of abstract algebra in any university. For a one-term course with four lectures every week, we suggest to teach the whole content of the first chapter, the first seven sections of the second chapter and the first four sections of the fourth chapter. For a one-term course with three lectures every week, we suggest to teach the first six sections of the first chapter, the first seven sections of the second chapter and the first four sections of the fourth chapter. The whole content can be entirely taught for a two-term course in abstract algebra.

There are many well-designed exercises in this book which are divided into two groups in every section, namely, basic exercises and hard exercises. Basic exercises are selected in accordance with the contents of the section, and they show our basic requirement for readers. In general, an average student can finish most of the basic exercises. On the other hand, hard exercises are selected to expand the scope of knowledge related to the content and to give readers a chance to train their skills and techniques in solving related algebraic problems. We don't require readers to finish all the exercises in this part, since it would be a very hard task even for an excellent student. Besides, we pose many thinking problems in the content, aiming either to acquiring a better understanding of the content or to introducing some important related results. In general, a reader can solve most of the thinking problems in this book. However, we must point out that, there are a few hard thinking problems in this book which only have reference value for readers.

This book is written based on our experience in teaching the course *Abstract Algebra* for many years, and it has been used for several years as a textbook for the S.S. Chern class of Nankai University. In the process of composition, many students and colleagues gave us much valuable suggestions for improving the presentation, and we have made several revisions based on their advises. Nevertheless, errors or mistakes are very likely to exist in this book. Comments and suggestions from readers, either for pointing out mistakes to the authors, or for the improvement of the book, are very welcome.

About the Authors

Shaoqiang Deng is currently a professor of mathematics at Nankai University and is the head of the Department of Mathematics. He received his Bachelor's degree from Xiangtan University in 1988, and a Ph.D. from Nankai University in 1998. His research fields include Lie groups, Lie algebras, and differential geometry.

Fuhai Zhu currently works as a professor of mathematics at Nanjing University. He received his Ph.D. from Nankai University in 2002, worked as a postdoctoral fellow at Peking University from 2002 to 2004, and then joined Nankai University until he moved to Nanjing University in 2019. His research interests are Lie groups, Lie algebras, and their representation theory.

Contents

Chapter 1

Groups

Group theory germinated in the research on number theory at the beginning of the 19th century. Lagrange, Raffini, and Abel used the idea of permutations of roots to study the solutions of algebraic equations. Galois found the necessary and sufficient condition for solving algebraic equations by radicals, which heralded the creation of group theory. In the past several decades, the most important achievement in group theory was the classification of finite simple groups, which was the joint work of more than 100 group theorists. Of course, the research work in group theory is far from complete. Groups are becoming more and more widely used. For example, in the second half of the 18th century, Klein applied group theory to study geometry. At the same time, Lie put forward the concept of Lie groups to solve differential equations. Group theory has many applications in other disciplines, such as physics, chemistry, etc. In this chapter, we will introduce the basic theory of groups and study the basic ideas and tools of group classification.

1.1 Semigroups and groups

The main purpose of abstract algebra is to study different kinds of algebraic structures. An **algebraic structure** is a non-empty set endowed with one or more operations and some identities those operations must satisfy.

Let us begin with some familiar examples. As usual, we denote by $\mathbb{N}, \mathbb{Z}, \mathbb{Q}, \mathbb{R}, \mathbb{C}$ the set of natural numbers (including 0), integers, rational numbers, real numbers, and complex numbers, respectively. The set $\mathbb{N}$ has two operations: the usual addition and multiplication. Both of them satisfy the same properties: commutativity and associativity. Furthermore, there is a special number 0 (or 1, resp.) such that for any $a \in \mathbb{N}$, $0 + a = a$ (or $1 \cdot a = a$, resp.). We call 0 (or 1, resp.) the unit for the addition (or the multiplication, resp.). Similarly, the set $\mathbb{Z}$ is endowed with addition and multiplication, but it has one more property: for any $a \in \mathbb{Z}$ there exists a (unique) $-a \in \mathbb{Z}$ such that $a + (-a) = (-a) + a = 0$. Therefore, one may introduce subtraction: $a - b = a + (-b)$. Similarly, for the set $\mathbb{Q}, \mathbb{R}$, or $\mathbb{C}$, there exists a unique number a^{-1} for each non-zero number a such that $aa^{-1} = a^{-1}a = 1$. Then one may introduce division: a/b ($b \neq 0$) is defined as ab^{-1}.

We may find more interesting examples with one or more operations in linear algebra, say, the set $\mathbb{F}[x]$ of polynomials, the n-dimensional vector space $\mathbb{F}^n$, the set $\mathbb{F}^{m \times n}$ of $m \times n$ matrices, the set $\mathrm{GL}(n, \mathbb{F})$ of $n \times n$ invertible matrices, the set $\mathrm{O}(n)$ of real orthogonal matrices, etc. Here, $\mathbb{F}$ is a **number field**, that is, a subset of $\mathbb{C}$ including 1, closed under addition, subtraction, multiplication, and division. All the above mathematical objects share some common properties. To be more precise, we need some notation and terminologies. The Cartesian product of two sets A, B, denoted by $A \times B$, is the set of all ordered pairs (a, b), where $a \in A, b \in B$, that is,

$$A \times B = \{(a, b)\,|\, a \in A, b \in B\}.$$

A **binary operation** on A, B is a map from $A \times B$ to a set S, or more precise, given any $a \in A, b \in B$, there is a unique element $c \in S$ corresponding to the ordered pair (a, b). Here, c is called the **product** of a, b, denoted by $a * b$, or ab if there is no danger of confusion. For example, the scalar product in any vector space is a binary operation, which will be discussed in Chapter 3 as a special case of modules.

In this chapter, we will focus on the case when $S = A = B$. A binary operation on a non-empty set S is said to be **associative** if

$$(ab)c = a(bc), \quad \forall a, b, c \in S.$$

In this case, S is called a **semigroup**. If, furthermore, the binary operation is commutative, i.e.,

$$ab = ba, \quad \forall a, b \in S,$$

then S is called a **commutative semigroup**. An element $e \in S$ is called a **left** (or **right**) **identity** if $ea = a$ (or $ae = a$) for all a in S. If e is both a left and right identity, then it is called an **identity**. A semigroup with an identity e is called a **monoid**. The identity of a monoid is unique, for if e' is another identity, then we have $e = ee' = e'$.

Problem 1.1.1. Find a semigroup S such that S has a left identity but does not have a right identity.

All the examples mentioned at the beginning of this section are semigroups. For example, the set of positive integers $\mathbb{N}^*$ is a monoid with multiplication and a semigroup with addition. We left it to the reader to check which of the other examples are monoids.

To construct a semigroup, we need an operation satisfying the associative law. A natural example of such operations is the following.

Example 1.1.2. Let $M(X)$ be the set of transformations of a non-empty set X. Then $M(X)$ is a monoid whose operation is the composition of transformations, and the identity is the identity transformation id_X.

If a non-empty set X is endowed with some additional structure, then the subset of $M(X)$ consisting of elements preserve the structure is more interesting. The following is a typical example.

Example 1.1.3. Let V a vector space, and $\mathrm{End}\, V$ the set of all linear transformations of V. Then $\mathrm{End}\, V$ has a natural addition and multiplication, which define two monoid structures on $\mathrm{End}\, V$.

We still have some very special examples.

Example 1.1.4. Let X be a non-empty set and let $P(X)$ be the **power set** of S, i.e., the set of all subsets of S. Then $\{P(X); \cup\}$ is a monoid, and the identity is the empty set $\varnothing$. Furthermore, $\{P(X); \cap\}$ is also a monoid, X itself being the identity. Here, $\cup$ and $\cap$ denote the operation of the union and intersection of sets, respectively.

In a monoid, different elements have different properties. For example, in the set $\mathbb{F}^{n \times n}$ of $n \times n$ matrices, some matrices are invertible, while others are singular. Invertible matrices have better properties and are easier to handle. Thus, we introduce the following definition.

Definition 1.1.5. Let S be a monoid with identity e. For $a \in S$, if there exists $b \in S$ such that $ba = e$ (or $ab = e$, resp.), then b is called a **left** (or **right**, resp.) **inverse** of a. Furthermore, b is called an **inverse** of a if b is both a left and right inverse of a, i.e., $ab = ba = e$. In this case, a is called **invertible**.

Problem 1.1.6. Give an example of a monoid S such that there exists $a \in S$ which has a left inverse but not a right inverse.

Problem 1.1.7. If S is a monoid, and an element a in S has a left inverse b and a right inverse c, is a invertible?

Proposition 1.1.8. *Let S be a monoid with identity e. Then any invertible element $a \in S$ has a unique inverse, usually denoted by a^{-1}.*

Proof. Let e be the identity of the monoid, and let b, b' be two inverses of a. Then $b = be = b(ab') = (ba)b' = b'$.   □

Now we are ready to introduce the most important concept in this chapter.

Definition 1.1.9. A monoid G is called a **group** if every element in G has an inverse. A group G is called **abelian** (or **commutative**) if the operation on G is commutative. A group with finite elements is called a **finite group**; otherwise, it is called an **infinite group**. For a finite group G, the cardinality of G (the number of elements in G) is called the **order** of G. In particular, a group with only one element is called a **trivial group**.

When studying solutions of algebraic equations, Galois considered some special permutations of all roots of a given algebraic equation; those permutations form a group and determine whether the equation is solvable by radicals. We will discuss Galois theory in Chapter 4. After more than one century's development, group theory now has extensive applications to almost all branches of mathematics, say, algebra, analysis, geometry, topology, etc.

The following result is frequently used to check whether an algebraic system is a group.

Theorem 1.1.10. *Let G be a semigroup. Then G is a group if and only if the following conditions are satisfied:*

(1) *There is a left identity e in G, i.e., for any $a \in G$, $ea = a$;*
(2) *Every element a in G has a left inverse, i.e., there exists $b \in G$ such that $ba = e$.*

Proof. It is easy to see that a group always satisfies the two conditions. For the converse, we need to show that e is an identity and b is an inverse of a.

Let c be a left inverse of b. Then we have $a = (cb)a = c(ba) = ce$. Therefore, $ab = (ce)b = cb = e$, which implies that b is an inverse of a. Furthermore,

$$ae = a(ba) = (ab)a = a.$$

Thus, e is an identity of G. □

Problem 1.1.11. In the above theorem, if we replace "left identity" by "right identity", "left inverse" by "right inverse", does the conclusion still hold? Furthermore, if a semigroup G has a left identity, and every element in G has a right inverse, is G still a group?

Generally speaking, to determine the structure of a group, we need to determine the product of any two elements in the group. For a finite group $G = \{a_1, \ldots, a_n\}$ with a_1 the identity, we use the following table to give the product of any two elements in G:

	a_1	$\cdots$	a_n
a_1	$a_1 a_1$	$\cdots$	$a_1 a_n$
$\vdots$	$\vdots$		$\vdots$
a_n	$a_n a_1$	$\cdots$	$a_n a_n$

Such a table is called the **multiplicative table** of G. Multiplicative tables can be defined for all finite semigroups or finite sets with binary operations.

In the following, several simple but frequently used properties of groups will be introduced. We will see that some techniques used in matrix theory can be applied in group theory.

Lemma 1.1.12. *A group G satisfies the **left (right) cancellation law**, i.e., for any $a, b, c \in G$, $ab = ac$ ($ba = ca$) implies $b = c$.*

Proof. Since $ab = ac$, one has $a^{-1}(ab) = a^{-1}(ac)$. By the associativity we have $(a^{-1}a)b = (a^{-1}a)c$, i.e., $b = c$. Hence, the left cancellation law holds. Similarly, one can prove the right cancellation law. □

Proposition 1.1.13. *Let G be a semigroup. Then G is a group if and only if for $a, b \in G$, the solutions to $ax = b$ and $xa = b$ exist.*

Proof.   If G is a group, it is quick to check that $a^{-1}b$ (ba^{-1}, resp.) is a solution to $ax = b$ ($xa = b$, resp.).

Conversely, by Theorem 1.1.10, we just need to prove that G has a left identity and has a left inverse for any element in G. Let $e \in G$ be a solution to $xa = a$, for some $a \in G$. Then for any $b \in G$, $ax = b$ has a solution $c \in G$. Hence, $eb = e(ac) = (ea)c = ac = b$, which implies that e is a left identity of G. Furthermore, any solution to the equation $xa = e$ is a left inverse of a. Thus, G is a group.   □

Proposition 1.1.14. *Let G be a finite semigroup satisfying left and right cancellation laws. Then G is a group.*

Proof.   Let $G = \{a_1, a_2, \ldots, a_n\}$. Since G is a semigroup, by the left cancellation law, $a_i a_1, a_i a_2, \ldots, a_i a_n$ must be a permutation of $a_1, a_2 \ldots, a_n$. Hence, for any $a_i, a_j \in G$, there exists $a_k \in G$ such that $a_i a_k = a_j$, i.e, the equation $a_i x = a_j$ has a solution in G. Similarly, $xa_i = a_j$ has a solution too. Therefore, G is a group by Proposition 1.1.13.   □

The idea of the proof is valuable. Any group has a binary operation $x \cdot y$. Given an element x, we get a transformation of G (called a left translation): $y \mapsto x \cdot y$, which is injective. The left translation is also surjective; the proof is left to the reader. Left translations will be helpful when we study group actions later. The reader should notice that Proposition 1.1.14 does not hold for infinite semigroups. Try to find counterexamples.

Lots of examples that we are familiar with are groups. For example, $\mathbb{Z}$, vector spaces, the set $\mathbb{F}^{n \times m}$ of $n \times m$ matrices are abelian groups under addition; the set $\mathrm{GL}(n, \mathbb{F})$ of $n \times n$ invertible matrices and the set $O(n)$ of real orthogonal matrices are groups under multiplication. In general, we may get a group from any monoid by the following result.

Proposition 1.1.15. *Denote by $U(S)$ the subset of all invertible elements in a monoid S. Then $U(S)$ is a group.*

Proof.   It is easy to see that $e \in U(S)$. If $a, b \in U(S)$, then $(b^{-1}a^{-1})(ab) = e$. Therefore, $ab \in U(S)$, and it follows that $U(S)$ is a monoid. Furthermore, for any $a \in U(S)$, one has that $a^{-1} \in U(S)$. Thus, $U(S)$ is a group. $\qquad\qquad\square$

For example, the set of integers is a monoid under multiplication, and the subset of invertible elements $\{1, -1\}$ is a group. The following examples are frequently used.

Example 1.1.16. (1) We have known that $M(X)$ (see Example 1.1.2) is a monoid. Therefore, $U(M(X))$ is a group and is usually denoted by S_X, called the **symmetric group** on X. In particular, if X is a finite set, say, $X = \{1, 2, \ldots, n\}$, then we write S_n for S_X, and call it the **symmetric group on n letters**.

(2) Let $\mathrm{GL}(V)$ be the set of invertible linear transformations on a vector space V. Then $\mathrm{GL}(V)$ is a group, called the **general linear group**, since it is the set of all invertible elements in the monoid $\mathrm{End}(V)$.

(3) Let $\mathbb{F}$ be a number field. Then $\mathbb{F}$ is an abelian group with addition and a monoid with multiplication. An element in $\mathbb{F}$ is invertible under multiplication if and only if it is non-zero. Thus, $\mathbb{F}^* = \mathbb{F} \setminus \{0\}$ is a group under multiplication.

Exercises

(1) Determine whether the operation $*$ for each of the following set G is a binary operation? If so, check whether $(G, *)$ is abelian, a semigroup, a monoid, or a group.

(i) $G = \mathbb{Z}$, $a * b = a - b$;
(ii) $G = \mathbb{Z}$, $a * b = a + b - ab$;
(iii) $G = \mathbb{Q} \setminus \{0, 1\}$, $a * b = ab$;
(iv) G is the set of all primitive polynomials in $\mathbb{Z}[x]$, $f(x) * g(x) = f(x)g(x)$;
(v) $G = \mathbb{N}^*$, $a * b = 2^{ab}$;
(vi) $G = \mathbb{N}^*$, $a * b = a^b$.

(2) If a semigroup S has a left identity and a right identity, is S a monoid?

(3) Define on $\mathbb{Z} \times \mathbb{Z}$ the following multiplication

$$(x_1, x_2)(y_1, y_2) = (x_1 y_1 + 2x_2 y_2, x_1 y_2 + x_2 y_1).$$

Show that $\mathbb{Z} \times \mathbb{Z}$ is an abelian monoid with the above multiplication.

(4) Let $M(\mathbb{N})$ be the monoid of all transformations of $\mathbb{N}$. Let $f \in M(\mathbb{N})$ be defined by

$$f(n) = n + 1, \quad \forall n \in \mathbb{N}.$$

Show that f has infinite many left inverses but no right inverses.

(5) Let "$\cdot$" and "$*$" be two operations on a set X, each having an identity. Assume that for any $a, b, c, d \in X$,

$$(a * b) \cdot (c * d) = (a \cdot c) * (b \cdot d).$$

Prove that the two operations are the same and $(X, \cdot)$ is an abelian monoid.

(6) List all the semigroups of two elements and find out all the monoids and groups.

(7) Show that a finite semigroup is a group if it has a left identity and satisfies the right cancellation law.

(8) Find a finite semigroup, which satisfies the right cancellation law, but is not a group.

(9) Find an infinite abelian monoid, which has an identity and satisfies the cancellation laws, but is not a group.

(10) Let $P(X)$ be the power set of a non-empty set X. Define an operation Δ (called **symmetric difference**) on $P(X)$ by

$$A \Delta B = (A \setminus B) \cup (B \setminus A), \quad \forall A, B \in P(X),$$

where $A \setminus B = \{x \in A | x \notin B\}$. Prove that $(P(X), \Delta)$ is a group. Furthermore, if $|X| = 2$, determine the multiplicative table of P_X. In this case, $P(X)$ is called a **Klein group**.

(11) Prove that $\left\{ e^{\frac{2k\sqrt{-1}\pi}{n}} = \cos\frac{2k\pi}{n} + \sqrt{-1}\sin\frac{2k\pi}{n} | k = 0, \ldots, n-1 \right\}$ is a group under multiplication.

(12) For $n \in \mathbb{N}$, set $\mathbb{Z}_n = \{\overline{0}, \overline{1}, \ldots, \overline{n-1}\}$. Define two operations on $\mathbb{Z}_n$ as follows:
$$\overline{a} \cdot \overline{b} = \overline{c}, \quad \text{where } ab \equiv c \pmod{n};$$
$$\overline{a} + \overline{b} = \overline{d}, \quad \text{where } a + b \equiv d \pmod{n}.$$

   (i) Prove that $\{\mathbb{Z}_n; \cdot\}$ is a commutative monoid, and $\{\mathbb{Z}_n; +\}$ is an abelian group.

(ii) Determine the multiplicative table of $\{\mathbb{Z}_4; \cdot\}$.

(iii) Let $\mathbb{Z}_n^* = \{\bar{a} \in \mathbb{Z}_n | (a, n) = 1\}$. Prove that $\{\mathbb{Z}_n^*; \cdot\}$ is a group.

(iv) Let p be a prime. Use group theory to prove **Wilson's theorem**: $(p - 1)! \equiv -1 \pmod{p}$.

(13) Denote by $\varphi(n)$ (the **Euler's φ function**) the number of positive integers which are no more than n and relatively prime to n.

(i) Let $n = p_1^{e_1} p_2^{e_2} \cdots p_r^{e_r}$, where p_i are different primes, $e_i \in \mathbb{N}^*$. Show that

$$\varphi(n) = n \left(1 - \frac{1}{p_1}\right) \left(1 - \frac{1}{p_2}\right) \cdots \left(1 - \frac{1}{p_r}\right).$$

(ii) Prove **Euler's theorem**:

$$a^{\varphi(n)} \equiv 1 \pmod{n}, \quad a \in \mathbb{N}, \quad (a, n) = 1.$$

In particular, if $n = p$ is a prime, then $a^{p-1} \equiv 1 \pmod{p}$ (**Fermat's little theorem**).

(14) Let $\mathrm{SL}(2, \mathbb{Z}_n) = \left\{ A = \begin{pmatrix} \bar{a} & \bar{b} \\ \bar{c} & \bar{d} \end{pmatrix} | \bar{a}, \bar{b}, \bar{c}, \bar{d} \in \mathbb{Z}_n, \bar{a} \cdot \bar{d} - \bar{b} \cdot \bar{c} = \bar{1} \right\}$. Show that $\mathrm{SL}(2, \mathbb{Z}_n)$ is a group under matrix multiplication. Determine A^{-1} for any $A \in \mathrm{SL}(2, \mathbb{Z}_n)$. Try to define $\mathrm{SL}(m, \mathbb{Z}_n)$ for any $m \in \mathbb{N}$.

(15) Assume that G is a semigroup and a transformation $a \mapsto a'$ of G satisfies

$$a'(ab) = b = (ba)a', \quad \forall a, b \in G.$$

Show that G is a group.

(16) Let $G = \{(a, b) | a, b \in \mathbb{R}, a \neq 0\}$ and define an operation on G by $(a, b)(c, d) = (ac, ad + b)$. Show that G is a group.

(17) For any $A \in \mathbb{R}^{n \times n}$, $\beta \in \mathbb{R}^n$, define a transformation $T_{(A,\beta)}$ of $\mathbb{R}^n$ by

$$T_{(A,\beta)}(\alpha) = A\alpha + \beta, \quad \forall \alpha \in \mathbb{R}^n.$$

(i) Show that $T_{(A,\beta)}$ is a bijection if A is invertible.

(ii) Show that $\mathrm{Aff}\,(n, \mathbb{R}) = \{T_{(A,\beta)}\,||A| \neq 0\}$ is a group under the composition of transformations. It is called the **affine transformation group** of $\mathbb{R}^n$.

(18) Let $\mathbb{F}$ be a number field. Find a subset G of $\mathbb{F}^{n \times n}$ such that G consists of singular matrices and G is a group under multiplication. Show that for any such group G, there exists a non-negative integer $k \leq n$ and an invertible matrix $T \in \mathbb{F}^{n \times n}$ such that for any $A \in G$, $TAT^{-1} = \begin{pmatrix} A_1 & \\ & 0 \end{pmatrix}$, where $A_1 \in \mathbb{F}^{k \times k}$ is invertible.

Exercises (hard)

(19) Let F be a figure in a 2 or 3-dimensional Euclidean space, and let G_F be the set of orthogonal transformations which map F into itself.

    (i) Show that G_F is a group under multiplication (the composition of maps), called the **symmetric group** of F;

    (ii) Determine the order of the symmetric group of a regular n-gon inscribed in the unit circle;

    (iii) Determine the orders of the symmetric groups of all regular polyhedrons inscribed in the unit ball.

(20) Let $G = \{x_1, x_2, \ldots, x_n\}$ be a finite group. The **group matrix** for G is an $n \times n$ matrix A whose (i, j)-entry is the product of x_i and x_j^{-1}. Thus, each row or column of A is a permutation of $x_1, x_2, \ldots, x_n$. The determinant of A is the **group determinant** of G.

    (i) Computer the group determinants $|A|$ for some groups of lower order.

    (ii) Factorize $|A|$ as a complex polynomial of n indeterminants, and check that the degree of each irreducible factor is equal to its multiplicity.

(21) Let S be the set of all **arithmetic functions**, i.e., functions from $\mathbb{N}^*$ to $\mathbb{C}$. For $f, g \in S$, the **Dirichlet convolution** $f * g$

is a new arithmetic function defined by:

$$f * g(n) = \sum_{d|n} f(d) g\left(\frac{n}{d}\right) = \sum_{ab=n} f(a)g(b).$$

(i) Show that $(S, *)$ is a commutative monoid and the identity is ε defined by $\varepsilon(n) = \delta_{1n}$, where δ is the Kronecker symbol.

(ii) Show that $f \in S$ is invertible if and only if $f(1) \neq 0$.

(iii) An arithmetic function f is called **multiplicative**, if $f(mn) = f(m)f(n)$ for all coprime positive integers m, n. Show that the set of all non-zero multiplicative functions is an abelian group under the operation $*$.

(iv) An arithmetic function f is called **completely multiplicative**, if $f(mn) = f(m)f(n)$ for all $m, n \in \mathbb{N}^*$. Is the set of non-zero completely multiplicative functions an abelian group under the operation $*$?

(22) Let $\mu : \mathbb{N}^* \to \mathbb{C}$ be the **Möbius function**, i.e., (a) $\mu(1) = 1$; (b) $\mu(n) = 0$ if n has a squared prime divisor; (c) $\mu(n) = (-1)^s$ if n is a product of s different primes.

(i) Show that μ is multiplicative, and $\mu(n) = \sum_{m \leq n, (m,n)=1} e^{\frac{2m\pi\sqrt{-1}}{n}}$;

(ii) Find the inverse of μ in S;

(iii) (Möbius inversion formula) For any $f \in S$, let $g(n) = \sum_{d|n} f(d)$. Show that $f(n) = \sum_{d|n} \mu\left(\frac{n}{d}\right) g(d)$.

1.2 Subgroups and cosets

In algebra, it is an effective method to study the properties of algebraic systems by studying subsystems and quotient systems, just like subspaces and quotient spaces in the theory of vector spaces. In this section, we will introduce the subsystems of the group: subgroups and some related notions.

Definition 1.2.1. A subset H of a group G is called a **subgroup**, denoted by $H < G$, if H is a group under the same operation of G.

Both $\{e\}$ and G are subgroups of G; we call them **trivial subgroups**, and other subgroups **non-trivial**. A subgroup H of G is called **proper** if $H \neq G$.

By definition, it is easy to see that the identity of a subgroup H is that of G, and the inverse of a in H is that of a in G. In general, we have the following criteria for subgroups.

Theorem 1.2.2. *Let H be a non-empty subset of a group G. Then the followings are equivalent*:

(1) H *is a subgroup of G*;
(2) *For any $a, b \in H$, $ab, a^{-1} \in H$*;
(3) *For any $a, b \in H$, $ab^{-1} \in H$*.

Proof. $(1) \Rightarrow (2)$. By definition.

$(2) \Rightarrow (3)$. If $a, b \in H$, then $b^{-1} \in H$, which implies $ab^{-1} \in H$.

$(3) \Rightarrow (1)$. For any $a \in H$, we have $e = aa^{-1} \in H$, which implies that $a^{-1} = ea^{-1} \in H$. Furthermore, if $a, b \in H$, then $a, b^{-1} \in H$, hence $ab = a(b^{-1})^{-1} \in H$. Therefore, H is closed under the operation of G containing the identity e, and each element in H has an inverse. Thus, H is a subgroup of G. $\square$

Let us look at some examples of subgroups.

Example 1.2.3. Let V be a vector space. Then V is an abelian group under the addition; any linear subspace of V is a subgroup.

Example 1.2.4. Let V be a vector space. Denote by $\mathrm{SL}(V)$ the set of linear transformations on V with determinant 1. Then $\mathrm{SL}(V)$ is a subgroup of $\mathrm{GL}(V)$, called the **special linear group** of V.

If V is a Euclidean space, then the set $\mathrm{O}(V)$ of orthogonal transformations of V is a subgroup of $\mathrm{GL}(V)$, called the **orthogonal group**. In particular, the set $\mathrm{SO}(V)$ of special orthogonal transformations of V is a subgroup of $\mathrm{GL}(V)$ and $\mathrm{O}(V)$, called the **special orthogonal group** of V.

If V is a unitary space, then the set $\mathrm{U}(V)$ of unitary transformations of V is a subgroup of $\mathrm{GL}(V)$, called the **unitary group** of V. In particular, the set $\mathrm{SU}(V)$ of unitary transformations of determinant 1 is a subgroup of $\mathrm{U}(V)$ and $\mathrm{GL}(V)$, called the **special unitary group** of V.

If V is endowed with a symplectic form, i.e., a non-degenerate skew-symmetric bilinear form, then $\mathrm{Sp}(V)$, the set of linear transformations of V preserving the symplectic form, is a subgroup of $\mathrm{GL}(V)$, called the **symplectic group** of V.

Problem 1.2.5. Show that $\mathrm{Sp}(V)$ is a subgroup of $\mathrm{SL}(V)$.

Let V be an n-dimensional vector space over a number field $\mathbb{F}$. Fix a basis of V. Then there is a natural 1-1 correspondence between $\mathrm{End}\,(V)$ and $\mathbb{F}^{n\times n}$. Hence, the groups in the above example can be represented by matrices, denoting the corresponding matrix groups as $\mathrm{GL}(n,\mathbb{F})$, $\mathrm{SL}(n,\mathbb{F})$, $\mathrm{O}(n)$, $\mathrm{SO}(n)$, $\mathrm{U}(n)$, $\mathrm{SU}(n)$ and $\mathrm{Sp}(n,\mathbb{F})$.

Besides the above examples, we have plenty of examples of subgroups in matrix groups.

Example 1.2.6. The following subsets of $n \times n$ matrices are subgroups of $\mathrm{GL}(n,\mathbb{F})$.

(1) The set T of all invertible diagonal matrices, or, more generally, the set of invertible block diagonal matrices of the same type.
(2) The set R of invertible upper triangular matrices, or, more generally, the set of invertible block upper triangular matrices of the same type.
(3) The set N of triangular matrices with diagonal elements being 1.

Example 1.2.7. Let $m \in \mathbb{N}$. Then $m\mathbb{Z} = \{mn | n \in \mathbb{Z}\}$ is a subgroup of $\mathbb{Z}$.

Problem 1.2.8. Show that any subgroup of $\mathbb{Z}$ must be one of $m\mathbb{Z}$, for some $m \in \mathbb{N}$.

Example 1.2.9. Denote by $\mathbb{R}^*$ the multiplicative group of all nonzero real numbers.. Then $\{1, -1\}$, $\mathbb{Q}^+$, $\mathbb{Q}^*$, $\mathbb{R}^+$ are subgroups of $\mathbb{R}^*$, but they are not subgroups of $\mathbb{R}$.

For abstract groups, we can also construct some subgroups. The following notation is frequently used. Let G be a group, $a \in G$. Set $a^0 = e$; for any $k \in \mathbb{N}^*$, set $a^k = a \cdot a^{k-1}$; $a^{-k} = (a^{-1})^k$ (If G is an additive group, a^n is usually denoted by na). It is easy to check that $a^m a^n = a^{m+n}$, for any $m, n \in \mathbb{Z}$. The subset $\langle a \rangle = \{a^n | n \in \mathbb{Z}\}$ is a subgroup of G, called the subgroup generated by a. The order of such a group is also called the **order** of a. If the order of a is finite, we call a an element of finite order; otherwise, we call a an element of infinite order.

More generally, similar to a subspace generated by a set of vectors, we consider subgroups generated by a collection of elements. Let S be a non-empty subset of a group G. Set $S^{-1} = \{a^{-1} | a \in S\}$ and

$$\langle S \rangle = \{x_1 \cdots x_m | m \in \mathbb{N}^*, x_1, \ldots, x_m \in S \cup S^{-1}\}.$$

By Theorem 1.2.2, it is easy to show that $\langle S \rangle$ is a subgroup of G, called the subgroup generated by S. If $\langle S \rangle = G$, then S is called a set of **generators** of G. If there exists a finite set of generators for G, we call G **finitely generated**. Finite groups are finitely generated, but finitely generated groups are not necessarily finite; say, $\mathbb{Z} = \langle 1 \rangle$ is an infinite group.

Problem 1.2.10. Show that $\langle S \rangle$ is the intersection of all subgroups of G containing S, and it is the smallest subgroup of G containing S.

In linear algebra, for a subspace W of a vector space V, we can introduce the congruence class $\alpha + W$ for any $\alpha \in V$ and the quotient space V/W. A similar idea may be applied to group theory.

Definition 1.2.11. Let H be a subgroup of G, $a \in G$. We call $aH = \{ah | h \in H\}$ (resp. $Ha = \{ha | h \in H\}$) the **left** (resp. **right**) **coset** of H, where a is a representative of the coset.

It is worth noting that $aH = Ha$ for any abelian group. However, as the following example shows, it is invalid for non-abelian groups.

Example 1.2.12. Let $G = \mathrm{GL}(2, \mathbb{C})$, and let T be the subgroup of G consisting of diagonal matrices. Then $aT \neq Ta$, for $a = \begin{pmatrix} 1 & 1 \\ 0 & 1 \end{pmatrix}$.

A natural question is: under what conditions are the left and right cosets of any element equal? This is closely related to the concept of quotient groups, which will be described in the following section.

The discussion of left cosets and right cosets is similar. Hence, we will focus on left cosets in the following. Given a left coset aH, the choice of its representative a is not unique. In fact, if $b \in aH$, then $bH \subseteq aH$, and there is $h \in H$ such that $b = ah$. Thus, $a = bh^{-1}$, i.e., $a \in bH$, and consequently, $aH = bH$. It follows that any element in aH is a representative. Now consider any two left cosets aH and bH.

If $c \in aH \cap bH$, then the above discussion tells us $aH = cH = bH$. Therefore, we have the following lemma.

Lemma 1.2.13. *Let H be a subgroup of G, $a, b \in G$. Then either $aH \cap bH = \varnothing$ or $aH = bH$, and $aH = bH$ if and only if $a^{-1}b \in H$.*

For any subgroup H of G, G is the disjoint union of different left cosets of H. In general, a **partition** of a set X is a set of non-empty subsets of X such that every element in X is in precisely one of these subsets; in other words, X is a disjoint union of these non-empty subsets. Therefore, the left cosets of H form a partition of G.

We can look at partitions from another viewpoint, which is more conducive to understanding the relationship between elements of the same subset in a partition.

Definition 1.2.14. Let A be a non-empty set, R a subset of $A \times A$. For $a, b \in A$, if $(a, b) \in R$, we say a has the relation R with b, denoted by aRb or $a \sim b$. We call R a **binary relation** on A.

The concept of binary relation can also be described as a property between two elements of a set A, such that any two elements in A either have the property R or do not have property R. If $a, b \in A$ have the property R, we say that a, b have a relation R, denoted by aRb. Considering the set

$$\{(a, b) | a, b \in A, aRb\},$$

we get a subset of $A \times A$, For example, ">", "<" and "=" are all relations of real numbers $\mathbb{R}$, and the reader can easily write down the subsets of $\mathbb{R} \times \mathbb{R}$ corresponding to these relations. The properties of different relations vary greatly. In order to construct and study quotient groups and other quotient algebra systems, we need relations with good properties.

Definition 1.2.15. A relation R on a set A is called an **equivalence relation**, if for any $a, b, c \in A$, we have the following:

(1) reflexivity: aRa;
(2) symmetry: $aRb \Rightarrow bRa$;
(3) transitivity: aRb and bRc imply aRc.

One can easily check that "=" on $\mathbb{R}$ is an equivalence relation, while ">" and "<" are not. There are many examples of equivalence relations in matrix theory; say, the equivalence, similarity, and congruence of $n \times n$ matrices over a number field $\mathbb{F}$ are all equivalence relations. Therefore, the same set may have several different equivalence relations. Take the equivalence of matrices as an example. Let $\mathbb{F}^{n \times n}$ be the set of all $n \times n$ matrices over a number field $\mathbb{F}$. It is known that two matrices are equivalent if and only if they have the same rank. Then $\mathbb{F}^{n \times n}$ is a disjoint union of $n + 1$ subsets $A_0, A_1, \ldots, A_n$, where A_i is the collection of $n \times n$ matrices with rank i. Thus, we get a partition of $\mathbb{F}^{n \times n}$. In generally, an equivalence relation on a set A gives a partition of A.

Definition 1.2.16. Let R be an equivalence relation on a set A. For any $a \in A$, the subset $\bar{a} = \{b \in A | bRa\}$ is called the **equivalence class** of a, and a is a **representative**.

By definition and the transitivity of equivalence relations, we have $\bar{a} = \bar{b}$ if aRb. Thus, any element in an equivalence class can be regarded as a representative. Furthermore, different equivalence classes are disjoint, which implies that the equivalence relation determines a partition of A. Conversely, given a partition of A, we may define a binary relation R: aRb if and only if a, b are in the same subset. By Definition 1.2.15, it is easy to check that R is an equivalence relation on A. Therefore, we have the following:

Theorem 1.2.17. *There is a 1-1 correspondence between partitions of A and equivalence relations on A.*

Equivalence relations may bring us much convenience. For example, the matrices of a linear transformation under different bases are similar. Hence, we can choose any matrix in the equivalence class to study, and the results will not be essentially different. Naturally, we can regard all the elements in the equivalence class as a whole, thus we have the following definition.

Definition 1.2.18. Let R be an equivalence relation on A. Then the set of all equivalence classes of A is called the **quotient set** of A relative to R, denoted by A/R.

Note that the equivalence class $\bar{a}$ is a subset of A, while it is an element of the quotient set A/R. For example, if V is a vector space and W is a subspace, then the element of the quotient space V/W is a subset of the form $\alpha + W$, for $\alpha \in V$. There is a natural quotient map from V onto V/W. Similarly, we have the following:

Definition 1.2.19. Let R be an equivalence relation on A. Then the map

$$\pi : A \to A/R, \quad \pi(a) = \bar{a}$$

is called the **canonical map** from A onto A/R.

Now, we are ready to use equivalence relations to describe cosets.

Theorem 1.2.20. *Let H be a subgroup of G. Then*

$$aRb \iff a^{-1}b \in H$$

defines an equivalence relation on G, and the equivalence class $\bar{a}$ is just the left coset aH.

Proof. First, we prove that R is an equivalence relation. For any $a \in G$, we have aRa since $a^{-1}a = e \in H$. If aRb, then $a^{-1}b \in H$, hence $b^{-1}a = (a^{-1}b)^{-1} \in H$, which implies bRa. If aRb and bRc, then $a^{-1}b, b^{-1}c \in H$. Hence, $a^{-1}c = (a^{-1}b)(b^{-1}c) \in H$, i.e., aRc.

Now for any $a \in G$, if $b \in \bar{a}$, then $a^{-1}b \in H$ and $b = a(a^{-1}b) \in aH$. Thus, $\bar{a} \subseteq aH$. Conversely, if $c \in aH$, then there exists $h \in H$ such that $c = ah$ and $a^{-1}c = h \in H$. Hence, $c \in \bar{a}$, which implies $aH \subseteq \bar{a}$. Therefore, $\bar{a} = aH$. $\qquad\square$

The quotient set defined by the above equivalence relation is called the **left coset space** of G relative to H, denoted by G/H. The cardinality $|G/H|$ of G/H is called the **index** of H in G, also denoted by $[G : H]$.

If the group G is abelian, the left coset of H is usually denoted by $a + H = \{a + h | h \in H\}$.

Example 1.2.21. $m\mathbb{Z}$ is a subgroup of $\mathbb{Z}$ with index $[\mathbb{Z} : m\mathbb{Z}] = m$ since

$$\mathbb{Z} = (0 + m\mathbb{Z}) \cup (1 + m\mathbb{Z}) \cup \cdots \cup ((m-1) + m\mathbb{Z}).$$

Example 1.2.22. The special orthogonal group $\mathrm{SO}(n)$ is a subgroup of $\mathrm{O}(n)$. Let $D = \mathrm{diag}\,(-1, 1, \ldots, 1)$. Then

$$\mathrm{O}(n) = \mathrm{SO}(n) \cup D\mathrm{SO}(n).$$

Consequently, $[\mathrm{O}(n) : \mathrm{SO}(n)] = 2$.

For finite groups, the following result is fundamental and valuable.

Theorem 1.2.23 (Lagrange). *Let G be a finite group, $H < G$. Then*

$$|G| = [G : H] \cdot |H|,$$

i.e., $|H|$ is a divisor of $|G|$. In particular, the order of any element in G is a divisor of $|G|$.

Proof. For any $a \in G$, consider the map

$$\phi : h \to ah, \quad \forall h \in H.$$

It is easy to see that ϕ is a bijection by the cancellation law. Thus, aH has exactly $|H|$ elements. Since G is the disjoint union of different left cosets of H, we have $|G| = [G : H] \cdot |H|$.   □

By the above theorem, one has $[G : H] = |G|/|H|$. Then, we have the following result.

Corollary 1.2.24. *Let G be a finite group, $H < K < G$. Then*

$$[G : H] = [G : K] \cdot [K : H].$$

Problem 1.2.25. By Lagrange's theorem, the order of a subgroup is the divisor of that of the group. What about the converse? More precisely, for any divisor m of $|G|$, is there a subgroup whose order is exactly m?

One needs to find more examples of groups to solve the above problem.

Exercises

(1) Show that the intersection of any number of subgroups of G is still a subgroup.
(2) Show that any group is not the union of two proper subgroups.
(3) Let H be a non-empty finite subset of a group G. Show that $H < G$ if and only if $ab \in H$, for any $a, b \in H$. Does the conclusion still hold if H consists of (possibly infinitely many) elements of finite order?

(4) Let G be a group such that $a^2 = e$ for any $a \in G$. Show that G is commutative.

(5) Find all subgroups of order 5 of S_5.

(6) Let G be a group, $a, b \in G$. Show that the following pairs of elements have the same order:

(i) a and a^{-1}; (ii) a and bab^{-1}; (iii) ab and ba.

(7) Let a, b be two elements in a group G of order m, n, respectively, and $ab = ba$.

(i) If $(m, n) = 1$, show that the order of ab is mn.

(ii) If $\langle a \rangle \cap \langle b \rangle = \{e\}$, show that the order of ab is $[m, n]$, where $[m, n]$ is the least common multiple of m, n.

(iii) If $\langle a \rangle \cap \langle b \rangle \neq \{e\}$, is the order of ab still $[m, n]$? Are there elements in G of order $[m, n]$?

(8) Let $a \in G$ be of order d, $k \in \mathbb{N}$.

(i) Show that the order of a^k is $d/(d, k)$, where (d, k) is the greatest common divisor of d, k.

(ii) Show that the order of a^k is d if and only if $(d, k) = 1$.

(9) Let G be an abelian group, and let $n \in \mathbb{N}^*$ be the maximum of orders of elements in G. Show that the order of every element in G is a divisor of n.

(10) Let G be an abelian group, $n \in \mathbb{N}^*$. Show that $\{a \in G | a^n = e\}$ is a subgroup of G.

(11) Let G be a finite group, $k \in \mathbb{N}$, $k > 2$. Show that the number of elements in G of order k must be even.

(12) Let $\mathrm{SL}(n, \mathbb{Z})$ be the set of $n \times n$ integer matrices with determinant 1.

(i) Show that $\mathrm{SL}(n, \mathbb{Z})$ is a subgroup of $\mathrm{GL}(n, \mathbb{R})$.

(ii) Set $R = \begin{pmatrix} 0 & -1 \\ 1 & 1 \end{pmatrix}$, $S = \begin{pmatrix} 0 & -1 \\ 1 & 0 \end{pmatrix}$, $T = \begin{pmatrix} 1 & 1 \\ 0 & 1 \end{pmatrix}$, $U = \begin{pmatrix} 1 & 0 \\ 1 & 1 \end{pmatrix}$. Show that any two of R, S, T, U are generators of $\mathrm{SL}(2, \mathbb{Z})$.

(13) Determine which of the following relations are equivalence ones.

(i) In $\mathbb{R}$, $xRy \iff |x - y| \leq 3$;

(ii) In $\mathbb{R}$, $xRy \iff |x| = |y|$;

(iii) In $\mathbb{Z}$, $xRy \iff x - y$ is odd;

(iv) In $\mathbb{C}^{n \times n}$, $ARB \iff$ there exists $P, Q \in \mathbb{C}^{n \times n}$ such that $A = PBQ$.

(14) Find some examples of relations that satisfy any two but not the other of the three conditions in the definition of equivalence relations, proving that the three conditions are independent.

(15) Let H, K be two subgroups of G. Set

$$HK = \{hk | h \in H, k \in K\}.$$

(i) If H, K are finite groups, prove $|HK| = |H||K|/|H \cap K|$;

(ii) Show that HK is a subgroup of G if and only if $HK = KH$.

(16) Let G be a group of odd order. Show that for any $a \in G$, there is a unique $b \in G$ such that $a = b^2$.

(17) Let A be a subgroup of G, and let R be a set of representatives of right cosets of A, i.e., the intersection of R and each right coset has exactly one element. Show that R^{-1} is a set of representatives of left cosets of A. Is R a set of representatives of left cosets of A?

(18) Let H_1, H_2 be subgroups of a finite group G. Prove:

$$[G : H_1 \cap H_2] \leq [G : H_1][G : H_2].$$

If, furthermore, $[G : H_1]$ and $[G : H_2]$ are coprime, then $[G : H_1 \cap H_2] = [G : H_1][G : H_2]$ and $G = H_1 H_2$.

Exercises (hard)

(19) Let G be a group of even order. Show that G has an element of order 2. More generally, if $n | |G|$, does G contain elements of order n?

(20) Let $G = \{A_1, A_2, \ldots, A_m\}$ be a subset of $\mathrm{GL}(n, \mathbb{R})$ such that $A_i A_j \in G$ for any $A_i, A_j \in G$. Show that G is a subgroup of $\mathrm{GL}(n, \mathbb{R})$. Furthermore, if $\sum_{i=1}^{m} \mathrm{tr}\, A_i = 0$, then $\sum_{i=1}^{m} A_i = 0$.

(21) Let A, B be subgroups of G.

(i) Show that $g(A \cap B) = gA \cap gB$, for any $g \in G$.

(ii) If both $[G : A]$ and $[G : B]$ are finite, then $[G : A \cap B]$ is finite and

$$[G : A \cap B] \leq [G : A][G : B].$$

(22) Let H, K be subgroups of G. For each $g \in G$, the (H, K)-**double coset** of g is the set

$$HgK = \{hgk | h \in H, k \in K\}.$$

(i) Show that $|HgK| = |H|[K : g^{-1}Hg \cap K] = |K|[H : H \cap gKg^{-1}]$.

(ii) Show that G is the disjoint union of its double cosets, which defines an equivalence relation on G.

(23) Let H be a subgroup of G of index n. Prove that there exist $g_1, g_2, \ldots, g_n \in G$ such that

$$G = \bigcup_{i=1}^{n} g_i H = \bigcup_{i=1}^{n} Hg_i.$$

1.3 Normal subgroups and quotient groups

In the last section, we defined the left coset space G/H for any subgroup H of a group G. A natural idea is whether we can define an operation on G/H such that it becomes a group. For example, for any vector space V and a subspace W of V, the quotient space V/W has a natural vector space structure and is naturally an abelian group. Similarly, a natural way to define multiplication on G/H is to define the product of coset aH and bH to be abH. Is this well-defined? That is to say, is this definition independent of the choice of representatives of the cosets? Or, more essentially, is the product of two left cosets of H still a left coset of H? If $aH \cdot bH = abH$, then $H \cdot bH = bH$, hence, we have $Hb = bH$. In other words, the necessary condition for defining a natural multiplication on G/H is $bH = Hb$ for any $b \in G$. It is not always the case, as Example 1.2.12 shows. Therefore, we need the following definition.

Definition 1.3.1. A subgroup H of G is said to be **normal**, denoted by $H \lhd G$, if

$$ghg^{-1} \in H, \quad \forall g \in G, h \in H.$$

Normal subgroups are very special subgroups. If H is a subgroup of G, $a \in H$, then we know that $a, a^2, a^3, \ldots$ and $a^{-1}, a^{-2}, \ldots$ are elements in H. If H is normal, then we may get more elements in H, say, $gag^{-1} \in H$, for any $a \in H$, $g \in G$. We say two elements $a, b \in G$ are **conjugate**, if there exists $g \in G$ such that $b = gag^{-1}$. Denote by C_a all elements in G which are conjugate with a, called the **conjugacy class** of a.

A non-trivial group G has at least two normal subgroups: G and $\{e\}$, called **trivial normal subgroups**. If a non-trivial group has only trivial normal subgroups, it is called a **simple group**. Simple groups play the same role as prime numbers and irreducible polynomials, as we shall see soon. Classification of finite simple groups was a massive mathematical project in the 20th century, finished in 2004. The final proof consists of more than ten thousand pages in about 500 journal papers.

Any subgroup of an abelian group is normal. Hence, simple abelian groups are precisely those of prime order by Lagrange's theorem. For non-abelian groups, determining whether a group is simple or, more generally, finding a non-trivial normal subgroup of a group is challenging.

We can also look at normal subgroups from another point of view. Given a subgroup H of G, we define an equivalence relation on G by $aRb \Leftrightarrow a^{-1}b \in H$ (see Theorem 1.2.20). If we want to define a natural multiplication on G/H, then the equivalence relation must satisfy the following condition: $aRb, cRd \Rightarrow acRbd$. In this case, G/H can inherit the operation on G. For this reason, we need better equivalence relations–congruence relations.

Definition 1.3.2. Let a set A be endowed with a binary operation $*$. A **congruence relation** R on A is an equivalence relation such that

$$aRb, cRd \implies (a * c)R(b * d), \quad \forall a, b, c, d \in A.$$

The equivalence class $\bar{a}$ for each $a \in A$ is called the **congruence class** of a.

Example 1.3.3. For any non-zero integer m, define a binary relation R on $\mathbb{Z}$ by

$$aRb \iff m|(a - b).$$

Then R is a congruence relation on $\mathbb{Z}$ with respect to the addition and multiplication. This relation R is also called congruence relation modulo m, aRb is written as $a \equiv b \pmod{m}$, and we say that a, b are congruent modulo m.

The following result tells us that normal subgroups are what we need to define congruence relations in a group.

Theorem 1.3.4. *Let G be a group, $H < G$. Then the following assertions are equivalent:*

(1) $H \lhd G$;
(2) $gH = Hg$, *for any* $g \in G$;
(3) $aH \cdot bH = abH$, *for any* $a, b \in G$. *Here,*

$$aH \cdot bH = \{ah_1bh_2 | h_1, h_2 \in H\}.$$

Proof. (1) $\Rightarrow$ (2). Since $H \lhd G$, for $g \in G$, $h \in H$, we have

$$gh = ghg^{-1}g \in Hg; \quad hg = gg^{-1}hg \in gH.$$

Hence, $gH = Hg$.

(2) $\Rightarrow$ (3). From (2), we have $Hb = bH$. Hence,

$$aH \cdot bH = a(Hb)H = a(bH)H = (ab)HH = abH.$$

(3) $\Rightarrow$ (1). Since $H < G$, for any $g \in G$, $h \in H$, we have

$$ghg^{-1} = ghg^{-1}e \in gH \cdot g^{-1}H = gg^{-1}H = eH = H.$$

Thus, $H \lhd G$. $\qquad\square$

By Theorem 1.3.4, the product of two left cosets of a subgroup H is still a left coset of H if and only if H is normal in G. We can define a multiplication on G/H by

$$aH \cdot bH = abH. \tag{1.1}$$

Theorem 1.3.5. *Let G be a group, $H \lhd G$. Then G/H is a group under the multiplication defined above. We call G/H the* **quotient group** *or* **factor group** *of G by H.*

Proof.   Since H is a normal subgroup of G, the definition of the multiplication is well-defined. For any $aH, bH, cH \in G/H$, we have

$$(aH \cdot bH) \cdot cH = abH \cdot cH = (abc)H = aH \cdot (bc)H = aH \cdot (bH \cdot cH).$$

Thus, the operation is associative. One can see that $eH = H$ is a left identity since $eH \cdot aH = (ea)H = aH$ for any $aH \in G/H$. Furthermore, every element $aH \in G/H$ has a left inverse $a^{-1}H$ since $a^{-1}H \cdot aH = (a^{-1}a)H = eH$. Therefore, the left coset space G/H is a group. $\qquad\square$

Example 1.3.6. Since $\{\mathbb{Z}; +\}$ is an abelian group, any subgroup $m\mathbb{Z}$ is normal in $\mathbb{Z}$. Hence, we have the quotient group $\mathbb{Z}/m\mathbb{Z}$, and

$$\mathbb{Z}/m\mathbb{Z} = \begin{cases} \mathbb{Z}, & m = 0, \\ \{\overline{0}, \overline{1}, \dots, \overline{m-1}\}, & m \neq 0. \end{cases}$$

To be more precise, for $m > 0$, we have

$$\overline{r}_1 + \overline{r}_2 = \overline{r_1 + r_2} = \overline{r},$$

where r is defined by $r_1 + r_2 = qm + r$, $0 \leq r < m$. Such group is usually denoted by $\mathbb{Z}_m$, see Exercise 12 of Section 1.1.

Example 1.3.7. The special linear group $\mathrm{SL}(n, \mathbb{R})$ is a normal subgroup of $\mathrm{GL}(n, \mathbb{R})$, since for any $A \in \mathrm{GL}(n, \mathbb{R})$, $B \in \mathrm{SL}(n, \mathbb{R})$, $|ABA^{-1}| = |A||B||A^{-1}| = |B| = 1$, which implies that $ABA^{-1} \in \mathrm{SL}(n, \mathbb{R})$. Similarly, one can prove that $\mathrm{SO}(n)$ is a normal subgroup of $\mathrm{O}(n)$.

In general, we have the following result.

Lemma 1.3.8. *Let H be a subgroup of G, N a normal subgroup of G. Then $H \cap N$ is normal in H.*

Proof.   It is easy to see that $H \cap N$ is a subgroup of G and H. For any $h \in H$, $n \in H \cap N$, we have $hnh^{-1} \in H$. Furthermore, $hnh^{-1} \in N$ since $N \lhd G$. Thus, $hnh^{-1} \in H \cap N$. Hence, $H \cap N$ is a normal subgroup of H. $\qquad\square$

Recall that if W, W' are subspaces of a vector space V, we say V is the direct sum of W, W', denoted by $W \oplus W' = V$, if $W + W' = V$ and $W \cap W' = \{0\}$. Similarly, let G be a group, and let H, N be subgroups with N normal. We call G the **semidirect product** of

H and N if $HN = G$ and $H \cap N = \{e\}$, denoted by $G = H \ltimes N$. If, furthermore, H is also a normal subgroup, then G is called the **direct product** of H and N, denoted by $G = H \times N$. Vector spaces are abelian groups. Hence, every subspace is normal and direct sums of subspaces are exactly the direct product of normal subgroups. In general, we can define the direct product of any two groups; see Exercise 3 in this section.

Similar to the definition of the quotient map from a vector space V to the quotient space V/W, we introduce the following definition.

Definition 1.3.9. Let H be a normal subgroup of G. The map $\pi\colon G \to G/H$ defined by

$$\pi(g) = gH, \quad \forall g \in G$$

is called the **canonical homomorphism** from G onto G/H.

For any $g, h \in G$, we have

$$\pi(gh) = ghH = gH \cdot hH = \pi(g)\pi(h).$$

Thus, the natural homomorphism π preserves the multiplications of the two groups. In the study of algebraic systems, maps that preserve the operations of algebraic systems are helpful to understand the algebraic structures by establishing the connection between two different objects in the same algebraic system. In group theory, the maps that preserve group multiplication, that is, group homomorphisms, are the main object to be studied in the following section.

Exercises

(1) Let A, B be normal subgroups of G. For any $a \in A$, $b \in B$, show that $aba^{-1}b^{-1} \in A \cap B$. In particular, if $A \cap B = \{e\}$, then $ab = ba$.

(2) Let H be a subgroup of G. Then H is normal if and only if H is the union of some conjugacy classes of G.

(3) Let H, K be two groups. Define a multiplication on $H \times K = \{(h, k) | h \in H, k \in K\}$ by

$$(h_1, k_1)(h_2, k_2) = (h_1 h_2, k_1 k_2).$$

Show that $H \times K$ is a group, called the **direct product** of H and K. Let $H_1 = \{(h, e_K) | h \in H\}$ and $K_1 = \{(e_H, k) | k \in K\}$, where e_H, e_K are the identities of H, K respectively. Show that H_1, K_1 are normal subgroups of $H \times K$.

(4) Show that the conjugacy of elements in a group is an equivalence relation, and conjugacy classes form a partition of the group.

(5) If H is a normal subgroup of G, and K is a normal subgroup of H, is K a normal subgroup in G?

(6) Let H, K be two subgroups of G. Show that if one of H, K is normal in G, the HK is a subgroup of G (see Exercise 15 of Section 1.2); if H, K are normal subgroups of G, then $HK \lhd G$.

(7) If a group G has only one subgroup H of order m, show that $H \lhd G$.

(8) Let H be a subgroup of G. The subset $N_G(H) = \{g \in G | gHg^{-1} = H\}$ of G is called the **normalizer** of H in G. Show that
   (i) $N_G(H) < G$ and $H \lhd N_G(H)$;
   (ii) $H \lhd G$ if and only if $N_G(H) = G$.

(9) Show that any subgroup of index 2 is normal.

(10) Let H be a normal subgroup of a group G, $|H| = n$, $[G : H] = m$, and $(m, n) = 1$. Show that H is the only subgroup of G of order n.

Exercises (hard)

(11) Let H be a subgroup of G.
   (i) For any $g \in G$, show that $gHg^{-1} = \{gxg^{-1} | x \in H\}$ is also a subgroup of G, and it is call a **conjugate subgroup** of H.
   (ii) Suppose $[G : H] = n$. Show that the number of conjugate subgroups of H is a divisor of n.

(12) Let H be a proper subgroup of a finite group G. Show that $G \neq \bigcup_{g \in G} gHg^{-1}$. Does the conclusion still hold if G is infinite?

(13) Show that SO(3) is a simple group, while SO(4) is not simple.

(14) Determine all normal subgroups of $SL(n, \mathbb{R})$.

1.4 Homomorphisms and isomorphisms

When studying algebraic systems, we must consider the relationship between different research objects. For example, linear maps (and transformations) are introduced for vector spaces; these maps preserve the additions and scalar multiplications of vector spaces, with which we establish the relationship between different vector spaces. Some linear maps are bijective, i.e., linear isomorphisms, which tell us that some vector spaces, although looking pretty different, are essentially the same. Vector spaces are abelian groups. Hence, linear maps are maps between abelian groups. Similarly, we introduce the following definition.

Definition 1.4.1. Let G, G' be two groups. A map $f : G \to G'$ is called a **group homomorphism** if

$$f(ab) = f(a)f(b), \quad \forall a, b \in G. \tag{1.2}$$

If $f(a) = e'$ for any $a \in G$, then f is a **trivial homomorphism**. If f is injective (resp. surjective), then we call it a **monomorphism** (resp. **epimorphism**); if f is bijective, we call it an **isomorphism**, and we say that the groups G and G' are **isomorphic**, denoted by $G \simeq G'$. In particular, a homomorphism (an isomorphism, resp.) from a group G to itself is called an **endomorphism** (an **automorphism**, resp.).

Taking $a = b = e$ in (1.2), we have $f(e) = e'$; similarly, we have $f(a^{-1}) = f(a)^{-1}$ by taking $b = a^{-1}$.

Problem 1.4.2. If the map $a \to a^2$ is an endomorphism of a group G, then what conclusion can you get?

The following proposition is easy to check; we leave it to the reader.

Proposition 1.4.3. *If* $f : G_1 \to G_2$, $g : G_2 \to G_3$ *are group homomorphisms (monomorphisms, epimorphisms, or isomorphisms), so is* $gf : G_1 \to G_3$. *In particular, if* f *is an isomorphism, so is* f^{-1}.

Denote by End (G) the set of endomorphisms of G, and by Aut (G) that of automorphisms of G. Then End (G) is a monoid, and Aut (G)

is a group by the above proposition. We call $\text{Aut}(G)$ the **automorphism group** of G.

Let's look at some examples.

Example 1.4.4. The exponential map $\exp : \mathbb{R} \to \mathbb{R}^+$ is an isomorphism of groups, where $\mathbb{R}$ is an additive group, and $\mathbb{R}^+$ is a multiplicative group.

Example 1.4.5. Let V be an n-dimensional vector space over a number field $\mathbb{F}$, and $\mathbb{F}^n$ the n-dimensional column space over $\mathbb{F}$, both of which are abelian groups under addition. Fix a basis $\alpha_1, \alpha_2, \ldots, \alpha_n$ of V. Define a map

$$\varphi : V \to \mathbb{F}^n,$$

such that for any $\alpha \in V$, $\varphi(\alpha)$ is the coordinate of α with respect to the basis. Then φ is an isomorphism between abelian groups V and $\mathbb{F}^n$.

Example 1.4.6. Let V be an n-dimensional vector space over a number field $\mathbb{F}$, $\alpha_1, \alpha_2, \ldots, \alpha_n$ a basis of V. For any invertible linear transformation $\mathscr{A}$, there corresponds to an invertible matrix A relative to the basis. Then we have an isomorphism

$$\varphi : \text{GL}(V) \to \text{GL}(n, \mathbb{F}), \quad \varphi(\mathscr{A}) = A.$$

If $\beta_1, \beta_2, \ldots, \beta_n$ is also a basis of V, B is the matrix of $\mathscr{A}$ relative to this basis, then we have another isomorphism

$$\psi : \text{GL}(V) \to \text{GL}(n, \mathbb{F}), \quad \psi(\mathscr{A}) = B.$$

Let T be the transition matrix from $\beta_1, \beta_2, \ldots, \beta_n$ to $\alpha_1, \alpha_2, \ldots, \alpha_n$. Then $B = TAT^{-1}$. Thus, we get a map from $\text{GL}(n, \mathbb{F})$ to itself:

$$\text{Ad}_T : \text{GL}(n, \mathbb{F}) \to \text{GL}(n, \mathbb{F}), \quad A \mapsto \text{Ad}_T(A) = TAT^{-1}.$$

One may check that Ad_T is an automorphism of $\text{GL}(n, \mathbb{F})$.

This example inspires us to come to a more general conclusion.

Proposition 1.4.7. *Let G be a group, $a \in G$. Define a map $\text{Ad}_a : G \to G$ by*

$$\text{Ad}_a(g) = aga^{-1}, \quad \forall g \in G.$$

*Then $\text{Ad}_a \in \text{Aut}(G)$, called an **inner automorphism** by a. The set $\text{Inn}(G) = \{\text{Ad}_a | a \in G\}$ is a normal subgroup of $\text{Aut}(G)$,* called the **inner automorphism group** of G. The quotient group $\text{Aut}\,(G)/\text{Inn}\,(G)$ is called the **outer automorphism group** of G, denoted by $\text{Out}\,G$.

Proof. For any $a, b, c \in G$, we have

$$\text{Ad}_a\text{Ad}_b(c) = a(bcb^{-1})a^{-1} = (ab)c(ab)^{-1} = \text{Ad}_{ab}(c).$$

Thus,

$$\text{Ad}_a\text{Ad}_b = \text{Ad}_{ab}, \tag{1.3}$$

which implies that $\text{Ad}_{a^{-1}}\text{Ad}_a = \text{Ad}_a\text{Ad}_{a^{-1}} = \text{Ad}_e = \text{id}_G$. Hence, Ad_a is a bijection with the inverse $\text{Ad}_{a^{-1}}$. For any $g, h \in G$, one has

$$\text{Ad}_a(gh) = agha^{-1} = aga^{-1}aha^{-1} = \text{Ad}_a(g)\text{Ad}_a(h).$$

Therefore, $\text{Ad}_a \in \text{Aut}\,(G)$.

Furthermore, $\text{Ad}_a(\text{Ad}_b)^{-1} = \text{Ad}_a\text{Ad}_{b^{-1}} = \text{Ad}_{ab^{-1}} \in \text{Inn}\,(G)$. Thus, $\text{Inn}\,(G) < \text{Aut}\,(G)$.

For any $\varphi \in \text{Aut}\,(G)$, $a, g \in G$, we have

$$\varphi\text{Ad}_a\varphi^{-1}(g) = \varphi(a\varphi^{-1}(g)a^{-1}) = \varphi(a)g\varphi(a)^{-1} = \text{Ad}_{\varphi(a)}(g).$$

Therefore, $\varphi\text{Ad}_a\varphi^{-1} = \text{Ad}_{\varphi(a)} \in \text{Inn}\,(G)$, which implies that $\text{Inn}\,(G) \lhd \text{Aut}\,(G)$. $\square$

Problem 1.4.8. Find a group G such that $\text{Inn}\,(G) \neq \text{Aut}\,(G)$.

Besides the inner automorphism Ad_a, we have the following transformations on a group G.

Definition 1.4.9. For any $a \in G$, define transformations L_a, R_a on G by

$$L_a(g) = ag, \quad R_a(g) = ga, \quad \forall g \in G,$$

called the **left** and **right translations** of G by a, respectively.

It is quick to check that $L_aL_b = L_{ab}, R_aR_b = R_{ba}, \text{Ad}_a = L_aR_{a^{-1}}$. Since $L_aL_{a^{-1}} = L_{a^{-1}}L_a = \text{id}_G$, L_a is a bijection. Similarly, R_a is a bijection. Hence,

$$L_G = \{L_a | a \in G\}, \quad R_G = \{R_a | a \in G\}$$

are subgroups of S_G. Thus, we have the following result.

Theorem 1.4.10 (Cayley). *Any group G is isomorphic to a subgroup of the symmetric group S_G of G.*

Proof.   Define a map $L : G \to L_G$ by $L(a) = L_a$. Then L is an epimorphism. If $L_a = \mathrm{id}_G$, then for any $b \in G$, we have $L_a(b) = b$, which implies that $a = e$. Hence, L is injective, and consequently, $G \simeq L_G < S_G$.   □

A subgroup of the symmetric group S_X of a set X is called a **transformation group**. Thus, Cayley's theorem says that every group G is isomorphic to a transformation group of G. This result shows the importance of transformation groups in group theory. It is crucial to establish the relationship between an abstract group and a transformation group of some set X, especially when we want to understand the structure of finite groups.

Let's begin with the general properties of group homomorphisms. It seems that isomorphisms are better than homomorphisms, but we will see that one may get more information from the latter. Recall that when studying a linear map, the kernel and image of the map are crucial. Similarly, we have the following definition.

Definition 1.4.11. Let $f : G \to G'$ be a homomorphism of groups. The **kernel** and **image** of f are defined to be

$$\mathrm{Ker}\, f = \{a \in G | f(a) = e'\}, \quad \mathrm{Im}\, f = f(G) = \{f(a) | a \in G\}.$$

If there exist $a, b \in G$ such that $f(a) = f(b)$, then $f(b)^{-1} f(a) = f(b^{-1}a) = e' = f(e)$, i.e., $b^{-1}a \in \mathrm{Ker}\, f$. Thus, we get the following result.

Lemma 1.4.12. *Let $f : G \to G'$ be a homomorphism. Then f is injective if and only if* $\mathrm{Ker}\, f = \{e\}$.

One knows that the kernel and image of a linear map are closed under addition and scalar multiplication; hence they are subspaces. Do the kernel and image of a group homomorphism have similar properties? The answer is affirmative.

Lemma 1.4.13. *If $f : G \to G'$ is a homomorphism, then $\mathrm{Im}\, f < G'$, $\mathrm{Ker}\, f \lhd G$.*

Proof.   For $f(a), f(b) \in \mathrm{Im}\, f$, we have $f(a)f(b)^{-1} = f(a)f(b^{-1}) = f(ab^{-1}) \in \mathrm{Im}\, f$, hence $\mathrm{Im}\, f < G'$. For any $a, b \in \mathrm{Ker}\, f$,

$f(ab^{-1}) = f(a)f(b)^{-1} = e'$, thus $ab^{-1} \in \operatorname{Ker} f$, which implies $\operatorname{Ker} f < G$. Furthermore, for any $g \in G$, $a \in \operatorname{Ker} f$, we have $f(gag^{-1}) = f(g)f(a)f(g)^{-1} = e'$. Thus, $gag^{-1} \in \operatorname{Ker} f$, and consequently, $\operatorname{Ker} f \lhd G$. $\qquad\square$

A vector space is an abelian group, and every subspace is normal; hence there is no significant difference between the kernel and the image of a linear map. But groups are not necessarily abelian. The previous lemma tells us that the kernel has the better property: it is normal. Conversely, any normal group H of G is the kernel of the canonical homomorphism $\pi : G \to G/H$. Thus, we have the following.

Corollary 1.4.14. *A subgroup H of G is normal if and only if H is the kernel of some homomorphism with G as the domain.*

Therefore, constructing homomorphisms is an efficient way to find normal subgroups. Let's look at some examples.

Example 1.4.15. Let G be a group, $a \in G$. Since $a^m a^n = a^{m+n}$, the map

$$f : \mathbb{Z} \to G, \quad f(n) = a^n \qquad (1.4)$$

is a homomorphism. The image is the subgroup $\langle a \rangle = \{a^n | n \in \mathbb{Z}\}$ of G generated by a, and the kernel $\operatorname{Ker} f$ must be $m\mathbb{Z}$, for some $m \in \mathbb{N}$. If $m = 0$, then the homomorphism is injective, and a is of infinite order; if $m > 0$, then the order of a is m. Therefore, the order of elements in a group can be described by homomorphisms, and we will use similar ideas later.

Example 1.4.16. Let V be an n-dimensional vector space over a number field $\mathbb{F}$. Then the map

$$\det : \operatorname{GL}(V) \to \mathbb{F}^*, \quad \mathscr{A} \mapsto \det(\mathscr{A})$$

is an epimorphism with $\operatorname{Ker} \det = \operatorname{SL}(V)$.

Example 1.4.17. Let V be an n-dimensional vector space over a number field $\mathbb{F}$, $\varepsilon_1, \varepsilon_2, \ldots, \varepsilon_n$ a basis of V. For any $\sigma \in S_n$, there exists a unique linear transformation π_σ on V satisfying

$$\pi_\sigma(\varepsilon_i) = \varepsilon_{\sigma(i)}.$$

Thus, we get a map

$$\pi : S_n \to \mathrm{GL}(V), \quad \sigma \mapsto \pi_\sigma.$$

One can check that π is a monomorphism. Combining with the above example, we get a homomorphism

$$\det \circ \pi : S_n \to \mathbb{F}^*.$$

For $n > 1$, the image of this homomorphism is $\{1, -1\}$, and the kernel of it is denoted by A_n. We will discuss S_n and A_n in Section 1.6.

Example 1.4.18. Let G be a group, $g \in G$. Then Ad_g is an inner automorphism of G. Define a map $\mathrm{Ad} : G \to \mathrm{Inn}\,(G)$ by

$$\mathrm{Ad}(a) = \mathrm{Ad}_a, \quad \forall a \in G.$$

By (1.3) we have $\mathrm{Ad}(ab) = \mathrm{Ad}(a)\mathrm{Ad}(b)$, hence Ad is an epimorphism.

Now for any $a \in \mathrm{Ker}\,\mathrm{Ad}$, $\mathrm{Ad}(a) = \mathrm{Ad}_a = \mathrm{id}$, i.e., $aga^{-1} = g$ for any $g \in G$, hence

$$\mathrm{Ker}\,\mathrm{Ad} = \{a \in G | ag = ga, \forall g \in G\}.$$

We call $\mathrm{Ker}\,\mathrm{Ad}$ the **center** of G, usually denoted by $C(G)$, which is a normal group of G.

Now we are ready to prove the following **Fundamental Theorem of Homomorphisms** of groups.

Theorem 1.4.19. *Let $f : G \to G'$ be an epimorphism. Then $G/\mathrm{Ker}\,f \simeq G'$.*

Proof. Set $N = \mathrm{Ker}\,f$. We must define an isomorphism $\overline{f} : G/N \to G'$. A natural way is to define $\overline{f}(gN) = f(g)$.

First, we need to check that $\overline{f}$ is well-defined. If $gN = hN$, then $g^{-1}h \in N$, and $e' = f(g^{-1}h) = f(g^{-1})f(h)$. Hence, $f(g) = f(h)$, which implies that $\overline{f}$ is well-defined since it is independent of the choice of representatives in left cosets.

It is easy to see that $\overline{f}$ is surjective. Furthermore, if $\overline{f}(gN) = \overline{f}(hN)$, then $f(g) = f(h)$. Thus, $g^{-1}h \in N$, which implies $gN = hN$. Hence, $\overline{f}$ is injective, and consequently, a bijection.

We need to check that $\overline{f}$ is a homomorphism. Note that f is a homomorphism. Then for any $gN, hN \in G/N$, we have

$$\overline{f}(gN \cdot hN) = \overline{f}(ghN) = f(gh) = f(g)f(h) = \overline{f}(gN) \cdot \overline{f}(hN),$$

hence $\overline{f}$ is a homomorphism. $\qquad\qquad\square$

If $f : G \to G'$ is not surjective, we may replace G' by the image $f(G)$. Thus, we have the following result.

Corollary 1.4.20. *Let $f : G \to G'$ be a homomorphism. Then $G/\mathrm{Ker}\, f \simeq f(G)$.*

To summarize, the image of a homomorphism $f : G \to G'$ is isomorphic to some quotient group of G. A quotient group of G is uniquely determined by some normal subgroup of G, which is the kernel of some homomorphism. Therefore, homomorphisms are bridges connecting different groups, subgroups, normal subgroups, and quotient groups. Homomorphisms can tell us more, as the following result shows.

Theorem 1.4.21. *Let $f : G \to G'$ be an epimorphism with $N = \mathrm{Ker}\, f$.*

(1) *f establishes a bijection between the set of subgroups of G containing N and that of subgroups of G'.*
(2) *f is a bijection between normal subgroups.*
(3) *If $H \lhd G$, $N \subseteq H$, then $G/H \simeq G'/f(H)$.*

Proof. (1) Let S be the set of subgroups of G containing N, S' the set of subgroups of G'. For any $K \in S$, $f(K)$ is a subgroup of G', hence f defines a map from S to S'. We will show that f is a bijection.

For any $H' \in S'$, considering the **preimage** of H':

$$f^{-1}(H') = \{g \in G | f(g) \in H'\},$$

we may check that $f^{-1}(H')$ is a subgroup of G containing N, and $f(f^{-1}(H')) = H'$. Thus, $f : S \to S'$ is surjective. Now for $H_1, H_2 \in S$ such that $f(H_1) = f(H_2)$, we need to show that $H_1 = H_2$. For any $h_1 \in H_1$, $f(h_1) \in f(H_1) = f(H_2)$, hence there exists $h_2 \in H_2$ such that $f(h_1) = f(h_2)$. It follows that $h_2^{-1}h_1 \in N \subseteq H_2$. Thus, $h_1 \in H_2$, i.e., $H_1 \subseteq H_2$. Similarly, $H_2 \subseteq H_1$. Therefore, $H_1 = H_2$, i.e., f is injective. Hence, f is bijective.

(2) Let $H \in S$ be normal, and $H' = f(H)$. For any $h' = f(h) \in H'$, $g' = f(g) \in G'$, where $h \in H$, $g \in G$, we have

$$g'h'(g')^{-1} = f(g)f(h)f(g)^{-1} = f(ghg^{-1}).$$

Since H is normal, $ghg^{-1} \in H$, hence $g'h'(g')^{-1} \in H'$. It follows that H' is normal.

Conversely, if $f(H)$ is a normal subgroup of G', then for any $f(g) \in G'$, $f(g)f(H)f(g)^{-1} = f(gHg^{-1}) = f(H)$. Since $N \subseteq H$, we have $N \subseteq gHg^{-1}$. Thus, $gHg^{-1} = H$ by (1), which implies that H is normal in G.

(3) By (2), $f(H)$ is a normal subgroup of G. Let $\pi : G' \to G'/f(H)$ be the canonical homomorphism. Then $\pi \circ f$ is an epimorphism from G onto $G'/f(H)$. Hence,

$$\text{Ker}\,(\pi \circ f) = (\pi \circ f)^{-1}(e'f(H)) = f^{-1}(\pi^{-1}(e'f(H))) = f^{-1}(f(H)).$$

Since $N \subseteq H$, one has $\text{Ker}\,(\pi \circ f) = H$ by (1). By the Fundamental Theorem of Homomorphisms, we have $G/H \simeq G'/f(H)$. $\square$

Applying the above theorem to the canonical homomorphism, we have the following.

Corollary 1.4.22. *Let G be a group, $N \lhd G$, and π the canonical homomorphism from G onto G/N. Then π establishes a bijection between the set of subgroups of G containing N and that of subgroups of G/N, and it maps normal groups to normal subgroups. Furthermore, if $H \lhd G$, $N \subseteq H$, then $G/H \simeq (G/N)/(H/N)$.*

In Theorem 1.4.21, we only consider subgroups containing $\text{Ker}\,f$. Similar results can be obtained for any subgroup. By Corollary 1.4.20, the image of a homomorphism is isomorphic to a quotient group. Thus, we consider the canonical homomorphisms from groups to their quotient groups. The readers are encouraged to rewrite the statement for general homomorphisms.

Theorem 1.4.23. *Let H, N be subgroups of G with N normal, and let $\pi : G \to G/N$ be the canonical homomorphism. Then*

(1) *HN is a subgroup of G and $HN = \pi^{-1}(\pi(H))$, i.e., HN is the preimage of $\pi(H)$;*
(2) *$\text{Ker}\,(\pi|_H) = H \cap N$. In particular, $(H \cap N) \lhd H$;*
(3) *$HN/N \simeq H/(H \cap N)$.*

Proof.   (1) For any $h_1, h_2 \in H$, $n_1, n_2 \in N$, we have

$$h_1 n_1 (h_2 n_2)^{-1} = h_1 n_1 n_2^{-1} h_2^{-1} = h_1 h_2^{-1} h_2 n_1 n_2^{-1} h_2^{-1}.$$

Since $N \lhd G$ and $H < G$, one has $h_2 n_1 n_2^{-1} h_2^{-1} \in N$ and $h_1 h_2^{-1} \in H$, which implies $h_1 n_1 (h_2 n_2)^{-1} \in HN$. It follows that HN is a subgroup of G by Theorem 1.2.2. Furthermore, $N \subseteq HN$, and

$$\pi(HN) = \{hnN|h \in H, n \in N\} = \{hN|h \in H\} = \pi(H).$$

By Theorem 1.4.21 (1), $HN = \pi^{-1}(\pi(H))$.

(2) Considering the homomorphism $\pi|_H : H \to G/N$, one has

$$\mathrm{Ker}\,(\pi|_H) = \{h \in H|\pi|_H(h) = \pi(N)\} = H \cap N.$$

(3) By (1), we have $\pi(H) = \pi(HN) = HN/N$, i.e., π is an epimorphism from H onto HN/N. Thus, $H/\mathrm{Ker}\,(\pi|_H) \simeq HN/N$ by the Fundamental Theorem of Homomorphisms of groups. By (2) one has $\mathrm{Ker}\,(\pi|_H) = H \cap N$. Therefore, $HN/N \simeq H/(H \cap N)$. $\qquad\square$

Exercises

(1) Show that a group of order 6 must be isomorphic to either $\mathbb{Z}_6$ or S_3.

(2) Let exp be a map from $\{\mathbb{R}; +\}$ to $\{\mathbb{R}^+; \cdot\}$ such that $\exp(x) = e^x$, $\forall x \in \mathbb{R}$. Show that exp is a group isomorphism.

(3) Let G be a group. Define $R : G \to S_G$ by $R(a) = R_a$, where R_a is the right translation of G by a. Show that R is an **anti-isomorphism**, i.e., $R(ab) = R(b)R(a)$.

(4) Determine the automorphism group $\mathrm{Aut}(K_4)$ of K_4.

(5) Let G be a finite abelian group, $k \in \mathbb{N}$. Show that $\varphi_k(g) = g^k$ is an endomorphism of G, and φ_k is an automorphism of G if and only if $(k, |G|) = 1$.

(6) Let $\phi : \{\mathbb{R}^*; \cdot\} \to \{\mathbb{R}^+; \cdot\}$ be defined by $\phi(x) = |x|$, for any $x \in \mathbb{R}$. Is ϕ an epimorphism? If so, determine the kernel of ϕ.

(7) Show that $f(x) = \cos x + \sqrt{-1}\sin x$ is a homomorphism from $\{\mathbb{R}; +\}$ to $\{\mathbb{C}^*; \cdot\}$. Determine $\mathrm{Ker}\, f$.

(8) Let G, H be groups, $\varphi : G \to \mathrm{Aut}\,(H)$ a homomorphism. Define a multiplication on the set $\{(g, h)|g \in G, h \in H\}$ by

$$(g_1, h_1)(g_2, h_2) = (g_1g_2, \varphi(g_2^{-1})(h_1)h_2),$$

Show that the set is a group under the multiplication, called a **semidirect product** of G and H and denoted by $G \ltimes H$. Furthermore, show that $G_1 = \{(g, e_H)|g \in G\}$ is a subgroup of $G \ltimes H$, and $H_1 = \{(e_G, h)|h \in H\}$ is a normal subgroup of $G \ltimes H$, where e_G, e_H are the identities of G, H, respectively.

(9) Let f, g be homomorphisms from G to H, $D = \{x \in G | f(x) = g(x)\}$. Show that $D < G$.

(10) Let $f : G \to G'$ be a homomorphism, $a \in G$. If f is an isomorphism, show that a and $f(a)$ have the same order. If f is not an isomorphism, does the conclusion still hold?

(11) Let $G = \{(a, b) | a, b \in \mathbb{R}, a \neq 0\}$ be the group in Exercise 16 of Section 1.1. Show that $K = \{(1, b) | b \in \mathbb{R}\}$ is a normal subgroup of G and $G/K \simeq \mathbb{R}^*$, where $\mathbb{R}^*$ is the set of non-zero real numbers.

(12) Let Aff $(n, \mathbb{R})$ be the affine transformation group (see Exercise 17 of Section 1.1). Set $H = \{T_{(I_n, \beta)} | \beta \in \mathbb{R}^n\}$.

    (i) Show that $H \lhd$ Aff $(n, \mathbb{R})$ and Aff $(n, \mathbb{R})/H \simeq \mathrm{GL}(n, \mathbb{R})$.

    (ii) Show that Aff $(1, \mathbb{R})$ is isomorphic to the group G in the above exercise.

(13) Give the definition of isomorphisms of monoids. Define an operation on $\mathbb{Z}$ by:

$$a * b = a + b - ab, \quad \forall a, b \in \mathbb{Z}.$$

Show that $\{\mathbb{Z}; *\}$ is a monoid, and $\{\mathbb{Z}; *\}$ is isomorphic to the monoid $\{\mathbb{Z}; \cdot\}$.

(14) Find a subgroup N of $\{\mathbb{C}^*; \cdot\}$ such that $\{\mathbb{C}^*; \cdot\}/N \simeq \{\mathbb{R}^+; \cdot\}$.

(15) (**Legendre symbol**) Let p be an odd prime. For any $a \in \mathbb{Z}$, define

$$\left(\frac{a}{p}\right) = \begin{cases} 1, & \text{if } a \equiv b^2 \pmod{p} \text{ for some } b \in \mathbb{Z} \text{ and } p \nmid a, \\ -1, & \text{if } a \not\equiv b^2 \pmod{p} \text{ for any } b \in \mathbb{Z}, \\ 0, & \text{if } p \mid a. \end{cases}$$

    (i) Show that $\varphi(a) = \left(\frac{a}{p}\right)$ is an epimorphism from $\mathbb{Z}_p^*$ to $\{1, -1\}$.

    (ii) Show $f_2 : a \mapsto a^2$ is an endomorphism of $\mathbb{Z}_p^*$, and $\mathbb{Z}_p^*/\mathrm{Im}\, f_2 \simeq \{1, -1\}$. Denote by ψ the isomorphism from $\mathbb{Z}_p^*/\mathrm{Im}\, f_2$ onto $\{1, -1\}$. Show that $\varphi = \psi \circ \pi$, where $\pi : \mathbb{Z}_p^* \to \mathbb{Z}_p^*/\mathrm{Im}\, f_2$ is the canonical homomorphism.

Exercises (hard)

(17) Show that the image of any homomorphism $\chi : \mathrm{SL}(2, \mathbb{Z}) \to \mathbb{C}^*$ lies in the set of 12th roots of unity. Determine all these homomorphisms.

(18) Set $V = \left\{ \begin{pmatrix} \alpha & \beta \\ -\bar{\beta} & \bar{\alpha} \end{pmatrix} \middle| \alpha, \beta \in \mathbb{C} \right\}$. Define an inner product on V by

$$(A, B) = \frac{1}{2}\mathrm{tr}\,(A\overline{B}'), \quad \forall A, B \in V.$$

Then V is a 4-dimensional Euclidean space with an orthonormal basis

$$\mathbf{1} = \begin{pmatrix} 1 & 0 \\ 0 & 1 \end{pmatrix}, \quad \mathbf{i} = \begin{pmatrix} \sqrt{-1} & 0 \\ 0 & -\sqrt{-1} \end{pmatrix}, \quad \mathbf{j} = \begin{pmatrix} 0 & 1 \\ -1 & 0 \end{pmatrix},$$

$$\mathbf{k} = \begin{pmatrix} 0 & \sqrt{-1} \\ \sqrt{-1} & 0 \end{pmatrix}.$$

For any $q \in \mathrm{SU}(2)$, define a linear transformation on V by $\varphi_q(A) = qAq^{-1}$.

    (i) Show that φ_q is an orthogonal transformation, and $W = L(\mathbf{i}, \mathbf{j}, \mathbf{k})$ is a φ_q-invariant subspace.

    (ii) Assume that $\Phi(q)$ is the matrix of $\varphi_q|_W$ related to the basis $\mathbf{i}, \mathbf{j}, \mathbf{k}$. Show that $\Phi(q) \in \mathrm{SO}(3)$ and $\Phi : \mathrm{SU}(2) \to \mathrm{SO}(3)$ is an epimorphism.

(19) Let $\mathbb{F}$ be a number field. Determine the monoid of epimorphisms and the automorphism group of $\mathrm{GL}(n, \mathbb{F})$.

(20) Let σ be an automorphism of G having no fixed points, i.e., $\sigma(g) \neq g$ for any $g \neq e$.

    (i) Show that $f : g \to \sigma(g)g^{-1}$ is injective.

    (ii) If G is finite, show that every element in G has the form $\sigma(g)g^{-1}$, for some $g \in G$.

    (iii) If G is finite and $\sigma^2 = \mathrm{id}_G$, show that G is an abelian group of odd order.

(21) For any $A = (a_{ij}) \in \mathrm{SL}(n, \mathbb{Z})$, define $\varphi(A) = \overline{A} \in \mathrm{SL}(n, \mathbb{Z}_m)$, where $\overline{A} = (\bar{b}_{ij})$ such that $a_{ij} \equiv b_{ij} \pmod{m}$. Show that $\varphi : \mathrm{SL}(n, \mathbb{Z}) \to \mathrm{SL}(n, \mathbb{Z}_m)$ is an epimorphism.

1.5 Cyclic groups

In Section 1.2, we introduced the notion of generators of a group. In this section, we will discuss a special case, $G = \langle a \rangle$, i.e., the group G is generated by a single element a; we call G a **cyclic group**, and call a a **generator** of G. Since every element in a cyclic group $\langle a \rangle$ is of the form a^m, for some $m \in \mathbb{Z}$, a cyclic group is always abelian.

Example 1.5.1. The groups $\{\mathbb{Z}_n; +\}$, $n \geq 2$, are all cyclic groups with a generator $\bar{1}$. Note that $\mathbb{Z}_1$ is the trivial group.

Example 1.5.2. Let $U_n = \{z \in \mathbb{C} | z^n = 1\}$. It is a cyclic group with multiplication, each primitive nth root of unity being a generator. In particular, $U_2 = \{1, -1\}$ has a generator -1; $U_3 = \{1, \omega, \omega^2\}$ has generators ω or ω^2; $U_4 = \{1, \sqrt{-1}, -1, -\sqrt{-1}\}$ has generators $\pm\sqrt{-1}$.

Example 1.5.3. Let E_{ij} be the $n \times n$ matrix with 1 in the (i, j)-entry and 0 otherwise. Set $J = E_{12} + E_{23} + \cdots + E_{n-1,n} + E_{n1}$. Then $J^n = I_n$, and $\langle J \rangle = \{I_n, J, J^2, \ldots, J^{n-1}\}$ is a cyclic group of order n. Furthermore, $a_0 I_n + a_1 J + \cdots + a_{n-1} J^{n-1}$ is the group matrix of $\langle J \rangle$.

Example 1.5.4. Let G be a group of prime order. Then for any $a \in G$, $a \neq e$, $\langle a \rangle$ is a subgroup of G. By Lagrange's theorem, the order of $\langle a \rangle$ is a divisor of $|G|$, hence, we have $G = \langle a \rangle$, that is to say, G is a cyclic group.

A fundamental problem in group theory is to classify groups up to isomorphism. For cyclic groups, this question is easy to answer. Suppose $G = \langle a \rangle$ is a cyclic group. The map $f(n) = a^n$ (see (1.4)) is an epimorphism from $\mathbb{Z}$ onto G, which implies that $G \simeq \mathbb{Z}/\mathrm{Ker}\, f$ by the Fundamental Theorem of Homomorphism. Thus, we have the following result.

Theorem 1.5.5. *Let G be a cyclic group. If $|G|$ is infinite, then $G \simeq \{\mathbb{Z}; +\}$; if $|G| = n$, then $G \simeq \{\mathbb{Z}_n; +\}$. Therefore, two cyclic groups are isomorphic if and only if they have the same order.*

The theorem (and its proof) is quite simple, but it is crucial because it classifies cyclic groups up to isomorphism. Thus, the examples $\{\mathbb{Z}_n; +\}$, U_n, and $\langle J \rangle$ mentioned above have no essential difference, or we say that they are different representations of the same abstract group. This theorem begins the epic task of classifying all groups, especially finite simple groups. One should also note that it is a little display of the Fundamental Theorem of Homomorphisms, which will be used frequently in studying group structure.

In the following, we will focus on subgroups of cyclic groups. Since cyclic groups are abelian, their subgroups are normal. Lagrange's theorem (Theorem 1.2.23) tells us that if H is a subgroup of a finite group G, then $|H|$ is a divisor of $|G|$. The natural question is, for any divisor m of $|G|$, is there a subgroup H of order m? In general, the answer is negative.

Problem 1.5.6. Find a group G such that for some divisor m of $|G|$, there is no subgroup of G of order m.

But for cyclic groups, the answer is affirmative. Let's look at the structure of subgroups of cyclic groups.

Lemma 1.5.7. *Let $G = \langle a \rangle$ be a cyclic group. Then any subgroup H of G is of the form $\langle a^l \rangle$, $l \in \mathbb{N}$. Consequently, H is a cyclic group.*

Proof. Let H be a non-trivial subgroup of G, $l = \min\{m \in \mathbb{N}^* | a^m \in H\}$. Then $a^l \in H$. Thus, $\langle a^l \rangle \subseteq H$. Conversely, if $a^m \in H$, then we have unique q, r such that $m = ql + r$, where $0 \le r < l$. Thus, $a^r = a^{m-ql} = a^m \cdot (a^l)^{-q} \in H$, which implies $r = 0$ by the choice of l and $m = lq$. It follows that $H \subseteq \langle a^l \rangle$, and consequently, $H = \langle a^l \rangle$ is a cyclic group. Since the trivial groups $\{e\}$ and G are generated by a^0, a respectively, any subgroup of G is of the form $\langle a^l \rangle$, for some $l \in \mathbb{N}$. $\qquad\square$

If $G = \langle a \rangle$ is a cyclic group of order n, then one may choose the generator of each subgroup of G as a^l such that $l | n$. In this case, the order of the subgroup $\langle a^l \rangle$ is n/l. Thus, we have the following result.

Theorem 1.5.8. *Let G be a cyclic group of order n, $k | n$. Then there exists a unique subgroup of G of order k.*

The converse statement of the above result still holds; see Exercises 15 and 16 in this section.

Now, we can get the first class of finite simple groups.

Theorem 1.5.9. *Let G be an abelian group. Then G is simple if and only if G is a (cyclic) group of prime order.*

Proof.   Since G is an abelian group, for any $a \in G$, $a \neq e$, $\langle a \rangle$ is a normal group of G, which implies that $G = \langle a \rangle$ if G is simple. If G is an infinite group, then $\langle a^2 \rangle$ is a non-trivial normal subgroup of G; if $|G| = n$ and k is a non-trivial divisor of n, then $\langle a^k \rangle$ is a non-trivial normal subgroup of G. Therefore, an abelian simple group must be a cyclic group of prime order.

Conversely, if $|G| = p$ is a prime, then for any $a \in G$, $a \neq e$, $|a|$ is a divisor of p, which implies that $|a| = p$ and $\langle a \rangle = G$. Hence, the only subgroups of G are the trivial ones. Thus, G is simple.   □

Exercises

(1) Show that up to isomorphism, there are two groups of order 4: the cyclic group of order 4 and the Klein group K_4.

(2) Determine the group determinant of the cyclic group of order n (see Exercise 20 of Section 1.1).

(3) Let m, n be coprime positive integers. Show that $\mathbb{Z}_m \times \mathbb{Z}_n \simeq \mathbb{Z}_{mn}$.

(4) Determine the monoid of the homomorphisms and the automorphism group of a cyclic group.

(5) If any proper subgroup of a group G is cyclic, is G cyclic?

(6) Let $G = \langle a \rangle$ be a cyclic group, $f : G \to K$ a homomorphism. Show that the image $f(G)$ is a cyclic group with a generator $f(a)$.

(7) Let $G = \langle a \rangle$ be an infinite cyclic group, $H = \langle b \rangle$ a cyclic group of order m. Determine all generators of G and H, respectively.

(8) Let G, H be cyclic groups of order m, n, respectively. Show that there is an epimorphism from G to H if and only if $n | m$.

(9) Let G be an abelian group of order n, $K = \{k \in \mathbb{N}^* | a^k = e, \forall a \in G\}$. Show that G is a cyclic group if and only if $n = \min K$.

(10) Show that every finite generated subgroup of $\{\mathbb{Q}; +\}$ is cyclic.

(11) Let $\mathbb{F}$ be a number field. Show that every finite subgroup of $\mathbb{F}^*$ is cyclic.

(12) Show that a group G has only a finite number of subgroups if and only if G is finite.

(13) Let G be a cyclic group of order n, $m | n$. Show that the number of elements in G of order m is $\varphi(m)$, where $\varphi(m)$ is the Euler's function. Furthermore, show that $\sum_{m | n} \varphi(m) = n$.

Exercises (hard)

(14) Assume that the map $\sigma(x) = x^3$ is a monomorphism from G to itself. Show that G is an abelian group.

(15) Prove the converse of Theorem 1.5.8, i.e., if G is a group of order n, and for any divisor m of n, there is a unique subgroup of G of order m, then G is a cyclic group.

(16) Assume that different subgroups of G has different orders. Show that G is a cyclic group.

(17) Let G be an abelian group of order $p_1^{l_1} p_2^{l_2} \cdots p_k^{l_k}$, where $p_1, p_2, \ldots, p_k$ are different primes, $l_1, l_2, \ldots, l_k \in \mathbb{N}^*$. Show that for any $i = 1, 2, \ldots, k$, $G_i = \{x \in G | x^{p_i^{l_i}} = e\}$ is a subgroup of G with $|G_i| = p_i^{l_i}$, and for any $x \in G$, there exist unique $x_i \in G_i$ such that $x = x_1 x_2 \cdots x_k$.

(18) Let G be an abelian group of p^n, where p is a prime, $n \in \mathbb{N}^*$. Show that there exist cyclic subgroups G_i, $i = 1, 2, \ldots, s$, of G such that for $x \in G$, there exist unique $x_i \in G_i$ such that $x = x_1 x_2 \cdots x_s$. Furthermore, let $|G_i| = p^{n_i}$ and $n_1 \leq n_2 \leq \cdots \leq n_s$. Show that $n_1, n_2, \ldots, n_s$ is a partition of n, which is uniquely determined by G.

1.6 Symmetric groups and alternating groups

In the last section, we studied the simplest groups–cyclic groups. In this section, we will focus on the symmetric group S_n and its subgroup A_n. Such groups are significant in that every finite group is isomorphic to a subgroup of the symmetric group S_G of G by Cayley's theorem. In the early history of group theory, the main problem was to study symmetric groups.

Every element $\sigma \in S_n$ is a bijection from $\{1, 2, \ldots, n\}$ to itself. Set $\sigma(k) = i_k$. Then we may write σ as

$$\sigma = \begin{pmatrix} 1 & 2 & \ldots & n \\ i_1 & i_2 & \ldots & i_n \end{pmatrix}.$$

Here, $i_1, i_2, \ldots, i_n$ is a permutation of $1, 2, \ldots, n$. There is a 1-1 correspondence between permutations of $1, 2, \ldots, n$ and elements in S_n. In the rest of the book, we identify permutations of $1, 2, \ldots, n$ with elements in S_n. Therefore, the order of S_n is $n!$

If $j_1, j_2, \ldots, j_n$ is a permutation of $1, 2, \ldots, n$, we may also write $\sigma \in S_n$ as

$$\sigma = \begin{pmatrix} j_1 & j_2 & \cdots & j_n \\ \sigma(j_1) & \sigma(j_2) & \cdots & \sigma(j_n) \end{pmatrix}.$$

Thus, there are $n!$ different ways to describe σ.

For any $\sigma \in S_n$, $i_1 \in \{1, 2, \ldots, n\}$, set

$$i_2 = \sigma(i_1), i_3 = \sigma(i_2) = \sigma^2(i_1), \ldots.$$

There must be some $r \in \mathbb{N}$ such that $i_1, i_2, \ldots, i_r$ are different, while $i_{r+1} = i_j$ for some $j \in \{1, 2, \ldots, r\}$. Thus, $\sigma^r(i_1) = \sigma^{j-1}(i_1)$, which implies $\sigma^{r-j+1}(i_1) = i_1$. Hence, $i_{r-j+2} = i_1$. We have $j = 1$ by the choice of r.

Definition 1.6.1. If $\{i_1, i_2, \ldots, i_r\} \subseteq \{1, 2, \ldots, n\}$, and $\sigma \in S_n$ satisfies

$$\sigma(i_1) = i_2, \ \ \sigma(i_2) = i_3, \ \ \cdots, \ \ \sigma(i_{r-1}) = i_r, \ \ \sigma(i_r) = i_1,$$
$$\sigma(k) = k, \quad \forall k \notin \{i_1, i_2, \ldots, i_r\},$$

then σ is a **cycle** of **length** r, or an r-cycle, denoted by $\sigma = (i_1 i_2 \cdots i_r)$. In particular, a 2-cycle (ij) is called a **transposition**, and a 1-cycle is the identity.

The order of an r-cycle is r. Every r-cycle can be written as

$$\sigma = (i_1 i_2 \cdots i_r) = (i_2 i_3 \cdots i_r i_1) = \cdots = (i_r i_1 \cdots i_{r-1}).$$

Definition 1.6.2. Two cyclics $(i_1 i_2 \ldots i_r), (j_1 j_2 \ldots j_s)$ in S_n are called **disjoint**, if $\{i_1, i_2, \ldots, i_k\} \cap \{j_1, j_2, \ldots, j_s\} = \varnothing$.

By the definition we have the following.

Proposition 1.6.3. *If σ, τ are disjoint cycles, then $\sigma\tau = \tau\sigma$.*

Now we are ready to prove the central theorem in this section.

Theorem 1.6.4. *Every permutation in S_n is a product of disjoint cycles.*

Proof.   For $\sigma = (1)$, there is nothing to prove. Assume that $\sigma \neq (1)$. Then for any $a_1 \in \{1, 2, \ldots, n\}$, consider the sequence

$$a_1 = \sigma^0(a_1), \sigma(a_1), \sigma^2(a_1), \ldots,$$

where σ^0 is the identity (1). Let m_1 be the smallest number such that $\sigma^{m_1}(a_1) = \sigma^k(a_1)$ for some natural number $k < m_1$. Then $k = 0$, otherwise $\sigma^{k-1}(a_1) = \sigma^{m_1-1}(a_1)$, which contradicts the choice of m_1. Thus, $\sigma_1^m(a_1) = a_1$. Considering the m_1-cycle

$$\sigma_1 = (a_1 \ \sigma(a_1) \ \cdots \ \sigma^{m_1-1}(a_1)),$$

we have that $\sigma_1^{-1}\sigma$ fixes each of $a_1, \sigma(a_1), \ldots, \sigma^{m_1-1}(a_1)$.

If there is some $a_2 \notin \{a_1, \sigma(a_1), \ldots, \sigma^{m_1-1}(a_1)\}$ (note that $\sigma_1(a_2) = a_2$) such that $\sigma_1^{-1}\sigma(a_2) \neq a_2$, then by the same procedure we get another cycle

$$\sigma_2 = (a_2 \ \sigma(a_2) \ \cdots \ \sigma^{m_2-1}(a_2)).$$

Then the cycles σ_1, σ_2 are disjoint, and the permutation $\sigma_2^{-1}\sigma_1^{-1}\sigma$ fixes each elements in $\{a_1, \sigma(a_1), \ldots, \sigma^{m_1-1}(a_1), a_2, \sigma(a_2), \ldots, \sigma^{m_2-1}(a_2)\}$. Finally, we can get disjoint cycles $\sigma_1, \sigma_2, \ldots, \sigma_k$ such that $\sigma_i = (a_i, \sigma(a_i), \ldots, \sigma^{m_i-1}(a_i))$ $(m_i > 1)$ and $\sigma_k^{-1} \cdots \sigma_2^{-1}\sigma_1^{-1}\sigma$ fixes every element, i.e., $\sigma_k^{-1} \cdots \sigma_2^{-1}\sigma_1^{-1}\sigma = (1)$. It follows that $\sigma = \sigma_1\sigma_2 \cdots \sigma_k$.

If $\sigma = \tau_1\tau_2 \cdots \tau_l$, where τ_i are disjoint cycles of length > 1, then for each a_i there exists a unique j_i such that $\tau_{j_i}(a_i) \neq a_i$ and $\tau_m(a_i) = a_i$ for $m \neq j_i$. Noticing that $\tau_{j_i}^t(a_i) = \sigma^t(a_i)$ and τ_{j_i} is a cycle, one can easily see that $\tau_{j_i} = \sigma_i$. Then by induction, we have $l = k$, and there is a permutation $(j_1, j_2, \ldots, j_k)$ such that $\tau_{j_i} = \sigma_i$. $\qquad\square$

Example 1.6.5. $\begin{pmatrix} 1 & 2 & 3 & 4 & 5 & 6 & 7 \\ 1 & 7 & 5 & 2 & 3 & 6 & 4 \end{pmatrix} = (1)(274)(35)(6) = (274)(35) = (35)(274)$. The order of this permutation is $[2, 3] = 6$. We omit (1) in the above expression since it is the identity.

Problem 1.6.6. Let p be a prime. Determine the number of elements of order p in S_p.

Furthermore, we may decompose each r-cycle as the product of transpositions.

Lemma 1.6.7. *Any r-cycle can be written as a product of $r - 1$ transpositions. More precisely,*

$$(i_1 i_2 \cdots i_r) = (i_1 i_r)(i_1 i_{r-1}) \cdots (i_1 i_3)(i_1 i_2).$$

A 1-cycle can be regarded as the product of 0 transpositions.

By Theorem 1.6.4, we have the following.

Proposition 1.6.8. *Any permutation is a product of some transpositions.*

In general, the ways of writing permutations as products of transpositions are not unique. Applying the homomorphism $\det \circ \pi$ in Example 1.4.17, we have, for any $\sigma \in S_n$, $\det \circ \pi(\sigma) = \pm 1$, which is called the **sign** of the permutation σ, denoted by $\mathrm{Sgn}\,\sigma$. A permutation is **even** if its sign is 1, and **odd** if its sign is -1. A transposition is an odd permutation. Writing a permutation σ as the product of transpositions $\sigma = \tau_1 \tau_2 \cdots \tau_k$, one gets that σ is odd (or even) if k is odd (or even). Furthermore, we have that the product of two even (or odd) permutations is even, and the product of an even permutation and an odd one is odd. By Example 1.4.17 again, we have that A_n, the set of all even permutations, is a normal subgroup of S_n and is called the **alternating group**. For $n \geq 2$, $|A_n| = \frac{n!}{2}$. A_1, A_2 are trivial, A_3 is a cyclic group of order 3, and the order of A_4 is 12. One may check that $\{(1), (12)(34), (13)(24), (14)(23)\}$, which is isomorphic to the Klein group, is a normal subgroup of A_4. If $n \geq 5$, A_n are simple groups, which was first proved by French mathematician Galois in 1831. Thus, we get the first class of non-abelian finite simple groups, opening the prelude of the classification of finite simple groups.

The standard way to prove the simplicity of a group G is the following. First, find a set of generators A of G such that any two elements in A are conjugate; then show that every non-trivial normal subgroup N of G contains an element in A. Hence, N contains every element in A and, consequently, $N = G$. We will use this strategy to prove the following result.

Theorem 1.6.9. *A_n is simple for $n \geq 5$.*

Proof.   Firstly, we claim that A_n $(n \geq 3)$ is generated by all 3-cycles. To prove this, for different i, j, k, l, we have

$$(ij)(ik) = (jik), \quad (ij)(kl) = (ij)(jk)(jk)(kl) = (ijk)(jkl).$$

Thus, the product of two different transpositions is a product of 3-cycles. Since every element in A_n is the product of an even number of transpositions, A_n is generated by all 3-cycles.

Secondly, we claim that the subset of 3-cycles in A_n $(n \geq 5)$ is a conjugacy class. Since for any 3-cycle (ijk), there exists $\sigma \in S_n$ such that $\sigma(i) = 1, \sigma(j) = 2, \sigma(k) = 3$, hence $\sigma(ijk)\sigma^{-1} = (123)$. If σ is an odd permutation, then $\tau = (45)\sigma \in A_n$, and $\tau(ijk)\tau^{-1} = (123)$, which shows that every 3-cycle is conjugate with (123).

Lastly, we prove that A_n has no non-trivial normal subgroups. Based on the above discussion, we need to show that if $N \lhd A_n$, $N \neq \{e\}$, then N contains a 3-cycle.

For any $\sigma \in N$, $\sigma \neq (1)$, decompose σ as the product of disjoint cycles. We deal with it case by case.

(1) If the decomposition of σ contains an r-cycle, $r \geq 4$, then, without loss of generality (why?), we assume $\sigma = (12 \cdots r)\tau$, where $\tau \in S_{\{r+1, r+2, \ldots, n\}}$. Since N is normal, $(123)\sigma(123)^{-1} \in N$, and consequently, $(123)\sigma(123)^{-1}\sigma^{-1} \in N$. On the other hand,

$$(123)\sigma(123)^{-1}\sigma^{-1} = (123)(\sigma(1)\sigma(3)\sigma(2)) = (123)(324) = (124),$$

which implies that $(124) \in N$, i.e., N contains a 3-cycle.

(2) If the decomposition of σ contains no cycles of length greater than 3 and contains two or more 3-cycles, say, $\sigma = (123)(456)\tau$, then $(124)\sigma(124)^{-1}\sigma^{-1} = (124)(253) = (12534)$. Thus, N contains a 3-cycle by (1).

(3) If the decomposition of σ contains only one 3-cycle, say $\sigma = (123)\tau$, where τ commutes with (123) and $\tau^2 = (1)$, then $\sigma^2 = (132) \in N$.

(4) If σ is the product of disjoint transpositions, say, $\sigma = (12)(34)\tau$, then $(123)\sigma(123)^{-1}\sigma^{-1} = (123)(241) = (13)(24) \in N$. Furthermore, we have

$$(135)(13)(24)(153)(13)(24) = (135)(351) = (153) \in N.$$

Therefore, the normal subgroup N contains a 3-cycle. Hence, $N = A_n$. $\square$

In the proof, one fact we frequently used is the following result which will be helpful for further discussion.

Proposition 1.6.10. *Let N be a normal subgroup of G, $\sigma \in N$, then $\delta\sigma\delta^{-1}\sigma^{-1} \in N$, for any $\delta \in G$.*

Exercises

(1) Show that $S_3 \simeq \text{Aut } S_3 = \text{Inn } S_3$.
(2) Determine the multiplicative table of S_3. Compute the group determinant D_G of S_3, and check that $D_G = f_1 f_2 g^2$, where

f_1, f_2 are linear factors, and g is an irreducible factor of degree two.

(3) Determine the cardinality of each conjugacy class of S_5, and show that A_5 is the only non-trivial normal subgroups of S_5.

(4) Show that $\sigma(i_1 i_2 \ldots i_r)\sigma^{-1} = (\sigma(i_1)\sigma(i_2)\ldots\sigma(i_r))$, for any $\sigma \in S_n$. Show that two elements in S_n are conjugate if and only if they have the same length.

(5) Show that $S_n = \langle\{(12), (13), \ldots, (1n)\}\rangle = \langle\{(12), (12\ldots n)\}\rangle$.

(6) Show that $A_n = \langle\{(123), (124), \ldots, (12n)\}\rangle$.

(7) Show that non-trivial normal subgroups of S_n are A_n or K_4 (when $n = 4$).

(8) Show that $A_n = \langle\{(ab)(cd)|a, b, c, d \text{ are different}\}\rangle$ for $n \geq 5$.

(9) Show that there are 1-1 correspondences among conjugacy classes of S_n, partitions of n, and equivalence classes of nilpotent matrices of order n.

(10) Show that the center $C(S_n)$ of S_n is trivial for $n \geq 3$.

(11) Let G be a permutation group containing an odd permutation. Show that G has a normal subgroup of index 2.

(12) Let $|G| = 2n$ with n odd. Show that G has a normal subgroup of index 2.

Exercises (hard)

(13) Let S_∞ be the symmetry group of $\mathbb{N}$. Determine all non-trivial normal subgroups of S_∞.

(14) Determine the symmetry groups of all regular n-gons and convex regular polyhedrons.

(15) Show that $\text{Aut}(S_n) \simeq S_n$ for $n \neq 2, 6$. Determine $\text{Aut}(S_2)$ and $\text{Aut}(S_6)$.

1.7   Group extensions and Jordan–Hölder theorem

We have discussed two classes of groups: the cyclic groups and the symmetric groups. Now we will begin with the study of the general theory of groups. Fix a group homomorphism $\mu : G \to B$. By replacing B by $\text{Im}\,\mu$, we may assume that μ is an epimorphism. Then $N = \text{Ker}\,\mu$ is a normal subgroup of G, and we have the following sequence of group homomorphisms

$$N \xrightarrow{\lambda} G \xrightarrow{\mu} B,$$

where λ is the natural embedding of N into G. Such a sequence has a special property: $\operatorname{Im} \lambda = \operatorname{Ker} \mu$. Hence, we introduce the following definition.

Definition 1.7.1. Let G, A, B be groups. A sequence of group homomorphisms

$$A \xrightarrow{\lambda} G \xrightarrow{\mu} B \tag{1.5}$$

is called **exact** at G if $\operatorname{Im} \lambda = \operatorname{Ker} \mu$. Furthermore, if λ is injective and μ is surjective, i.e., if the sequence

$$1 \longrightarrow A \xrightarrow{\lambda} G \xrightarrow{\mu} B \longrightarrow 1 \tag{1.6}$$

is exact at A, G, B, then (1.6) is called a **short exact sequence**, and G is an **extension** of B by A. In this case, $N = \operatorname{Im} \lambda = \operatorname{Ker} \mu$, called the **kernel** of the extension, is a normal subgroup of G with $N \simeq A$ and $G/N \simeq B$.

The followings are some familiar examples of group extensions.

G	N	A	G/N	B
V	W	W	V/W	V/W
$\mathbb{Z}_4 = \langle a \rangle$	$\{e, a^2\}$	$\mathbb{Z}/2\mathbb{Z}$	$\{\bar{1}, \bar{a}\}$	$\mathbb{Z}/2\mathbb{Z}$
$K_4 = \{e, a, b, c\}$	$\{e, a\}$	$\mathbb{Z}/2\mathbb{Z}$	$\{\bar{1}, \bar{b}\}$	$\mathbb{Z}/2\mathbb{Z}$
S_3	A_3	$\mathbb{Z}/3\mathbb{Z}$	S_3/A_3	$\mathbb{Z}/2\mathbb{Z}$
$\mathbb{Z}$	$2\mathbb{Z}$	$\mathbb{Z}$	$\mathbb{Z}/2\mathbb{Z}$	$\mathbb{Z}/2\mathbb{Z}$
$O(n)$	$SO(n)$	$SO(n)$	$O(n)/SO(n)$	$\mathbb{Z}/2\mathbb{Z}$

Here, V is a vector space, W is a subspace of V, and $\mathbb{Z}_4$ is a cyclic group of order 4.

The first example tells us that a vector space V, when regarded as an additive group, is an extension of the quotient space V/W by a subspace W. In the second and third examples, both the cyclic group $\mathbb{Z}_4$ and the Klein group K_4 are extensions of $\mathbb{Z}_2$ by $\mathbb{Z}_2$, which shows that extensions of the same groups may be different. In the fourth example, S_3 is an extension of $\mathbb{Z}_2$ by $\mathbb{Z}_3$. Hence, complicated groups may be obtained by extensions of simpler ones. In the fifth example, $\mathbb{Z}$ is the extension of $\mathbb{Z}_2$ by $\mathbb{Z}(\cong 2\mathbb{Z})$. Thus, we may not get new groups by group extensions. In the last example, $O(n)$ is not isomorphic to $SO(n)$ (why?).

In linear algebra, subspaces and quotient spaces play crucial roles in the study of vector spaces and linear transformations. Normal subgroups and quotient groups play similar roles in group theory. If a group G is an extension of B by A, it means that G can be decomposed; or we may say that G is composed of A and B. The above examples show that the extensions of B by A are not unique. Hence, a natural question in group extensions is to determine all different extensions of B by A. Before going deeper, we first study some general properties of extensions.

Theorem 1.7.2. *Let A, B, G, G' are groups.*

(1) *If G is an extension of B by A, $G \simeq G'$, then G' is also an extension of B by A.*
(2) *If both G and G' are extensions of B by A, and there is a homomorphism $f : G \to G'$ such that the following diagram commutes (i.e., $\lambda' = f \circ \lambda$, $\mu' \circ f = \mu$), then f is a group isomorphism. We call G and G' **equivalent extensions** of B by A.*

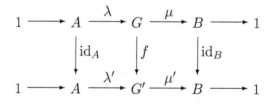

Proof. (1) By the short exact sequence $1 \longrightarrow A \xrightarrow{\lambda} G \xrightarrow{\mu} B \longrightarrow 1$ and the group isomorphism $f : G \to G'$, define $\lambda' = f\lambda$, $\mu' = \mu f^{-1}$. Then λ' is a monomorphism, and μ' is an epimorphism. Furthermore, $\operatorname{Ker} \mu' = \operatorname{Ker}(\mu f^{-1}) = f(\operatorname{Ker} \mu) = f(\lambda(A)) = \lambda'(A)$, which implies that $1 \longrightarrow A \xrightarrow{\lambda'} G' \xrightarrow{\mu'} B \longrightarrow 1$ is a short exact sequence, i.e., G' is an extension of B by A.

(2) Need to prove that f is bijective. If $f(x) = e$, then $\mu(x) = \mu' f(x) = e$. Hence, $x \in \operatorname{Ker} \mu = \operatorname{Im} \lambda$. Thus, there exists $y \in A$ such that $x = \lambda(y)$. Therefore, $\lambda'(y) = f\lambda(y) = f(x) = e$. Since λ' is injective, we have $y = e$, and consequently $x = e$, i.e., f is injective.

For any $z \in G'$, since μ is surjective, there exists $x \in G$ such that $\mu'(z) = \mu(x) = \mu' f(x)$. Hence, $z^{-1} f(x) \in \operatorname{Ker} \mu' = \operatorname{Im} \lambda'$. Then there exists $y \in A$ such that $z^{-1} f(x) = \lambda'(y) = f(\lambda(y))$. Thus, $z = f(x\lambda(y)^{-1}) \in \operatorname{Im} f$, which implies that f is surjective. $\qquad \square$

For any subspace W of a finite-dimensional vector space V, there exists a subspace U of V such that $V = W \oplus U$. Then U is isomorphic to V/W as vector spaces. We may apply such an idea to group extensions and get the following result.

Theorem 1.7.3. *Let G be an extension of B by A with the kernel N and the corresponding short exact sequence is* (1.6).

(1) *If there exists $H < G$ with $G = HN$, $H \cap N = \{e\}$, then $\mu|_H$ is an isomorphism between H and B. In this case, $\nu = (\mu|_H)^{-1}$ is a homomorphism from B to G such that $\mu\nu = \operatorname{id}_B$.*
(2) *If there exists a homomorphism $\mu : B \to G$ such that $\mu\nu = \operatorname{id}_B$, then $H = \nu(B) < G$, $G = HN$, and $H \cap N = \{e\}$.*

Proof. (1) Since $\mu|_H$ is a natural homomorphism from H to B, we need prove that it is a bijective. If $h \in H$ and $\mu(h) = e$, then $h \in H \cap N = \{e\}$; hence $\mu|_H$ is injective. For any $b \in B$, there exists $g \in G$ such that $\mu(g) = b$. Since $G = HN$, there exist $h \in H, n \in N$ such that $g = hn$. Hence, $b = \mu(g) = \mu(hn) = \mu(h)\mu(n) = \mu(h)$, proving that $\mu|_H$ is surjective.

(2) Since H is a subgroup of G and $\nu : B \to H$ is an isomorphism. If $g \in H \cap N$, then there exists $b \in B$ such that $g = \nu(b)$. Since $\mu\nu = \operatorname{id}_B$, $b = \mu\nu(b) = \mu(g) = e$. Thus, $H \cap N = \{e\}$. Now assume $g \in G$. Then there exists $h \in H$ such that $\mu(g) = \mu(h)$. Thus, $\mu(h^{-1}g) = e$ and $n = h^{-1}g \in N$, which implies $g = hn$. Therefore, $G = HN$. $\qquad \square$

Definition 1.7.4. Let G be an extension of B by A with kernel N. The corresponding short exact sequence is (1.6). If there is a subgroup $H < G$ such that $G = HN$, $H \cap N = \{e\}$, then we call this extension a **non-essential extension** and G a **semidirect product** of H and N, denoted by $G = H \ltimes N$.

Furthermore, if $H \triangleleft G$, then G a direct product of H and N, and we call such extension a **trivial extension**, denoted by $G = H \times N$. If $N \subseteq C(G)$, then the extension is called a **central extension**.

Example 1.7.5.

(1) $\mathbb{Z}$ is an extension of $\mathbb{Z}/2\mathbb{Z}$ by $\mathbb{Z}$, but it is not a non-essential extension.
(2) If $n \geq 2$, S_n is a non-essential extension of $\mathbb{Z}/2\mathbb{Z}$ by A_n.
(3) $O(n)$ is a non-essential extension of $\mathbb{Z}/2\mathbb{Z}$ by $SO(n)$. Furthermore, if n is odd, the extension is trivial.
(4) A cyclic group of order 15 is a trivial extension of $\mathbb{Z}/3\mathbb{Z}$ by $\mathbb{Z}/5\mathbb{Z}$.

The following lemma is similar to the direct sum decomposition of vector spaces.

Lemma 1.7.6. *Let A, B be subgroups of G, $G = AB$. Then the following assertions are equivalent:*

(1) $A \cap B = \{e\}$;
(2) *For any $g \in G$, the decomposition $g = ab$ of g is unique, where $a \in A, b \in B$;*
(3) *The decomposition of the identity is unique.*

For normal subgroups, we have the following.

Lemma 1.7.7. *Let A, B be subgroups of G, $G = AB$, and $A \cap B = \{e\}$. Then A, B are normal subgroups of G if and only if $ab = ba$, for any $a \in A, b \in B$.*

Proof. If both A and B are normal subgroups of G, then for any $a \in A, b \in B$, $a^{-1}b^{-1}ab = a^{-1}(b^{-1}ab) \in A$, and $a^{-1}b^{-1}ab = (a^{-1}b^{-1}a)b \in B$. Thus, $a^{-1}b^{-1}ab = e$, i.e., $ab = ba$. Conversely, for any $g \in G$, there exist $a \in A, b \in B$ such that $g = ab$. For any $x \in A$, we have $gxg^{-1} = abxb^{-1}a^{-1} = axa^{-1} \in A$, since $xb = bx$. Therefore, $A \triangleleft G$. Similarly, $B \triangleleft G$. □

Theorem 1.7.8. *Let A, B be two groups. Then the trivial extension G of B by A is unique up to isomorphism.*

Proof. Existence. Let $G = \{(a, b) | a \in A, b \in B\}$ and define on G a product by

$$(a_1, b_1) \circ (a_2, b_2) = (a_1 a_2, b_1 b_2), \quad a_1, a_2 \in A, \quad b_1, b_2 \in B.$$

Then it is quick to check that G is a group, and $A_1 = \{(a, e_B) | a \in A\}$, $B_1 = \{(e_A, b) | b \in B\}$ are normal subgroups of G such that $G = A_1 B_1$.

Uniqueness. Let G' be a trivial extension of B by A. Then $G' = A'B'$, where A', B' are normal subgroups of G' isomorphic to A, B, respectively. Denote by λ (resp. μ) the isomorphism from A to A' (resp. from B to B'). Then it is easy to check that $f : G \to G'$, $f(a, b) = \lambda(a)\mu(b)$ is an isomorphism from G to G'. $\qquad\square$

In the rest of this section, we will use the idea of group extensions to study the structure of groups. Let A_1 be a non-trivial normal subgroup of G, $B = G/A_1$. Then there is a short exact sequence

$$1 \longrightarrow A_1 \xrightarrow{\lambda} G \xrightarrow{\mu} B \longrightarrow 1.$$

If B has a non-trivial normal subgroup B_1, and let $B_2 = B/B_1$, then we have another short exact sequence

$$1 \longrightarrow B_1 \xrightarrow{\lambda_1} B \xrightarrow{\mu_1} B_2 \longrightarrow 1.$$

Consider the composition of the maps:

$$G \xrightarrow{\mu} B \xrightarrow{\mu_1} B_2.$$

Then we get an epimorphism $\mu_1 \circ \mu : G \to B_2$ with kernel $A_0 = (\mu_1 \circ \mu)^{-1}(1) = \mu^{-1}(B_1)$ which is a normal subgroup of G containing A_1, and $A_0/A_1 \simeq B_1$, $G/A_0 \simeq B_2$. Furthermore, if A_2 is a normal subgroup of A_1, then we have a sequence

$$G \supset A_0 \supset A_1 \supset A_2 \supset \cdots.$$

For convenience, we introduce the following definition.

Definition 1.7.9. A sequence of subgroups

$$G = G_1 \rhd G_2 \rhd \cdots \rhd G_t \rhd G_{t+1} = \{e\} \tag{1.7}$$

is called a **subnormal series** for G, and the quotient groups G_i/G_{i+1} are called the **factors** of the series. We do not require that G_i be normal subgroups of G. It $G_i \neq G_{i+1}$ for $i = 1, 2, \ldots, t$, then t is called the **length** of the series. A subnormal series

$$G = G_1' \rhd G_2' \rhd \cdots \rhd G_s' \rhd G_{s+1}' = \{e\} \tag{1.8}$$

is called a **refinement** of (1.7), if each subgroup G_i in (1.7) occurs in (1.8). If, in addition, each G_i is normal in G, then the series is called a **normal series**. If each factor G_i/G_{i+1} is simple in a subnormal series for G, then the subnormal series is called a **composition series** for G, and the factors G_i/G_{i+1} are called **composition factors**. Furthermore, if a composition series is also a normal series, it is called a **principal series**.

If (1.7) is a composition series for G, then G_{t-1} is the extension of the simple group G_{t-1}/G_t by the simple group G_t, G_{t-2} is the extension of the simple group G_{t-2}/G_{t-1} by $G_{t-1}, \ldots, G$ is the extension of the simple group G/G_2 by G_2. That is to say, G can be obtained by a series of extensions of simple groups. Let's look at some examples of composition series or principal series.

Example 1.7.10.

(1) If $n = 3$ or $n \geq 5$, $S_n \triangleright A_n \triangleright \{e\}$ is a principal series for S_n.
(2) The sequence $S_4 \triangleright A_4 \triangleright K_4 \triangleright \langle (12)(34) \rangle \triangleright \{e\}$ is a composition series for S_4, but not a principal series.
(3) $G = \mathbb{Z}_{15} \triangleright \mathbb{Z}_3 \triangleright \{e\}$ and $G = \mathbb{Z}_{15} \triangleright \mathbb{Z}_5 \triangleright \{e\}$ are two different principal series for G.

Problem 1.7.11. Show that the group $\mathbb{Z}$ has no composition series.

The composition series of a group may not exist in general but exist for finite groups. The above examples show that the composition series of $\mathbb{Z}_{15}$ are not unique. But the composition factors of the two composition series are the same. This is not a coincidence. We introduce the following notion.

Definition 1.7.12. Two subnormal series

$$G = G_1 \triangleright G_2 \triangleright \cdots \triangleright G_{t+1} = \{e\},$$
$$G = H_1 \triangleright H_2 \triangleright \cdots \triangleright H_{s+1} = \{e\}$$

for a group G are called **equivalent**, if $s = t$ and there exists a permutation $i \mapsto i'$ of $1, 2, \ldots, s$ such that $G_i/G_{i+1} \simeq H_{i'}/H_{i'+1}$.

Note that the definition does not require that $G_i/G_{i+1} \simeq H_i/H_{i+1}$. For example, in Example 1.7.10, the two principal series for $\mathbb{Z}_{15}$ are equivalent. The following theorem tells us that the composition series are equivalent if they exist.

Theorem 1.7.13 (Jordan–Hölder). *Assume that a group G has composition series. Then any two composition series for G are equivalent. In particular, any two principal series for G are equivalent.*

Proof. Let

$$G = G_1 \triangleright G_2 \triangleright \cdots \triangleright G_{t+1} = \{e\},$$
$$G = H_1 \triangleright H_2 \triangleright \cdots \triangleright H_{s+1} = \{e\}$$

be two composition series for G. Then we have a subnormal series for each G_i

$$G_i = G_i \cap H_1 \triangleright G_i \cap H_2 \triangleright \cdots \triangleright G_i \cap H_{s+1} = \{e\}.$$

Consider the sequence of groups

$$G_i = (G_i \cap H_1)G_{i+1} \supseteq (G_i \cap H_2)G_{i+1} \supseteq \cdots \supseteq (G_i \cap H_{s+1})G_{i+1}$$
$$= G_{i+1}.$$

Since G_i/G_{i+1} is simple, there exists a unique i' such that $G_i = (G_i \cap H_{i'})G_{i+1}$ and $(G_i \cap H_{i'+1})G_{i+1} = G_{i+1}$. Thus, we have

$$
\begin{aligned}
G_i/G_{i+1} &\simeq (G_i \cap H_{i'})G_{i+1}/(G_i \cap H_{i'+1})G_{i+1} \\
&\simeq G_i \cap H_{i'}/(G_i \cap H_{i'} \cap (G_i \cap H_{i'+1})G_{i+1}) \\
&= G_i \cap H_{i'}/(H_{i'} \cap (G_i \cap H_{i'+1})G_{i+1}) \\
&= G_i \cap H_{i'}/(H_{i'} \cap G_{i+1})(G_i \cap H_{i'+1}).
\end{aligned}
$$

Applying the same procedure for the pair $H_{i'} \supset H_{i'+1}$, we have

$$H_{i'}/H_{i'+1} \simeq G_k \cap H_{i'}/(H_{i'} \cap G_{k+1})(G_k \cap H_{i'+1}),$$

for some unique k, and we must have $k = i$. $\qquad\square$

Remark 1.7.14. It is well known that an organic molecule is usually composed of several atoms of carbon, hydrogen, and oxygen, and there are two different organic molecules with the same number of atoms of carbon, hydrogen, and oxygen. Groups and simple groups are the organic molecules and atoms in group theory. Jordan–Hölder theorem tells us that the composition factors (if they exist) of a group are unique up to isomorphism, although different extensions of the same simple groups may result in quite different groups. It is a challenging problem to determine all different extensions of given simple groups.

Exercises

(1) Show that any group of order 15 is a trivial extension of $\mathbb{Z}_3$ and $\mathbb{Z}_5$.

(2) Give equivalent refinements for the two subnormal series $\mathbb{Z} \rhd 20\mathbb{Z} \rhd 60\mathbb{Z} \rhd \{0\}$ and $\mathbb{Z} \rhd 49\mathbb{Z} \rhd 245\mathbb{Z} \rhd \{0\}$ for $\mathbb{Z}$.

(3) Find all composition series for $\mathbb{Z}_{60}$, and check that they are equivalent.

(4) Let p be the smallest prime divisor of the order of a group G, N a normal subgroup of order p. Show that $N \subseteq C(G)$.

(5) Let A, B be cyclic groups of order m, n respectively, $(m, n) = 1$. Determine all non-equivalent extensions of A by B (In general, a group is called a **metacyclic group** if it is an extension of a cyclic group by another cyclic group).

(6) A group G is called **complete** if $C(G) = \{e\}$ and $\mathrm{Aut}\, G = \mathrm{Inn}\, G$. Show that the extension of any group by a complete group is trivial.

(7) Let A, B be normal groups of G and $G = AB$. Show that $G/(A \cap B) = A/(A \cap B) \times B/(A \cap B)$.

(8) Let A, B be normal groups of G and $A \cap B = \{e\}$. Show that G is isomorphic to a subgroup of $(G/A) \times (G/B)$.

(9) Let Y be a subset of a set X. Show that $P(X) \simeq P(Y) \times P(X \setminus Y)$ (see Exercise 10 of Section 1.1).

(10) Assume that a group G has composition series. Show that any normal subgroup N of G and the quotient group G/N have composition series.

Exercises (hard)

(11) Let $G = G_1 \rhd G_2 \rhd \cdots \rhd G_t = \{e\}$ and $G = H_1 \rhd H_2 \rhd \cdots \rhd H_s = \{e\}$ be two composition series for G, and $G_2 \neq H_2$. Set $K_3 = G_2 \cap H_2 \lhd G$. Take any composition series $K_3 \rhd K_4 \rhd \cdots \rhd K_r = \{e\}$ for K_3. Show that

(i) $G = G_1 \rhd G_2 \rhd K_3 \rhd \cdots \rhd K_r = \{e\}$ and $G = H_1 \rhd H_2 \rhd K_3 \rhd \cdots \rhd K_r = \{e\}$ are composition series for G;

(ii) all the four composition series mentioned above are equivalent to each other, which gives another proof of Jordan–Hölder theorem.

1.8 Solvable and nilpotent groups

The Jordan–Hölder theorem states that the composition series of a group must be unique up to isomorphism if it exists. In particular, the composition series of a finite group does exist. How to find a composition series? In general, this question is not easy to answer. This section will focus on a class of groups whose composition factors are all cyclic groups of prime order. Such groups will be crucial in studying the solvability of algebraic equations. Suppose G has a composition series

$$G = G_1 \triangleright G_2 \triangleright \cdots \triangleright G_{t+1} = \{e\},$$

and each composition factor is a cyclic group of prime order. To construct this series, we first need to find G_2 such that G/G_2 is a cyclic group of prime order. Consider the canonical homomorphism

$$\pi : G \to G/G_2.$$

For any $g, h \in G$, one has $\pi(g)\pi(h) = \pi(h)\pi(g)$. Hence,

$$\pi(e) = \pi(g^{-1})\pi(h^{-1})\pi(g)\pi(h) = \pi(g^{-1}h^{-1}gh).$$

It follows that

$$g^{-1}h^{-1}gh \in \operatorname{Ker} \pi.$$

Thus, we introduce the following definition.

Definition 1.8.1. For $g, h \in G$, we call $g^{-1}h^{-1}gh$ the **commutator** of g, h, denoted by $[g, h]$.

If H, K are subgroups of G, we call

$$[H, K] = \langle \{[h, k] | h \in H, k \in K\} \rangle$$

the **commutator subgroup** of H, K. In particular, $[G, G]$ is the commutator subgroup of G.

Problem 1.8.2. Determine the commutator subgroup of S_3.

Note that the set of all commutators may not be a subgroup of G since the product of two commutators may not be a commutator (see Exercise 12 in this section). In 1952, Ore proved that every element in A_n is a commutator and conjectured that every element in a finite non-abelian simple group is a commutator. The conjecture was verified case by case and was finally proved in 2010 (see the survey paper (Malle, 2014)).

The following results can be easily derived from the definition.

Lemma 1.8.3. (1) $[g,h]^{-1} = [h,g]$, *and* $[g,h] = 1$ *if and only if* $gh = hg$.

(2) *Let* $\varphi : G \to H$ *be a homomorphism,* $g, h \in G$. *Then* $\varphi([g,h]) = [\varphi(g), \varphi(h)]$. *In particular, if* $\varphi \in \mathrm{Aut}\,(G)$, *then* $\varphi([g,h]) = [\varphi(g), \varphi(h)]$.

Lemma 1.8.4. *Let* H, K *be normal groups of* G. *Then* $[H, K] \lhd G$ *and* $[H, K] \subseteq H \cap K$. *In particular,* $[H, H] \lhd G$, $[G, G] \lhd G$.

Proof. Take $\varphi = \mathrm{Ad}_g$, for $g \in G$, in Lemma 1.8.3 (2). Then for any $h \in H, k \in K$, $\mathrm{Ad}_g([h,k]) = [\mathrm{Ad}_g(h), \mathrm{Ad}_g(k)]$. Since H, K are normal groups of G, $\mathrm{Ad}_g(h) \in H$, $\mathrm{Ad}_g(k) \in K$, we have $\mathrm{Ad}_g([h,k]) \in [H, K]$, and consequently, $[H, K]$ is a normal group of G. Furthermore, since $h^{-1}k^{-1}h \in K$, $k^{-1}hk \in H$, we have $[h, k] = h^{-1}k^{-1}hk \in H \cap K$, which implies $[H, K] \subseteq H \cap K$. □

Lemma 1.8.5. *Let* $N \lhd G$. *Then* G/N *is abelian if and only if* $[G, G] \subseteq N$. *In particular,* $G/[G, G]$ *is abelian.*

Proof. Let $\pi : G \to G/N$ be the canonical homomorphism. Then G/N is abelian if and only if $\pi(x)\pi(y) = \pi(y)\pi(x)$ for any $x, y \in G$, if and only if $\pi(x^{-1}y^{-1}xy) = \pi(e)$, if and only if $[x, y] \in \mathrm{Ker}\,\pi = N$, if and only if $[G, G] \subseteq N$. Thus, the first assertion follows, and the second one is a special case of the first. □

By the above lemma, the commutative subgroup of G is the smallest normal subgroup N of G such that G/N is abelian. Therefore, the subgroup G_2 we want to find at the beginning of this section must contain $[G, G]$.

Problem 1.8.6. Let N be a subgroup of G containing $[G, G]$. Is N normal?

By taking commutators, we can get several sequences of normal subgroups.

Definition 1.8.7. Let G be a group. Define $G^{(k)}$ and G^k for any $k \in \mathbb{N}$ as follows:

$$G^{(0)} = G, \ G^{(k+1)} = [G^{(k)}, G^{(k)}], \quad k \geq 0,$$
$$G^1 = G, \ G^{k+1} = [G, G^k], \qquad k \geq 1.$$

Then the following two sequences of normal subgroups

$$G = G^{(0)} \rhd G^{(1)} \rhd G^{(2)} \rhd \cdots,$$
$$G = G^1 \rhd G^2 \rhd G^3 \rhd \cdots$$

are call the **derived series** and **lower central series** of G, respectively.

For the derived series, if $G^{(k)} = G^{(k+1)}$ for some $k \in \mathbb{N}$, then $G^{(n)} = G^{(k)}$, for any $n \geq k$. A similar statement holds for the lower central series. In some cases, we have $G = G^{(1)} = G^2$, that is, there are no non-trivial normal subgroups in the derived series and the lower central series. But if there exists some t such that $G^{(t)} = \{e\}$ or $G^t = \{e\}$, then we get a sequence of non-trivial normal subgroups, which is also a normal series of G such that every factor is abelian. Thus, we introduce the following definitions.

Definition 1.8.8. A group G is called **solvable** if $G^{(t)} = \{e\}$ for some $t \in \mathbb{N}$. G is called **nilpotent** if $G^t = \{e\}$ for some $t \in \mathbb{N}$.

In 1902, the British mathematician William Burnside found an astonishing result: if the order of a group G is an odd number less than 40,000, then G is solvable! Based on this fact, the famous **Burnside's conjecture** was proposed: every finite group of odd order is solvable. The conjecture was of far-reaching influence and was proved by Feit and Thompson in 1962 in a journal paper of 255 pages.

In Galois theory (see Chapter 4), we will see that solvable groups have close relationship with the solvability of algebraic equations by radicals, where the term "solvable" arose.

Lemma 1.8.9. *Subgroups and quotient groups of solvable groups are solvable. Conversely, if N is a normal group of G, and both N and G/N are solvable, then G is solvable.*

Proof. Let G be a solvable group with $G^{(t)} = \{e\}$. If H is a subgroup of G, then $H^{(t)} \subseteq G^{(t)} = \{e\}$. Thus, H is solvable.

If N is a normal subgroup of G, $\pi : G \to G/N$ is the canonical homomorphism, then for any $g, h \in G$, $\pi([g,h]) = [\pi(g), \pi(h)]$. Hence, $\pi(G^{(k)}) = (\pi(G))^{(k)} = (G/N)^{(k)}$. Thus, G/N is solvable.

Conversely, since G/N is solvable, we have $(G/N)^{(k)} = \{\pi(e)\}$ for some $k \in \mathbb{N}$. Thus, $\pi(G^{(k)}) = (\pi(G))^{(k)} = (G/N)^{(k)} = \{\pi(e)\}$, which implies $G^{(k)} \subseteq \operatorname{Ker} \pi = N$. Hence, $G^{(k)}$ is solvable since it is a subgroup of the solvable group N. It follows that $(G^{(k)})^{(l)} = \{e\}$ for some $l \in \mathbb{N}$. Therefore, $G^{(k+l)} = (G^{(k)})^{(l)} = \{e\}$, i.e., G is solvable. $\square$

Corollary 1.8.10. *Let G, A, B be groups, and G an extension of B by A. Then G is solvable if and only if A, B are solvable.*

Now, we introduce some equivalent conditions for a finite group being solvable.

Theorem 1.8.11. *Let G be a finite group. Then the following assertions are equivalent:*

(1) *G is solvable;*
(2) *There is a normal series $G = G_1 \rhd G_2 \rhd \cdots \rhd G_{t+1} = \{e\}$ such that G_i/G_{i+1} $(i = 1, 2, \ldots, t)$ are abelian;*
(3) *There is a subnormal series $G = G_1' \rhd G_2' \rhd \cdots \rhd G_{s+1}' = \{e\}$ such that G_i'/G_{i+1}' $(i = 1, 2, \ldots, s)$ are abelian;*
(4) *There is a subnormal series $G = G_1'' \rhd G_2'' \rhd \cdots \rhd G_{r+1}'' = \{e\}$ such that G_i''/G_{i+1}'' $(i = 1, 2, \ldots, r)$ are groups of prime order.*

Proof. $(1) \Rightarrow (2)$ Since G is solvable, the derived series of G is a normal series whose factors are abelian groups.

$(2) \Rightarrow (3)$ Trivial.

$(3) \Rightarrow (4)$ Refine the subnormal series in (3) to a composition series, whose composition factors are abelian simple groups of prime order.

$(4) \Rightarrow (1)$ The groups G_r'' and G_{r-1}''/G_r'' are solvable since their order are prime numbers. By Lemma 1.8.9, G_{r-1}'' is solvable. Then, by induction, one may easily get that $G = G_1''$ is solvable. $\square$

By the definitions we have $G^{(k)} \subseteq G^{k+1}$. Thus, nilpotent groups must be solvable. Abelian groups are nilpotent, hence are solvable.

Problem 1.8.12. Show that nilpotent groups of order less than 8 are abelian. Find all non-abelian nilpotent groups of order 8.

Similar to Lemma 1.8.9, we have the following.

Lemma 1.8.13. *Subgroups and quotient groups of a nilpotent group are nilpotent.*

It is worth noting that if a normal subgroup N of G and the quotient group G/N are both nilpotent, G may not be nilpotent; in other words, the extension of one nilpotent group over another nilpotent group may not be nilpotent. For example, S_3 is not nilpotent, although it is the extension of $\mathbb{Z}_2$ over $\mathbb{Z}_3$. But we have the following result.

Lemma 1.8.14. *Let $N \subseteq C(G)$. If both N and G/N are nilpotent groups, then so is G. In other words, a central extension of a nilpotent group is nilpotent.*

Proof. Let $\pi : G \to G/N$ be the canonical homomorphism. Since G/N is nilpotent, $(G/N)^k = \{\pi(e)\}$ for some $k \in \mathbb{N}$. Thus, $\pi(G^k) = (\pi(G))^k = \{\pi(e)\}$, which implies that $G^k \subset \operatorname{Ker} \pi = N \subset C(G)$. Hence, $G^{k+1} = \{e\}$, i.e., G is nilpotent. $\qquad\square$

One may see from the lemma that the center of a nilpotent group plays an important role. Any non-trivial nilpotent group has a non-trivial center, which can be seen from the lower central series. To be more precise, let k be the smallest integer such that $G^k = \{e\}$. Then $G^{k-1} \neq \{e\}$ and $G^{k-1} \subseteq C(G)$.

Let $C(G)$ be the center of a nilpotent group G. Note that $G/C(G)$ is nilpotent; if it is non-trivial, then so is $C(G/C(G))$. Let $\pi : G \to G/C(G)$ be the canonical homomorphism. Then $\pi^{-1}(C(G/C(G)))$ is a normal subgroup of G containing $C(G)$ properly. Thus, we introduce the following series.

Definition 1.8.15. Let G be a group. Set $C_0(G) = \{e\}$, and

$$C_{k+1}(G) = \pi_k^{-1}(C(G/C_k(G))), \quad k \geq 0,$$

where $\pi_k : G \to G/C_k(G)$ is the canonical homomorphism. The series

$$C_0(G) \lhd C_1(G) \lhd C_2(G) \lhd \cdots$$

is called the **upper central series** of G.

Now, we are ready to prove the following criteria for nilpotent groups, which is similar to Theorem 1.8.11.

Theorem 1.8.16. *Let G be a group. Then the followings are equivalent:*

(1) *G is a nilpotent group;*
(2) *G has a normal series $G = G_1 \triangleright G_2 \triangleright \cdots \triangleright G_{t+1} = \{e\}$ such that $[G, G_i] \subseteq G_{i+1}$, for $i = 1, 2, \ldots, t$;*
(3) *G has a normal series $G = G'_1 \triangleright G'_2 \triangleright \cdots \triangleright G'_{s+1} = \{e\}$ such that $G'_i/G'_{i+1} \subseteq C(G/G'_{i+1})$, for $i = 1, 2, \ldots, s$;*
(4) *$C_r(G) = G$ for some $r \in \mathbb{N}$.*

Proof. (1) $\Rightarrow$ (2) Since G is a nilpotent group, the lower central series of G is a normal series satisfying the condition in (2).

(2) $\Rightarrow$ (3) Let $\pi : G \to G/G_{i+1}$ be the canonical homomorphism. Then by $[G, G_i] \subseteq G_{i+1}$ we have $\pi[G, G_i] = \{\pi(e)\}$. Hence, $[\pi(G), \pi(G_i)] = \{\pi(e)\}$, which implies $[G/G_{i+1}, G_i/G_{i+1}] = \{\pi(e)\}$. Therefore, $G_i/G_{i+1} \subseteq C(G/G_{i+1})$. Thus, we just let $G'_i = G_i$.

(3) $\Rightarrow$ (4) We prove it by induction. It is easy to see that $G'_{s+1} = \{e\} = C_0(G)$ and $G'_s \subseteq C_1(G)$. Assume that $G'_{s+1-k} \subseteq C_k(G)$. Then we need to prove $G'_{s-k} \subseteq C_{k+1}(G)$. Consider the following commutative diagram of canonical homomorphisms

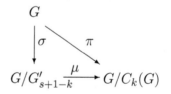

Since μ is surjective, we have

$$C(G/G'_{s+1-k}) \subseteq \mu^{-1}(C(G/C_k(G))).$$

By the fact $G'_{s-k}/G'_{s+1-k} \subseteq C(G/G'_{s+1-k})$, one has

$$G'_{s-k} \subseteq \sigma^{-1}(C(G/G'_{s+1-k})) \subseteq \sigma^{-1}\mu^{-1}(C(G/C_k(G)))$$
$$\subseteq \pi^{-1}(C(G/C_k(G))) = C_{k+1}(G).$$

Therefore, $C_s(G) \supseteq G'_1 = G$, i.e., $C_s(G) = G$.

$(4) \Rightarrow (1)$ Since $G = C_r(G)$, $G/C_{r-1}(G)$ is abelian, hence nilpotent. Assume that $G/C_{k+1}(G)$ is nilpotent. Then by the definition of $C_k(G)$ we have $C_{k+1}(G)/C_k(G) \subseteq C(G/C_k(G))$. By the short exact sequence

$$1 \to C_{k+1}/C_k(G) \to G/C_k(G) \to G/C_{k+1}(G) \to 1,$$

one gets that $G/C_k(G)$ is nilpotent since it is a central extension of the nilpotent group $G/C_{k+1}(G)$. Therefore, $G = G/C_0(G)$ is nilpotent by induction. □

Exercises

(1) Let H, K be subgroups of G.

    (i) Show that $[H, K] = \{e\}$ if and only if $H \subseteq C_G(K)$, if and only if $K \subseteq C_G(H)$.

    (ii) Show that $[H, K] \subseteq K$ if and only if $H \subseteq N_G(K)$.

    (iii) Suppose $H_1 < H$, $K_1 < K$. Show that $[H_1, K_1] \subseteq [H, K]$.

(2) Let G be a solvable group. Show that G has a composition series if and only if G is finite.

(3) Show that if H, K are solvable normal subgroup of G, then so is HK.

(4) Let R be a maximal solvable normal subgroup of G, H a solvable normal subgroup of G. Show that $H \subseteq R$ and G/R have no non-trivial solvable normal subgroups.

(5) Let R be a maximal solvable subgroup of G. Show that $N_G(R) = R$.

(6) Let G be a finite nilpotent group, H a proper subgroup of G. Show that $N_G(H) \neq H$.

(7) Let G be a finite group. Show that G is nilpotent if and only if for any proper normal subgroup H of G, $C(G/H)$ is non-trivial.

(8) Let $I^{(1)}(G) = \text{Inn}\,(G)$, $I^{(n)}(G) = \text{Inn}\,(I^{(n-1)}(G))$. Show that G is nilpotent if and only if $I^{(n)}(G)$ is trivial for some $n \in \mathbb{N}$.

(9) Let G be a finite nilpotent group.

    (i) If $\{e\} \neq N \triangleleft G$, show that $N \cap C(G) \neq \{e\}$.

    (ii) If G is a non-abelian group, N is a maximal normal group of G and N is abelian, show that $C_G(N) = N$.

(10) Let $a, b \in G$. If a, b commute with the commutator $[a, b]$, show that $[a^m, b^n] = [a, b]^{mn}$ for any $m, n \in \mathbb{N}$.

(11) Let A be a cyclic normal subgroup of G. Show that $ax = xa$ for any $a \in A$, $x \in G^{(1)}$.

Exercises (hard)

(12) Show that the commutator subgroup of $\mathrm{SL}(2, \mathbb{R})$ is itself, while $-I_2$ is not a commutator of $\mathrm{SL}(2, \mathbb{R})$.

(13) Let B_n be the group of $n \times n$ complex invertible upper-triangular matrices. Show that B_n is solvable. Furthermore, let G be a finite solvable group. Is G isomorphic to a subgroup of B_n for some n?

(14) Let N_n be the group of $n \times n$ complex invertible upper-triangular matrices with the diagonal entries all 1. Show that N_n is nilpotent. Furthermore, let G be a finite nilpotent group. Is G isomorphic to a subgroup of N_n for some n?

1.9   Group actions

We have seen that subgroups, especially normal subgroups of a group G are very important for studying the structure of G. However, it is not easy to find subgroups and normal subgroups. A good idea is to find group homomorphisms from G to a concrete group. The most natural choice for this group is the symmetric group S_X on a non-empty set X. Given a group homomorphism $\varphi : G \to S_X$. If φ is a monomorphism, then G is isomorphic to a subgroup of S_X; if φ is not a monomorphism, then $\operatorname{Ker} \varphi$ is a normal subgroup of G.

To establish a group homomorphism, we need to define a map $\pi : G \to S_X$, and then verify that π is a group homomorphism, that is, for any $g_1, g_2 \in G$, $\pi(g_1 g_2) = \pi(g_1)\pi(g_2)$. To verify this equation, we only need to apply the transformations on both sides of the equation to any element x in X so that

$$\pi(g_1 g_2)(x) = (\pi(g_1)\pi(g_2))(x) = \pi(g_1)(\pi(g_2)(x)). \qquad (1.9)$$

Now, we introduce the following definition.

Definition 1.9.1. An (left) **action** of a group G on a non-empty set X is a map

$$f : G \times X \to X, \quad (g, x) \to f(g, x)$$

such that for any $x \in X$, $g_1, g_2 \in G$, we have

$$f(e, x) = x, \tag{1.10}$$
$$f(g_1 g_2, x) = f(g_1, f(g_2, x)). \tag{1.11}$$

We also say that G acts on X. Usually, we simply write $g(x)$ or gx for $f(g, x)$.

Given an action f of G on X, we define a transformation $\pi(g)$, for any $g \in G$, on X by $\pi(g)(x) = f(g, x)$. By (1.11), we have

$$\pi(g_1 g_2) = \pi(g_1) \pi(g_2).$$

By (1.10), we have

$$\pi(g) \pi(g^{-1}) = \pi(g g^{-1}) = \pi(e) = \text{id}.$$

Thus, $\pi(g)$ is invertible and its inverse is $\pi(g^{-1})$. Hence, π is a group homomorphism, called the **action homomorphism** associated with the action of G on X. Conversely, Given a group homomorphism $\pi : G \to S_X$, one can define $f(g, x) = \pi(g)(x)$. Then it is easy to check that f is an action of G on X. Hence, we have the following.

Theorem 1.9.2. *There is a 1-1 correspondence between all actions of G on X and all homomorphisms from G to S_X.*

Therefore, group actions are equivalent forms of special group homomorphisms. Why should we focus on such cases? Group homomorphism is an algebraic term, while group action is more geometrical. One of the famous mathematical problems is the Erlangen program, which was put forward in 1872 by F. Klein, aiming to characterize geometry based on group actions. For example, Euclidean geometry focuses on invariants under the action of the isometry group; affine geometry considers invariants under the action of the affine groups. At the beginning of the 20th century, the French mathematician E. Cartan used Lie groups to study differential geometry. From these, we can see that the geometric meaning of group actions is powerful. Let us look at some examples first.

Example 1.9.3. Let G be a group and $X = G$. By Theorem 1.4.10 and Proposition 1.4.7, we get three different group homomorphisms from G to S_G, the corresponding actions $f : G \times G \to G$ being

(1) left translation action: $f(g, x) = L_g(x) = gx$;
(2) right translation action: $f(g, x) = R_{g^{-1}}(x) = xg^{-1}$;
(3) adjoint action: $f(g, x) = \mathrm{Ad}_g(x) = gxg^{-1}$.

The left translation action can be generalized to that on the left coset spaces.

Example 1.9.4. Let H be a subgroup of G. Define a map $f : G \times G/H \to G/H$ by

$$f(g, xH) = (gx)H, \quad \forall g, x \in G.$$

Then f is an action of G on G/H.

There are lots of examples of group actions in matrix theory.

Example 1.9.5. As transformation groups on $\mathbb{R}^n$, $\mathrm{GL}(n, \mathbb{R})$ and its subgroups $\mathrm{SL}(n, \mathbb{R})$, $\mathrm{O}(n)$, etc. act naturally on $\mathbb{R}^n$. Furthermore, let S^{n-1} be the set of unit vectors in $\mathbb{R}^n$. Then $\mathrm{O}(n)$ and its subgroup $\mathrm{SO}(n)$ act on S^{n-1} by matrix multiplication.

Example 1.9.6. In matrix theory, row/column transformations, equivalence, congruence, and similarity of matrices can all be viewed as group actions. Let $\mathbb{F}$ be a number field. For any $g \in \mathrm{GL}(n, \mathbb{F})$, $A \in \mathbb{F}^{n \times n}$, one may define various group actions as follows:

(1) Action by left multiplication: $f(g, A) = gA$;
(2) Action by right multiplication: $f(g, A) = Ag^{-1}$;
(3) Similarity: $f(g, A) = gAg^{-1}$;
(4) Congruence: $f(g, A) = gAg'$, where g' is the transpose of g.

Note that if A is symmetric (or skew-symmetric), so is gAg'. Let $\mathbb{S}$ (and $\mathbb{A}$, resp.) be the set of symmetric (and skew-symmetric, resp.) $n \times n$ matrices. The map $f(g, A) = gAg'$ is also an action of $\mathrm{GL}(n, \mathbb{F})$ on $\mathbb{S}$ (and $\mathbb{A}$, resp.).

The following concepts are frequently used when considering group actions.

Definition 1.9.7. Let a group G act on a set X. The action is called **transitive**, if for any $x, y \in X$, there exists $g \in G$ such that $gx = y$. In this case, X is called a **homogeneous space** of G. The action is called **effective** if $gx = x$ for all $x \in X$ implies that $g = e$. The action is called **trivial** if $gx = x$ for any $g \in G$, $x \in X$.

We left it to the reader to check whether the examples mentioned above are transitive or effective. Let $\pi : G \to S_X$ be the group homomorphism associated with a group action $f(g, x)$. Then the action is effective if and only if π is a monomorphism, and the action is trivial if and only if π is trivial, i.e., $\operatorname{Ker} \pi = G$.

Recall that invariant subspaces is crucial for the study of linear transformations. To explore group actions further, we can consider similar concepts: subsets of X invariant under the action of G. Therefore, we introduce the following definition.

Definition 1.9.8. Let a group G act on a set X, $x \in X$. The subset $O_x = \{gx | g \in G\}$ of X is called the **orbit** of x under G.

It is quick to check that $gO_x = O_x$, for any $g \in G$. Hence, the action can be restricted to O_x. If $|O_x| = 1$, then x is called a **fixed point** of G.

Proposition 1.9.9. (1) *For any* $x, y \in X$, *either* $O_x \cap O_y = \varnothing$ *or* $O_x = O_y$.

(2) *The action of G on O_x is transitive, and the action of G on X is transitive if and only if X has only one orbit.*

Proof. (1) If $z \in O_x$, then $O_z \subseteq O_x$, and there exists $g \in G$ such that $z = gx$. Hence, $x = g^{-1}z \in O_z$, which implies that $O_x \subseteq O_z$, and consequently, $O_x = O_z$. If $z \in O_x \cap O_y$, then $O_x = O_z = O_y$.

(2) For any $y, z \in O_x$, there exist $g, h \in G$ such that $y = gx$, $z = hx$. Hence, $z = hx = hg^{-1}y$, which implies that the action of G on O_x is transitive. The other assertion is trivial. $\quad\square$

Therefore, once we have an action of a group G on X, X is the disjoint union of orbits, or we may say that the set of orbits forms a partition of X. To study the action of G on X, it is necessary to consider that of G on each orbit.

For any $x \in X$, we have the following map

$$\varphi_x : G \to O_x, \quad \varphi_x(g) = gx, \tag{1.12}$$

which must be surjective. Denote by F_x the preimage of x, i.e.,

$$F_x = \{g \in G | gx = x\}.$$

It is quick to check that F_x is a subgroup of G, called the **isotropy subgroup** of x. For $y = gx \in O_x$,

$$F_y = \{h \in G | hy = y\} = \{h \in G | hgx = gx\} = \{h \in G | g^{-1}hgx = x\}.$$

Thus, we have the following result.

Lemma 1.9.10. *Suppose $y = gx \in O_x$. Then $F_y = gF_xg^{-1}$.*

Now, we look at some examples.

Example 1.9.11.

(1) Identify the surface of the Earth with the unit sphere $S^2 = \{x \in \mathbb{R}^3 || x | = 1\}$ in $\mathbb{R}^3$. Rotations of the Earth around the line through the north and south poles define an action of $\mathrm{SO}(2)$ on S^2, whose orbits are the circles of latitude. The isotropy subgroups of the north and south poles are $\mathrm{SO}(2)$, while the isotropy subgroups of other points are $\{e\}$.

(2) Considering the natural action of $\mathrm{GL}(n, \mathbb{R})$ on $\mathbb{R}^n$, we can get two orbits $\{0\}$ and $\mathbb{R}^n \setminus \{0\}$. The isotropy subgroup of 0 is $\mathrm{GL}(n, \mathbb{R})$, and the isotropy subgroup of $(1, 0, \ldots, 0)'$ is

$$\left\{ \begin{pmatrix} 1 & \alpha \\ 0 & A \end{pmatrix} \in \mathrm{GL}(n, \mathbb{R}) \middle| A \in \mathrm{GL}(n-1, \mathbb{R}), \alpha \in \mathbb{R}^{n-1} \right\}.$$

(3) For the natural action of $\mathrm{SO}(n)$ on $\mathbb{R}^n$, the orbits are $S_{n-1}(r) = \{x \in \mathbb{R}^n || x | = r\}$, $r \geq 0$. If $r > 0$, the isotropy subgroup of $(r, 0, \ldots, 0)'$ is $\{\mathrm{diag}\,(1, A) | A \in \mathrm{SO}(n-1)\}$.

(4) Regard S_{n-1} as a subgroup of S_n. The action of S_n on $\{1, 2, \ldots, n\}$ is transitive, the isotropy subgroup of n being S_{n-1}.

(5) For the natural action of $\mathrm{GL}(n, \mathbb{C})$ on $\mathbb{C}^{n \times n}$ by similarity of matrices, the orbit of J is $O_J = \{TJT^{-1} | T \in \mathrm{GL}(n, \mathbb{C})\}$, where J is one of the Jordan canonical form, and $F_J = \{A \in \mathrm{GL}(n, \mathbb{C}) | AJ = JA\}$.

(6) The congruence action of $\mathrm{GL}(n, \mathbb{R})$ on the set of real symmetric matrices is defined by $T \circ A = TAT'$, for $T \in \mathrm{GL}(n, \mathbb{R})$, A any symmetric matrix. Two matrices lie in the same orbit if and only if they have the same positive and negative indices of inertia (p, q), where $p, q \geq 0$, $p + q \leq n$. The representative element in the orbit $O_{p,q}$ is diag $(I_p, -I_q, 0)$. The isotropy subgroup of diag $(I_p, -I_{n-p})$ is denoted by $\mathrm{O}(p, n-p)$. In particular, the orbit $O_{n,0}$ is the set of positive definite matrices, and the isotropy subgroup of I_n is the orthogonal group $\mathrm{O}(n)$.

Now consider the map φ_x (see (1.12)), we have

$$\varphi_x^{-1}(gx) = \{h \in G | hx = gx\} = \{h \in G | g^{-1}hx = x\}.$$

Hence, $\varphi_x^{-1}(gx) = gF_x$, i.e., the preimage of gx is the left coset gF_x of F_x. Therefore, we get a bijection $\varphi : G/F_x \to O_x$. Each of G/F_x and O_x has a natural action of G, and

$$\varphi(s(gF_x)) = \varphi((sg)F_x) = (sg)x = s(gx) = s\varphi(gF_x), \quad \forall s, g \in G.$$

Thus, we have the following commutative diagram

$$
\begin{array}{ccc}
G/F_x & \xrightarrow{\varphi} & O_x \\
\downarrow{\scriptstyle s} & & \downarrow{\scriptstyle s} \\
G/F_x & \xrightarrow{\varphi} & O_x
\end{array}
$$

Now, we introduce the following definition.

Definition 1.9.12. The actions of G on two sets X, X' are called **equivalent** if there exists a bijection $\varphi : X \to X'$ such chat

$$s(\varphi(x)) = \varphi(s(x)), \quad \forall s \in G, x \in X.$$

This definition is natural, just like linear maps, which are not only a map between vector spaces but also preserve the additions and scalar multiplications. For the current situation, group actions play the same roles as the scalar multiplications of vector spaces. Therefore, the following result follows from the previous discussion on orbits.

Theorem 1.9.13. *Let G act transitively on X. Then the action of G on X is equivalent to that of G on G/F_x, for any $x \in X$.*

Corollary 1.9.14. $|O_x| = |G/F_x| = [G : F_x]$. *Hence,* $|O_x|\,\big|\,|G|$.

By Proposition 1.9.9 and Theorem 1.9.13, we know that if G acts on X, then X is a disjoint union of orbits, and the action of G on each orbit is equivalent to the left translation action of G on some left coset space. Therefore, the action of G on X may be regarded as that on the disjoint union of left coset spaces. Therefore, the study of homogeneous spaces is crucial. In differential geometry and Lie theory, homogeneous spaces are a class of important objects; see Exercises 4(i) and 5 in this section.

Example 1.9.15. Let G be a group. Consider the adjoint action of G on itself, the corresponding homomorphism being

$$\mathrm{Ad} : G \to S_G, \quad \mathrm{Ad}_x(g) = xgx^{-1}.$$

For any $g \in G$, the orbit of g under the adjoint action is the conjugacy class C_g of g. The isotropy subgroup at g is called the **centralizers** of g, denoted by $C_G(g)$ or $C(g)$. We call $\mathrm{Ker}\,\mathrm{Ad}$ the **center** of G, denoted by $C(G)$.

Problem 1.9.16. Determine all the conjugacy classes of S_n.

Example 1.9.17. Let $\sigma \in S_n$. Consider the action of $\langle \sigma \rangle$ on $\{1, 2, \ldots, n\}$, which is the restriction of the action of S_n. Then $\{1, 2, \ldots, n\}$ is a disjoint union of different orbits; each orbit is of the form

$$O_m = \{\sigma^k(m) | k \in \mathbb{N}\}.$$

Since O_m has only finite many elements, there exists a smallest positive integer r such that the restriction $\sigma^r|_{O_m}$ is the identity. Hence, σ can be decomposed as a product of disjoint cycles, each corresponding to an orbit of $\langle \sigma \rangle$.

Exercises

(1) Determine the number of 9-digit integers consisting of 1, 1, 1, 2, 2, 2, 2, 3, 3.

(2) Show that A_5 has no subgroups of indices 2, 3 or 4.

(3) Give a new proof of Exercise 15(i) of Section 1.2 via group actions.

(4) (i) **(Real projective space).** Denote by $P^{n-1}(\mathbb{R})$ the set of all 1-dimensional subspaces of $\mathbb{R}^n$. For any $\varphi \in \mathrm{GL}(n, \mathbb{R})$, $W \in P^{n-1}(\mathbb{R})$, $\varphi(W)$ is also a 1-dimensional subspace. Prove that $(\varphi, W) \mapsto \varphi(W)$ is a transitive action of $\mathrm{GL}(n, \mathbb{R})$ on $P^{n-1}(\mathbb{R})$, and determine the isotropy subgroups.

(ii) **(Grassmannian manifold).** Denote by $\mathrm{Gr}(k, n)$ the set of k-dimensional subspaces of $\mathbb{R}^n$. For any $\varphi \in \mathrm{GL}(n, \mathbb{R})$, $W \in \mathrm{Gr}(k, n)$, $\varphi(W)$ is also a k-dimensional subspace. Prove that this defines a transitive action of $\mathrm{GL}(n, \mathbb{R})$ on $\mathrm{Gr}(k, n)$, and determine the isotropy subgroups.

(iii) Consider the actions of $\mathrm{SL}(n, \mathbb{R})$, $\mathrm{O}(n)$, $\mathrm{SO}(n)$ on the above sets and check whether the actions are transitive.

(5) **(Stiefel manifold).** Denote by $V_k(\mathbb{R}^n)$ the set of all k-tuple orthogonal vectors in $\mathbb{R}^n$. For any $\varphi \in \mathrm{O}(n)$, $(\alpha_1, \alpha_2, \ldots, \alpha_k) \in V_k(\mathbb{R}^n)$, define

$$\varphi(\alpha_1, \alpha_2, \ldots, \alpha_k) = (\varphi(\alpha_1), \varphi(\alpha_2), \ldots, \varphi(\alpha_k)).$$

Prove that this is a transitive action of $\mathrm{O}(n)$ on $V_k(\mathbb{R}^n)$, and determine the isotropy subgroups.

(6) Generalize the above two exercises to the complex case.

(7) Suppose $\mathbb{E}$ is a number field, and $\mathbb{E}$ is finite-dimensional when regarded as a vector space over $\mathbb{Q}$. Set

$$\mathrm{Aut}\,(\mathbb{E}) = \{\varphi \in \mathrm{GL}(\mathbb{E}) | \varphi(xy) = \varphi(x)\varphi(y), \forall x, y \in \mathbb{E}\}.$$

Show that $\mathrm{Aut}\,(\mathbb{E})$ is a group. Let X be the set of all different complex roots of some polynomial $f(x) \in \mathbb{Q}[x]$. If $X \subseteq \mathbb{E}$, show that X is $\mathrm{Aut}\,(\mathbb{E})$-invariant. Then $\mathrm{Aut}\,(\mathbb{E})$ acts on X naturally. Show that this action is transitive if and only if $f(x)$ is irreducible over $\mathbb{Q}$.

(8) **(Poincaré upper-half plane).** Let $\mathbb{H} = \{z = x + y\sqrt{-1} \in \mathbb{C} | y > 0\}$. For any $g = \begin{pmatrix} a & b \\ c & d \end{pmatrix} \in \mathrm{SL}(2, \mathbb{R})$, $z \in \mathbb{H}$, define $g \circ z = \frac{az+b}{cz+d}$, called a **Möbius transformation** or **linear fractional transformation**. Show that this is an action of $\mathrm{SL}(2, \mathbb{R})$ on $\mathbb{H}$. Determine whether this action is transitive or effective, and determine the isotropy subgroup at $\sqrt{-1}$.

(9) Let G be a simple group, and let $H < G$ with $[G : H] \leq 4$. Show that $G \cong \mathbb{Z}_2$ or $\mathbb{Z}_3$.

(10) Let H be a subgroup of G with index n. Show that H contains a normal subgroup N of G such that $[G : N]|n!$

(11) Let G be a finite group, p the smallest prime divisor of $|G|$. Suppose H is a subgroup of G with $[G : H] = p$. Show that $H \lhd G$.

(12) Let $H < G$. Show that the number of conjugate subgroups of H is $[G : N_G(H)]$.

(13) Let H be a subgroup of G with finite index, and let $\pi : G \to S_{G/H}$ be the action homomorphism associated with the action of G on G/H. Show that the kernel of π is the intersection of all conjugate subgroups of H.

(14) Let B be the set of all invertible upper-triangular complex matrices, $G = \{\text{diag}\,(g, h)|g, h \in B\}$, $X = \text{GL}(n, \mathbb{C})$. Define
$$f : G \times X \to X, \quad (\text{diag}\,(g, h), x) \mapsto gxh^{-1}.$$
Show that f is a group action. Determine all the orbits of this action, and consequently, decompose $\text{GL}(n, \mathbb{C})$ as the disjoint union of different orbits (the double cosets).

(15) Explain the equivalence of matrices from the viewpoint of group actions and determine all the orbits.

(16) Let G be a group, X the set of all complex-valued functions $f : G \to \mathbb{C}$. Define
$$G \times X \to X, \quad (g, f) \mapsto g \cdot f,$$
by $(g \cdot f)(x) = f(gx)$. Is this map a group action?

(17) Show that the left and right translation actions of a group G on itself are equivalent.

Exercises (hard)

(18) **(Fermat's theorem on sums of two squares).** Let $p = 4k + 1$ be a prime, $X = \{(x, y, z) \in \mathbb{N}^3 | x^2 + 4yz = p\}$. Define two maps on X by
$$\varphi_1 : (x, y, z) \mapsto (x, z, y),$$

$$\varphi_2 : (x, y, z) \mapsto \begin{cases} (x + 2z, z, y - x - z), & x < y - z, \\ (2y - x, y, x - y + z), & y - z < x < 2y, \\ (x - 2y, x - y + z, y), & x > 2y. \end{cases}$$

Show that these two maps define two actions of $\mathbb{Z}_2$ on X. Determine the fixed points of the two actions, and show that $x^2 + y^2 = p$ has solutions of positive integers.

(19) **(Burnside's lemma).** Let a group G act on a set X, t the number of G-orbits in X. For any $g \in G$, Denote by $F(g)$ the number of fixed points of g in X, i.e., $F(g) = |\{x \in X | gx = x\}|$. Show that $\sum_{g \in G} F(g) = t|G|$.

(20) Determine all finite subgroups of SO(3), and compare with the symmetric groups of regular n-gons and regular polyhedron (see Exercise 19 of Section 1.1).

1.10   Sylow theorems

From the previous sections, we know that normal subgroups play an essential role in studying the structure of groups. Therefore, finding normal subgroups of a group becomes a significant problem. Generally speaking, it is not easy to find normal subgroups, so we settle for the more accessible work to find subgroups first. By Lagrange's theorem, if $H < G$, then $|H|$ is a divisor of $|G|$. Conversely, for a divisor k of $|G|$, there is not always a subgroup H of order k. However, the following result opens the door to finding subgroups for a given finite group.

Theorem 1.10.1 (Cauchy's theorem). *If p is a prime divisor of the order of G, then G has an element of order p.*

Proof.   Set $X = \{(a_1, a_2, \ldots, a_p) | a_i \in G, a_1 a_2 \cdots a_p = e\}$. Let $\sigma = (12 \cdots p) \in S_p$. Consider the action of the cyclic group $\langle \sigma \rangle$ on X defined by

$$\sigma \cdot (a_1, a_2, \ldots, a_p) = (a_{\sigma(1)}, a_{\sigma(2)}, \ldots, a_{\sigma(p)}).$$

Each orbit of the action has 1 or p elements. Denote by X_0 the set of fixed points, the union of orbits of 1 element. Since $p||X|$, we have $p||X_0|$. It is easy to see that elements in X_0 are of the form $(a, a, \ldots, a)$, for some $a \in G$ with $a^p = e$, and X_0 is not empty for $(e, e, \ldots, e) \in X_0$. Thus, X_0 contains some element $(a, a, \ldots, a)$ with $a \neq e$. Hence, a is an element of order p. □

By Cauchy's theorem, G contains a subgroup of order p if $p||G|$. More generally, if $|G| = p^l m$, $(p, m) = 1$, are there subgroups of G of order p^k, for $1 \leq k \leq l$? To answer this question, we need some preparation.

Definition 1.10.2. Let p be a prime. A finite group G is a p-**group** if $|G| = p^k$, for some positive integer k.

The order of a p-group G is so special that it must have some remarkable properties.

Proposition 1.10.3. *Let a p-group G act on a set X with $|X| = n$. Denote by X_0 the set of fixed points of G in X, $t = |X_0|$. Then $t \equiv n \pmod{p}$. In particular, if $(n, p) = 1$, then $t \geq 1$, i.e., G has fixed points in X.*

Proof. For the G action on X we have the following decomposition:

$$X = X_0 \cup O_1 \cup \cdots \cup O_k,$$

where O_i, $i = 1, \ldots, k$, are disjoint orbits with $|O_i| > 1$. Thus, $p||O_i|$, which implies that $p|(|X| - |X_0|)$, i.e., $t \equiv n \pmod{p}$. If $(n, p) = 1$, then $(t, p) = 1$. □

Corollary 1.10.4. *The center of a p-group G is also a p-group. Consequently, G is nilpotent.*

Proof. Consider the adjoint action of G on itself. Then the set of fixed points is exactly the center $C(G)$ of G. By Proposition 1.10.3, we have $p||C(G)|$ and $|C(G)| = p^l$ for some $l > 0$. Thus, $C(G)$ is a p-group. By induction, G is the central extension of a nilpotent group, hence nilpotent. □

The 19th century saw the significant contribution to group theory of three outstanding Norwegian mathematicians: Abel, Sylow, and Lie. In 1972, Sylow proved several powerful theorems about the structure of groups; more precisely, he revealed the existence of subgroups whose order are some powers of p with p a prime divisor of $|G|$. Therefore, the number-theoretic information of the order of G has deep connection with the group structure.

Theorem 1.10.5 (Sylow I). *Let $|G| = p^l m$, where p is a prime, $l \geq 1$, and $(p, m) = 1$. Then there exists a subgroup of G of order p^k, for any positive integer $k \leq l$.*

Proof. By Cauchy's theorem, G has a subgroup of order p. Assume that $l > 1$ and G has a subgroup H of order p^n, $0 < n < l$. Consider the left translation action of H on $X = G/H$, and Denote by X_0 the set of fixed points. If $gH \in X_0$, then $h(gH) = gH$, for any $h \in H$. Hence, $g^{-1}hg \in H$, i.e., $g \in N_G(H)$, where $N_G(H)$ is the normalizer of H in G. Conversely, if $g \in N_G(H)$, then $gH \in X_0$. Hence, $X_0 = N_G(H)/H$. Since $|G/H| = p^{l-n}m$ and $n < l$, we have $p \mid |G/H|$. Then by Proposition 1.10.3 we have $p \mid |X_0|$. Thus, $p \mid |N_G(H)/H|$. Noticing $H \triangleleft N_G(H)$, one has that $N_G(H)/H$ is a group. By Cauchy's theorem, $N_G(H)/H$ contains a subgroup of order p, which implies that $N_G(H)$ has a subgroup containing H of order p^{n+1} by the Fundamental Theorem of Homomorphisms. The assertion follows. $\square$

Definition 1.10.6. Let $|G| = p^l m$, p a prime, and $(p, m) = 1$. A subgroup of order p^l is called a **Sylow p-subgroup** of G.

The first Sylow's theorem shows the existence of Sylow p-subgroups. Is such a subgroup unique? If not, how many Sylow p-subgroup are there? Do different Sylow p-subgroups have any relation? We will answer these questions in the following.

Theorem 1.10.7 (Sylow II). *Let $|G| = p^l m$ with $(p, m) = 1$. Suppose P is a Sylow p-subgroup of G, and H is a subgroup of G of order p^k for some $k \leq l$. Then there exists $g \in G$ such that $H \subseteq gPg^{-1}$. In particular, Sylow p-subgroups of G are conjugate in G.*

Proof. Consider the left translation action of H on G/P. Since $p \nmid |G/P|$, by Proposition 1.10.3, one gets that H has fixed points. Let gP be a fixed point of H. Then $h(gP) = gP$, for any $h \in H$. Thus, $h \in gPg^{-1}$. It follows that $H \subseteq gPg^{-1}$. Furthermore, if H is a Sylow p-subgroup, then $|H| = |P|$, and consequently, $H = gPg^{-1}$. $\square$

Let X_p be the set of Sylow p-subgroups of G, $n_p = |X_p|$. Then, by the above theorem, there is a natural transitive G-action on X_p

by conjugation. For any $P \in X_p$, the isotropy subgroup of P is $F_P = \{g \in G | gPg^{-1} = P\} = N_G(P)$. Therefore, we have the following.

Lemma 1.10.8. *Let P be a Sylow p-subgroup of G. Then $n_p = [G : N_G(P)]$.*

The following result reveals more information on the number n_p.

Theorem 1.10.9 (Sylow III). *Let n_p be the number of Sylow p-subgroups of G. Then we have the following.*

(1) $n_p = 1$ *if and only if the Sylow p-subgroup of G is normal;*
(2) $n_p | m$, *and $n_p \equiv 1 (\mathrm{mod}\, p)$.*

Proof.   (1) A Sylow p-subgroup P of G is normal if and only if $G = N_G(P)$ if and only if $n_p = [G : N_G(P)] = 1$, by Lemma 1.10.8.

(2) By Lemma 1.10.8 again, we have $n_p = |G|/|N_G(P)|$. Thus, $|N_G(P)|/|P| \cdot n_p = |G|/|P| = m$. Since $P \lhd N_G(P)$, $|N_G(P)|/|P|$ is a positive integer, Hence, $n_p | m$. Now consider the action of P on X_p by conjugation. Then $P \in X_p$ is a fixed point of this action. Assume that $P_1 \in X_p$ is another fixed point. Then for any $p \in P$, $pP_1p^{-1} = P_1$, which implies that $P < N_G(P_1)$. Therefore, P, P_1 are Sylow p-subgroups of $N_G(P_1)$ with P_1 normal. By (1), we have that $N_G(P_1)$ has a unique Sylow p-subgroup, hence $P = P_1$, which implies that there is only one fixed point of the action of P on X_p. By Proposition 1.10.3, we have $n_p \equiv 1(\mathrm{mod}\, p)$.   □

Both Sylow's Theorems and their proofs are valuable. On the one hand, the proofs show us how to use various group actions to solve problems; on the other hand, Sylow's Theorems pave the way for us to find subgroups of a group and even help us find the normal subgroups in some cases. Let's look at some examples.

Example 1.10.10. If G is a group of order 12, then at least one of its Sylow subgroups is normal.

Proof.   By the third Sylow's theorem, $n_3 = 1$ or 4. If $n_3 = 1$, then the Sylow 3-subgroup is normal. If $n_3 = 4$, then G has eight elements of order 3 since the intersection of any two Sylow 3-subgroups contains only the identity. Therefore, the remaining four elements, except for the eight elements of order 3, form a Sylow 2-subgroup, which implies that $n_2 = 1$, and consequently, the Sylow 2-subgroup is normal.   □

Example 1.10.11. Show that any group of order 224 is not simple.

Proof. Since $224 = 2^5 \cdot 7$, we have $n_2 = 1$ or 7. If $n_2 = 1$, then the Sylow 2-subgroup is normal. If $n_2 = 7$, say, $X = \{P_1, P_2, \ldots, P_7\}$ is the set of all Sylow 2-subgroups of G, then G acts on X transitively. Thus, we get a group homomorphism $\pi : G \to S_7$ with $\mathrm{Ker}\,\pi \neq G$. Furthermore, $|S_7| = 2^4 \cdot 3^2 \cdot 5 \cdot 7$ is not a multiple of 224, we have $\mathrm{Ker}\,\pi \neq \{e\}$. Therefore, $\mathrm{Ker}\,\pi$ is a non-trivial normal subgroup of G. Hence, G is not simple. □

Problem 1.10.12. Use the same method to prove that for any prime p and positive integer m, $(p, m) = 1$, there exists $k \in \mathbb{N}$ such that any group of order $p^l m$ is not simple for any $l > k$.

Exercises

(1) Let G be a group of order $p^l m$, where p is a prime with $(p, m) = 1$. Set $X = \{A \subseteq G \mid |A| = p^k\}$. Then the left translation action of G on itself induces an action of G on X. For $A \in X$, F_A denotes the isotropy subgroup of A.

    (i) Show that A is a union of some right cosets of F_A, and consequently, F_A is a p-group and $|F_A| \leq p^k$.
    (ii) Show that there exists $A \in X$ such that $|F_A| = p^k$.

(2) Show that every group of order p^2 is abelian and classify such groups up to isomorphism.

(3) Determine the numbers of all Sylow subgroups of S_4.

(4) Let p be a prime. Determine the number of Sylow p-subgroup of S_p. Prove **Wilson's theorem:** $(p - 1)! \equiv 1 \pmod{p}$.

(5) Show that the number of non-normal subgroups of a p-group is a multiple of p.

(6) Let G be a p-group, N a normal subgroup of G of order p. Show that $N \subseteq C(G)$.

(7) Let $|G| = p^l m$ with p a prime, $l \geq 1$, $(p, m) = 1$. Assume that P is a subgroup of order p^k $(k < l)$. Show that P is a proper subgroup of its normalizer $N_G(P)$.

(8) Show that groups of order 56 or 72 are not simple.

(9) Assume that $|G| = p^l m$, where p is a prime, $l \geq 1$, $p > m > 1$. Show that G is not simple.

(10) Let $|G| = p^2 q$, where p, q are different primes. Show that G is solvable.

(11) Let p be the smallest prime divisor of the order of a finite group G, $H < G$ and $[G : H] = p$. Show that $H \lhd G$.

(12) (Frattini) Let $N \lhd G$, and let P be a Sylow subgroup of N. Show that $G = N_G(P)N$.

(13) Let P be a Sylow subgroup of G. Show that $N_G(N_G(P)) = N_G(P)$.

(14) Let H be a proper subgroup of a nilpotent group G. Show that the normalizer $N_G(H) \supsetneq H$.

(15) Let G be a finite nilpotent group. Show that every Sylow p-subgroup is normal in G. Consequently, a finite group is nilpotent if and only if it is a direct product of p-groups.

(16) Let P be a Sylow p-subgroup of G, and let H be a subgroup of G with $H \supseteq N_G(P)$. Show that $N_G(H) = H$.

(17) Let p, q be primes, $p < q$, $p \nmid (q - 1)$. Show that groups of order pq are cyclic.

(18) Let H be a normal subgroup of a finite group G, and let p be a prime divisor of $|G|$ with $p \nmid [G : H]$. Show that H contains all Sylow p-subgroups of G.

(19) Let G be a group of order $p^l m$ with p a prime, $(p, m) = 1$ and $m < 2p$. Show that G has a normal Sylow p-subgroup or a normal subgroup of order p^{l-1}.

Exercises (hard)

(20) Prove that every group of order less than 60 is solvable.

(21) Prove that a simple group of order 60 is isomorphic to A_5.

(22) **(Burnside's theorem).** Show that if $|G| = p^a q^b$ where p, q are primes, and $a, b \in \mathbb{N}$, then G is solvable.

1.11   Summary of the chapter

This chapter focuses on the basic theory of groups. We need systematic tools to study group structures as a brand-new mathematical object. By using a non-trivial normal subgroup N of a group G, we can get the quotient group G/N; in other words, G is the extension of G/N by N. Similarly, if N and G/N have non-trivial normal subgroups, they are extensions of smaller groups. For finite groups, the above process will terminate after finite steps with a series of groups without non-trivial normal subgroups, that is, finite simple groups. Just as molecules are made up of atoms, finite groups are obtained by a series of extensions of finite simple groups.

The Jordan-Hölder theorem tells us that the composition factor of a finite group is uniquely determined. Therefore, the classification of finite groups comes down to the extensions of groups and the classification of finite simple groups. To determine whether a group is simple or not, we need to find normal subgroups. Therefore, group homomorphisms, especially group actions, play a crucial role in finding normal subgroups. Based on this, Sylow obtained three crucial theorems. The further development of group theory is the representation theory, that is, group homomorphisms from groups to general linear groups.

Chapter 2

Rings

In this chapter, we will introduce the notion of a ring and study the fundamental properties. In view of the structure of algebraic systems, a group has only one binary operation, but a ring has two binary operations, and the two operations are related by distributive laws. Mathematical objects we routinely encounter usually have two binary operations, e.g., sets of numbers, sets of matrices, and sets of polynomials, etc. Therefore, it is important to study rings. The terminology *ring* was first posed by Hilbert, although many mathematicians had established a lot of beautiful results related to rings before him. Following Hilbert, Krull and Noether made great contributions to the study of rings. Ring theory was further developed by the work of Artin, Jacobson, etc. Many terminologies related to rings, such as divisor chain condition, maximal condition, and minimal condition, were first defined and studied by them. Moreover, many special rings that are studied extensively currently in ring theory were named after them (e.g., Noetherian rings, Artin rings, etc.). Ring theory is the foundation of commutative algebra, algebraic number theory and algebraic geometry. Thus, it is vital to grasp the content of this chapter for everyone who are interested in learning further algebraic courses.

2.1   The definition of a ring and fundamental properties

As we mentioned above, a ring is an algebraic system which has two binary operations, and the two operations are related by distributive laws. Now, we give the precise definition.

Definition 2.1.1. Let R be a non-empty set. Suppose there are two binary operations on R, $+$ and $\cdot$, called addition and multiplication respectively, that satisfy the following conditions:

(1) R is an abelian group with respect to the addition;
(2) R is a semigroup with respect to the multiplication;
(3) The left and right distributive laws hold for the two operations:

$$a(b+c) = ab + ac, \ (a+b)c = ac + bc, \quad \forall a, b, c \in R.$$

Then we say that $\{R; +, \cdot\}$ is a **ring**. Sometimes we simply say that R is a ring if the operations are clear.

The abelian group $\{R, +\}$ is usually called the **additive group** of the ring R, and the semigroup $\{R, \cdot\}$ is usually called the **multiplicative semigroup** of R.

Since $\{R, +\}$ is an abelian group, there exists a unique element in R which is the identity element with respect to $+$. This element is usually denoted as 0, and called the **zero** of R. Moreover, for any $a \in R$, there exists (unique) $a' \in R$ such that $a + a' = 0$. This element will be denoted as $-a$, and called the **negative element** of a.

Notice that, in the above definition of a ring, it is required that the set R be an abelian group with respect to the addition, but it is only required to be a semigroup with respect to the multiplication. A ring is called **unitary**, or a **ring with identity**, if it has an identity element with respect to the multiplication. It is called **commutative** if the multiplication is commutative. In the next sections, we will define other types of special rings. Here we pose the following.

Problem 2.1.2. If a ring is a group with respect to the multiplication, what conclusions can you draw?

We now give some examples of rings. These examples show that rings occur in various branches of mathematics.

Example 2.1.3. First of all, on the set $R = \{0\}$, we define $0 + 0 = 0$ and $0 \cdot 0 = 0$. Then R becomes a ring, called the **null ring**.

Example 2.1.4. Many sets of numbers in mathematics are rings with respect to the ordinary addition and multiplication. For example, any number field is a ring. Moreover, some sets of numbers are rings even if they are not number fields, e.g., $\mathbb{Z}$ (the set of integers) is a ring, called the ring of integers.

Furthermore, given $m \in \mathbb{Z}$, define

$$\mathbb{Z}[\sqrt{m}] = \{a + b\sqrt{m} \,|\, a, b \in \mathbb{Z}\}.$$

Then it is easy to show that $\mathbb{Z}[\sqrt{m}]$ is a ring. In particular, when $m = -1$ we have

$$\mathbb{Z}[\sqrt{-1}] = \{a + b\sqrt{-1} \,|\, a, b \in \mathbb{Z}\}.$$

This is a famous example of a ring, called **the ring of Gauss integers**.

Example 2.1.5. Two important objects we encounter in linear algebra, namely, the set of matrices and the set of polynomials, are rings. More precisely, let $\mathbb{F}$ be a number field, and $\mathbb{F}[x]$ the set of all polynomials of x with coefficients in $\mathbb{F}$. Then $\mathbb{F}[x]$ is a ring with respect to the addition and multiplication of polynomials. It is called the **polynomial ring** in one indeterminate over the number field $\mathbb{F}$, or simply the polynomial ring over $\mathbb{F}$. Similarly, denote by $\mathbb{F}^{n \times n}$ the set of all $n \times n$ matrices over $\mathbb{F}$. Then $\mathbb{F}^{n \times n}$ is a ring with respect to the addition and multiplication of matrices, called the ring of $n \times n$ matrices over $\mathbb{F}$.

Example 2.1.6. Many sets of functions in mathematical analysis are rings. Denote by $C(\mathbb{R})$ the set of all continuous real-valued functions on $\mathbb{R}$, and define addition and multiplication as:

$$(f + g)(x) = f(x) + g(x),$$
$$(fg)(x) = f(x)g(x), \quad x \in \mathbb{R}, f, g \in C(\mathbb{R}).$$

Then it is easy to check that $C(\mathbb{R})$ is a ring. Similarly, denote by $C^\infty(\mathbb{R})$ the set of all smooth functions on $\mathbb{R}$. Then $C^\infty(\mathbb{R})$ is a ring with respect to the above defined addition and multiplication.

The above rings can be generalized in various ways. For example, given a closed interval $[a, b] \subset \mathbb{R}$, denote by $C([a, b])$ the set of all continuous functions on $[a, b]$. Then $C([a, b])$ is a ring with the above addition and multiplication. Similarly, denote by $C^\infty(\mathbb{R}^n)$ the set of smooth real-valued functions on the Euclidean space $\mathbb{R}^n$. Then $C^\infty(\mathbb{R}^n)$ is a ring with respect to the addition and multiplication of functions. The study of the ring $C^\infty(\mathbb{R}^n)$ is very important in the field of differential geometry.

We now present some fundamental properties of a ring. Since a ring is an abelian group with respect to the addition, given $n \in \mathbb{Z}$ and $a \in R$, it makes sense to write na. Moreover, since a ring is associative with respect to the multiplication, given a positive integer k and $a \in R$, it also makes sense to write a^k. The following properties can be deduced directly from the definition of a ring.

(1) $(m + n)a = ma + na$, $m(-a) = -(ma)$, $(mn)a = m(na)$, $m(a + b) = ma + mb$, $\quad \forall a, b \in R, m, n \in \mathbb{Z}$.

(2) $a^m a^n = a^{m+n}$, $(a^m)^n = a^{mn}$, $\quad \forall m, n \in \mathbb{N}, a \in R$.

(3) $\left(\sum_{i=1}^n a_i\right)\left(\sum_{j=1}^m b_j\right) = \sum_{i=1}^n \sum_{j=1}^m a_i b_j$.

(4) For any $a, b \in R$, we have $a0 = 0a = 0$, here 0 denotes the zero of R; Moreover, $(-a)b = a(-b) = -(ab)$, $(-a)(-b) = ab$.

We will give the proof of (4), and leave the proofs of other assertions to readers. Since 0 is the zero element with respect to the addition, we have $0 + 0 = 0$. Then by the distributive laws, we get

$$0a = (0 + 0)a = 0a + 0a; \quad a0 = a(0 + 0) = a0 + a0.$$

This means that $0a, a0$ are identity elements with respect to the addition, hence $0a = a0 = 0$. On the other hand, since $(-a)b + ab = (-a + a)b = 0a = 0$, we have $(-a)b = -(ab)$. Similarly, $a(-b) = -(ab)$, $(-a)(-b) = ab$. This completes the proof of (4).

Let us make some observations to the above examples. In Example 2.1.4, by the property of numbers, we have $a \neq 0, b \neq 0 \Rightarrow ab \neq 0$. The same assertion is also valid for polynomials in Example 2.1.5. But in linear algebra, we have constructed some examples to show that the product of two non-zero matrices might be zero. On the other hand, consider $C([0, 1])$ in Example 2.1.6, and define

$$f(x) = \begin{cases} 0, & x \in [0, \frac{1}{2}]; \\ 2x - 1, & x \in (\frac{1}{2}, 1]. \end{cases}$$

$$g(x) = \begin{cases} 1 - 2x, & x \in [0, \frac{1}{2}]; \\ 0, & x \in (\frac{1}{2}, 1]. \end{cases}$$

Then f, g are non-zero elements in $C([0, 1])$, but $fg = 0$. If in a ring there exist non-zero elements whose product is zero, then the cancellation laws for the multiplication will no longer be valid. This observation leads to the following.

Definition 2.1.7. Let R be a ring, and $a, b \in R$. If $a \neq 0, b \neq 0$ but $ab = 0$, then a is called a **left zero divisor** of R, and b is called a **right zero divisor** of R. Left and right zero divisors will be simply referred to as **zero divisors**. If $ax = ay$ and $a \neq 0$ imply that $x = y$, then we say that the **left cancellation law** holds for R. Similarly, if $xa = ya$ and $a \neq 0$ imply that $x = y$, then we say that the **right cancellation law** holds for R.

We now prove the following.

Proposition 2.1.8. *Let R be a ring. Then there does not exist any zero divisor in R if and only if the left and right cancellation laws hold.*

Proof. *The only if part.* Suppose there does not exist any zero divisor in R. If $ax = ay$ and $a \neq 0$, then $a(x - y) = 0$. This implies that $x = y$, otherwise the non-zero element $x - y$ will be a right zero divisor, contradicting the assumption that R has no zero divisor. Hence, the left cancellation law holds. Similarly, the right cancellation law holds.

The if part. Suppose the left and right cancellation law hold for R. If $ax = 0$, and $a \neq 0$, then $ax = a0$. By the left cancellation, we have $x = 0$. This means that there does not exist any right zero divisor in R. Similarly, there does not exist any right zero divisor in R. $\square$

Now, we give an important definition.

Definition 2.1.9. If R is not the null ring, and has no zero divisor, then we say that R is a **ring without zero divisors**.

We stress here that a ring without zero divisors cannot be the null ring itself. Besides, the answer to Question 2.1.2 is that in this case the ring must be the null ring (the proof is left to readers), and this is the reason why in our definition of a ring, we do not require R to be a group with respect to the multiplication. Now, we prove the following.

Proposition 2.1.10. *Let R be a ring without zero divisors, and $R^* = R - \{0\}$. Then all the elements in R^* have the same order with respect to the addition. Moreover, if the above order is finite, then it must be a prime number.*

Proof. This will be proved in three steps.

1. If the order of any element in R^* is infinite, then the assertions hold.
2. Suppose there exists $a \in R^*$ with finite order n. Then for any $b \in R^*$, we have

$$(na)b = a(nb) = 0.$$

Since R has no zero divisor, and $a \neq 0$, we have $nb = 0$. This means that the order of b is a divisor of n. In particular, the order of b is also finite. Denote the order of b as m. Then the above argument shows that n is also a divisor of m. Thus, $m = n$. Hence, the order of any element of R^* is n.

3. Suppose the order of any element of R^* is n. We now show that n must be a prime number. Suppose conversely that this is not true. Then there exists positive integers n_1, n_2, such that $n_1 < n, n_2 < n$, and $n = n_1 n_2$. Since the order of a is equal to n, we have $n_1 a \neq 0$, and $n_2 a \neq 0$. On the other hand, we also have

$$(n_1 a)(n_2 a) = na^2 = (na)a = 0.$$

This is a contradiction with the assumption that R has no zero divisors. Hence, n must be a prime number. □

We remark here that the conclusion of the above proposition is interesting, in that the condition that R has no zero divisors is relevant to the multiplication, but we can deduce from this condition that all the elements of R^* have the same order with respect to the addition. The proposition reveals the fact that the distributive laws have important impact on the properties of rings. Now, we give the following.

Definition 2.1.11. Let R be a ring without zero divisors. If all elements in R^* have infinite order with respect to the addition, then we say that the **characteristic** of R is 0; If all elements in R^* have finite order p (where p is prime), then we say that the characteristic of R is p. The characteristic of R is denoted as $\operatorname{Ch} R$.

In the following we will present an important formula for commutative rings without zero divisors and with characteristic p. We first give the following.

Problem 2.1.12. Show that for any prime number p and positive integer k, with $1 \le k \le p-1$, $p \mid C_p^k$.

Proposition 2.1.13. *Let R be a commutative ring without zero divisors, and $\operatorname{Ch} R = p$, where p is a prime number. Then for any $a, b \in R$, we have*

$$(a+b)^p = a^p + b^p, \ (a-b)^p = a^p - b^p.$$

Proof. Since R is commutative, it is easy to prove inductively that

$$(a+b)^p = a^p + C_p^1 a^{p-1} b + \cdots + C_p^{p-1} a b^{p-1} + b^p.$$

Now, by the assertion of the above question, $p \mid C_p^k, k = 1, 2, \ldots, p-1$. Then by the assumption that $\operatorname{Ch} R = p$, we have $C_p^k a^{p-k} b^k = 0, k = 1, 2, \ldots, p-1$. Thus, $(a+b)^p = a^p + b^p$. Moreover, $(a-b)^p = (a+(-b))^p = a^p + (-b)^p = a^p + (-1)^p b^p$. Notice that, if $p \ne 2$, then p is odd. Moreover, if $p = 2$, then it follows from $2b^p = 0$ that $b^p = -b^p$. Therefore, $(a-b)^p = a^p - b^p$. $\qquad \square$

Corollary 2.1.14. *Let R be a commutative ring without zero divisors, and $\operatorname{Ch} R = p$, where p is a prime number. Then for any $a, b \in R$ and positive integer n, we have*

$$(a+b)^{p^n} = a^{p^n} + b^{p^n}, \ (a-b)^{p^n} = a^{p^n} - b^{p^n}.$$

Exercises

(1) Determine whether the following sets with the given additions and multiplications are rings:

    (i) $R = \{a + b\sqrt{m} \mid a, b \in \mathbb{Q}\}$, where m is an integer, and the addition and multiplication are the usual ones of numbers.

(ii) $R = \mathbb{Z}$, the addition is defined by

$$a \oplus b = a + b - 1, \quad a, b \in R,$$

and the multiplication is defined by

$$a \otimes b = a + b - ab, \quad a, b \in R.$$

(iii) $R = \mathbb{Z} \times \mathbb{Z}$, addition and multiplication are defined as

$$(a, b) + (c, d) = (a + c, b + d),$$
$$(a, b)(c, d) = (ac, bd), \quad \forall (a, b), (c, d) \in \mathbb{Z} \times \mathbb{Z}.$$

(iv) $\{R; +\}$ is an abelian group, and multiplication is $ab = 0, \forall a, b$.

(v) $\mathbb{F}$ is a number field, $R = \{A \in \mathbb{F}^{n \times n} | A = -A'\}$, and addition and multiplication are the usual ones for matrices.

(vi) $R = \{\left(\begin{smallmatrix} a & b \\ 0 & 0 \end{smallmatrix}\right) | a, b \in \mathbb{R}\}$, and addition and multiplication are the usual ones for matrices.

(2) Suppose R is a ring, and $|R|$ is prime. Show that R is commutative.

(3) Let $R = \{a, b, c\}$. Define $+$ and $\cdot$ by the following tables:

+	a	b	c
a	a	b	c
b	b	c	a
c	c	a	b

·	a	b	c
a	a	a	a
b	a	a	a
c	a	b	a

is $\{R; +, \cdot\}$ a ring?

(4) Construct an example to show that there exists a ring R with identity and an element $a \in R$ such that a has infinitely many right inverse elements.

(5) Let R be a ring with identity e, and $a \in R$. If there exits a positive integer m such that $a^m = 0$, then a is called nilpotent. Prove that if a is nilpotent, then $e + a$ is an invertible element.

(6) Let R be a ring with an identity, and $a, b \in R$. Suppose a is nilpotent and $a + b = ab$. Show that $ab = ba$.

(7) Let R be a ring, and $a \in R$. If $a \neq 0$ and $a^2 = a$, then a is called idempotent. Prove the following assertions:

   (i) If every non-zero element in R is idempotent, then R must be commutative;

   (ii) If R has no zero divisors and has one idempotent element, then the idempotent element of R is unique, and in this case R has an identity.

(8) Suppose the ring R has a unique left identity. Show that R has an identity.

(9) Let R be a ring without zero divisors, and e is a left (or right) identity element with respect to the multiplication. Prove that e must be an identity.

(10) Let R be a ring with identity e, and $u, v \in R$. Suppose $uvu = u$, $vu^2v = e$. Show that u, v are invertible and $v = u^{-1}$.

(11) Let R be a ring with identity, and $u, v \in R$. Suppose $uvu = u$, and v is the unique element satisfying the condition. Show that u, v are invertible and $v = u^{-1}$.

(12) Let R_1, R_2 be rings. Define addition and multiplication on the product set $R_1 \times R_2$ as the following:

$$(a_1, b_1) + (a_2, b_2) = (a_1 + a_2, b_1 + b_2);$$
$$(a_1, b_1)(a_2, b_2) = (a_1 a_2, b_1 b_2).$$

Show that $R_1 \times R_2$ is a ring with respect to the above operations, called the external direct product of the rings R_1, R_2, and denoted as $R_1 + R_2$. Prove the following assertions:

   (i) If R_1, R_2 are both rings with identity, then $R_1 + R_2$ is also a ring with identity;

   (ii) If R_1, R_2 are commutative, then $R_1 + R_2$ is also commutative. Furthermore, suppose R_1, R_2 are both rings without zero divisors. Is $R_1 + R_2$ necessarily a ring without zero divisors?

Exercises (hard)

(13) Let R be a ring with identity, and $u, v \in R$. Suppose $u^k v u^l = u^{k+l-1}$, where k, l are positive integers, and v is the unique element satisfying this condition. Show that u, v are invertible and $v = u^{-1}$.

(14) Suppose R is a non-empty set with addition and multiplication operations such that all the axioms of a ring are satisfied except for the community of the addition. Suppose there does not exist zero divisors for the multiplication. Show that R is a ring (i.e., the addition is commutative).

(15) (Hua) Let R be a ring with identity e, and $a, b \in R$. Show that $e - ab$ is invertible if and only if $e - ba$ is invertible.

(16) (Hua) Let R be a ring with identity e, and $a, b \in R$. Suppose $a, b, ab - e$ are invertible. Show that $a - b^{-1}$, $(a - b^{-1})^{-1} - a^{-1}$ are also invertible, and $((a - b^{-1})^{-1} - a^{-1})^{-1} = aba - a$.

(17) (Kaplansky) Let R be a ring with identity. Show that if an element in R has more than one right inverse, then it has infinitely many.

2.2 Ideals and quotient rings

In this section, we study subrings and quotient rings. As mentioned in Chapter 1, it is an important technique in algebra theory to study algebraic systems through subsystems and quotient systems. On one hand, sometimes we can get information of the algebraic system through the properties of subsystems and quotient systems. On the other hand, subsystems and quotient systems provide abundant explicit examples with special properties, which can enhance our understanding of related topics.

We first give the definition of a subring.

Definition 2.2.1. Let R be a ring and R_1 a non-empty subset of R. If R_1 is a ring with respect to the addition and multiplication of R, then R_1 is called a **subring** of R.

Notice that in the above definition, the addition and multiplication of a subring must coincide with that of the underlying ring. If on a non-empty subset R_1 of a ring R, we define new addition and multiplication, which are different from that of R, then R_1 can not be called a subring of R even if R_1 is a ring with respect to the new addition and multiplication. Readers are suggested to construct an example to show that this can actually happen. It is obvious that, if

$\mathbb{F}_1, \mathbb{F}_2$ are number fields such that $\mathbb{F}_1 \subseteq \mathbb{F}_2$, then $\mathbb{F}_1$ is a subring of $\mathbb{F}_2$. We now give several other examples of subrings.

Example 2.2.2. It is easy to check that, for any natural number m, the subset $m\mathbb{Z} = \{mn \,|\, n \in \mathbb{Z}\}$ of the ring $\mathbb{Z}$ is a subring. As an exercise, readers are suggested to prove that any subring of $\mathbb{Z}$ must be of this form.

Example 2.2.3. Let $\mathbb{F}$ be a number field, and $\mathbb{Z}^{n \times n}$ the subset of the ring $\mathbb{F}^{n \times n}$ consisting of the matrices whose entries are integers. Then it is easily seen that $\mathbb{Z}^{n \times n}$ is a subring of $\mathbb{F}^{n \times n}$. In the next chapter we will define the ring of matrices over an arbitrary ring, which can be viewed as a generalization of this example.

Example 2.2.4. In Example 2.1.6, $C^\infty(\mathbb{R}^n)$ is a subring of $C(\mathbb{R}^n)$.

In general, it is somehow complicated to prove directly that a subset of a ring is a subring. We now give a simple criterion for this.

Proposition 2.2.5. *Let R be a ring and R_1 a non-empty subset of R. Then R_1 is a subring of R if and only if for any $a, b \in R_1$, $a - b \in R_1$ and $ab \in R_1$.*

Proof. *The only if part.* Let R_1 be a subring of R, and $a, b \in R_1$. Since R_1 is a subgroup of R with respect to the addition of R, we have $a - b \in R_1$. Moreover, since R_1 is a semigroup with respect to the multiplication of R, it must be closed with respect to the multiplication. Thus, $ab \in R_1$.

The if part. If $a - b \in R_1$, $\forall a, b \in R_1$, then R_1 is a subgroup of R with respect to the addition, hence, R_1 is an abelian group with respect to the addition. Moreover, since $ab \in R_1$, $\forall a, b \in R_1$, R_1 is closed with respect to the multiplication. Since R is associative with respect to the multiplication, R_1 must also be associative, hence it is a semigroup with respect to the multiplication. Finally, since R satisfies the distributive laws with respect to the addition and multiplication, and R_1 is closed with respect to the operations, R_1 must satisfy the distributive laws. Therefore, R_1 is a ring with respect to the addition and multiplication of R, hence it is a subring of R. $\square$

Now, we consider quotient rings. Given a subring R_1 of R, R_1 is a normal subgroup of the additive group R with respect to the addition. Therefore, on the left coset space R/R_1 we can define an addition which makes R/R_1 into an abelian group. Our goal is to consider the condition under which there is a multiplication which makes the abelian group R/R_1 into a ring. A natural idea is to define

$$(a + R_1)(b + R_1) = ab + R_1, \quad a, b \in R. \tag{2.1}$$

Unfortunately, this is not always reasonable.

Problem 2.2.6. Construct an example to show that, there exists a ring R, a subring R_1 of R, and $a, b, a', b' \in R$, such that on the quotient group R/R_1, $a + R_1 = a' + R_1$, $b + R_1 = b' + R_1$, but $ab + R_1 \neq a'b' + R_1$.

Therefore, the operation in (2.1) is not always well-defined. The reason is that the condition R_1 is a subring of R cannot guarantee that the equivalent relation which defines the left coset space is a congruence relation with respect to the multiplication. Now, let us probe the appropriate condition. The equivalent relation defining the left coset R/R_1 space is $a \sim b \iff a - b \in R_1$. Therefore, to ensure the operation in (2.1) to be well-defined, we must have $a_1 \sim a_2$, $b_1 \sim b_2 \Rightarrow a_1 b_1 \sim a_2 b_2$, which is equivalent to

$$a_1 - a_2 \in R_1, b_1 - b_2 \in R_1 \Rightarrow a_1 b_1 - a_2 b_2 \in R_1.$$

Notice that

$$a_1 b_1 - a_2 b_2 = a_1 b_1 - a_1 b_2 + a_1 b_2 - a_2 b_2 = a_1 (b_1 - b_2) + (a_1 - a_2) b_2.$$

In the above equation, $a_1 - a_2 \in R_1$, $b_1 - b_2 \in R_1$, and a_1, b_2 can be arbitrary in R. Consequently, the condition for the operation (2.1) to be well-defined, is $ax \in R_1, ya \in R_1, \forall x, y \in R, a \in R_1$. This leads to the following definition:

Definition 2.2.7. Let R be a ring and I a subring of R. If I satisfies the condition $a \in I, x \in R \Rightarrow xa \in I$, then we say that I is a **left ideal** of R; Similarly, if I satisfies the condition $a \in I, y \in R \Rightarrow ay \in I$, then we say that I is a **right ideal** of R. If a subring is both a left and right ideal, then it is called a **two-sided ideal**.

In the literature, the word *ideal* generally means a two-sided ideal. In this book, unless otherwise stated, we will also use the term *ideal* to mean a two-sided ideal. It is clear that, for any ring R, the subrings $\{0\}$ and R are ideals. These are called **trivial ideals**.

Problem 2.2.8. Construct examples to show that, there exists a ring R and a left ideal I of R, which is not a right ideal. Similarly, there exists a ring and a right ideal, which is not a left ideal.

We now give some examples of ideals.

Example 2.2.9. As we mentioned above, any subring of $\mathbb{Z}$ must be of the form $m\mathbb{Z}$, $m \geq 0$. It is easy to check that $m\mathbb{Z}$ is a two-sided ideal of $\mathbb{Z}$. Therefore, $m\mathbb{Z}$, $m \geq 0$ are all the ideals of $\mathbb{Z}$.

Example 2.2.10. In Example 2.2.4, consider $C(\mathbb{R})$. Given $x_0 \in \mathbb{R}$, we set

$$Z_{x_0}(\mathbb{R}) = \{f \in C(\mathbb{R}) \mid f(x_0) = 0\}.$$

Then $Z_{x_0}(\mathbb{R})$ is a two-sided ideal of $C(\mathbb{R})$.

Another ideal of the ring of functions plays important roles in differential geometry. Given a point x in $\mathbb{R}^n$, define

$$O_x = \{f \in C^\infty(\mathbb{R}^n) \mid \exists \text{ a neighborhood } U \text{ of } x \text{ such that } f(y) = 0,$$
$$\forall y \in U\}.$$

Then is is easy to check that O_x is an ideal of $C^\infty(\mathbb{R}^n)$.

Combining Proposition 2.1.13 with the definition of ideals, we get the following.

Proposition 2.2.11. *Let R be a ring and I a non-empty subset of R. Then I is an ideal of R if and only if $a - b \in I$, and $ax, ya \in I$, $\forall a, b \in I, x, y \in R$.*

Problem 2.2.12. By the definition of a subring, it is easily seen that a subring of a subring of a ring is a subring. More precisely, if R_1 is a subring of a ring R, and R_2 is a subring of R_1, then R_2 is a subring of R. Please determine whether the above assertion holds for ideals, namely, if I_1 is an ideal of a ring R, and I_2 is an ideal of I_1, is I_2 necessarily an ideal of R?

Now, we introduce a method to construct ideals. As an easy exercise, one can prove that the intersection of a family of (finitely or infinitely many) ideals of a ring is also an ideal. Now, given a non-empty subset S of a ring R, there must be ideals of R which contains S as a subset (e.g., R itself is such an ideal). We define $\langle S \rangle$ to be the intersection of all such ideals. Then $\langle S \rangle$ is an ideal. We assert that $\langle S \rangle$ is the minimal ideal of R containing S. In fact, on one hand, $\langle S \rangle$ is an ideal containing S. On the other hand, since $\langle S \rangle$ is the intersection of all the ideals which contains S, any ideal of R containing S must contain $\langle S \rangle$. This proves our assertion. In the following, $I = \langle S \rangle$ will be called the ideal generated by S. Sometimes we also say that S is a set of generators of the ideal I.

We give some examples to illustrate the above definition.

Example 2.2.13. Let R be a commutative ring with identity. We now show that

$$\langle S \rangle = \left\{ \sum_{i=1}^{n} x_i a_i \mid n > 0, x_i \in R, a_i \in S, i = 1, 2, \ldots, n \right\}.$$

Denote the set in the right hand of the above equation as I. Then for any $a \in S$, $a = 1 \cdot a \in I$ (here 1 is the identity element of R), hence $S \subseteq I$. Moreover, by the criterion of Proposition 2.2.11, it is easy to check that I is an ideal of R. On the other hand, if I_1 is an ideal of R containing S, then for any $x_i \in R$, and $a_i \in S, 1 \leq i \leq n$, we have $x_i a_i \in I_1$. Thus, $\sum x_i a_i \in I_1$. Therefore, $I \subseteq I_1$, which implies that I is the minimal ideal containing S. Hence, $I = \langle S \rangle$.

Definition 2.2.14. Let I be an ideal of R. If there exists $a \in I$ such that $I = \langle a \rangle$, then we say that I is a **principal ideal** of R, and a is a **generator** of I.

Example 2.2.15. If R is a ring with identity, then

$$\langle a \rangle = \left\{ \sum_{i=1}^{m} x_i a y_i \mid x_i, y_i \in R, i = 1, 2, \ldots, m, \ m \geq 1 \right\};$$

If R is commutative, then

$$\langle a \rangle = \{ ra + na \mid r \in R, n \in \mathbb{Z} \};$$

If R is a commutative ring with identity, then

$$\langle a \rangle = \{ ra \mid r \in R \}.$$

The proofs are left as exercises.

Now, we can define quotient rings. As we showed above, to define a quotient ring, we must consider the quotient of a ring with respect to an ideal, rather than a subring.

Theorem 2.2.16. *Let R be a ring and I an ideal of R. Let R/I be the quotient group of the additive group $\{R; +\}$ with respect to the (normal) subgroup I, and denote the coset of $a \in R$ as $a + I$. Then R/I is a ring with respect to the addition of the quotient group and the multiplication defined by*

$$(a + I)(b + I) = ab + I. \tag{2.2}$$

*The ring R/I is called the **quotient ring** of R with respect to the ideal I.*

Proof. It is known that R/I is an abelian group with respect to the addition. By the observation before Definition 2.2.7, if I is an ideal, then the equivalent relation defining the quotient set R/I is congruent with respect to the multiplication of R. Therefore, the operation in (2.2) is well-defined. Since R satisfies the associative law with respect to the multiplication, it is easily seen that R/I also satisfies the associative law. Finally, for any $a, b, c \in R$, we have

$$\begin{aligned}
((a + I) + (b + I))(c + I) &= ((a + b) + I)(c + I) = (a + b)c + I \\
&= (ac + bc) + I = (ac + I) + (bc + I) \\
&= (a + I)(c + I) + (b + I)(c + I).
\end{aligned}$$

Similarly,

$$(a + I)((b + I) + (c + I)) = (a + I)(b + I) + (a + I)(c + I).$$

Thus, the distributive laws hold. This completes the proof of the theorem. □

By the above definition of quotient rings, it is easily seen that, if R is a commutative ring, then R/I is also commutative; If R is a ring with identity 1, then R/I is a ring with identity $1 + I$.

Example 2.2.17. Consider the ring of integers $\mathbb{Z}$. We have shown that any ideal of $\mathbb{Z}$ must be of the form $m\mathbb{Z}$, where $m \geq 0$. Therefore, we get a series of quotient rings $\mathbb{Z}_m = \mathbb{Z}/m\mathbb{Z}$. When $m > 0$, the ring $\mathbb{Z}_m$ is called the **residue ring** of $\mathbb{Z}$ modulo m.

Example 2.2.18. In Example 2.2.10, the quotient ring of $C^\infty(\mathbb{R}^n)$ with respect to O_x is usually denoted as $\mathcal{F}_x(\mathbb{R}^n)$. This quotient ring plays a very important role in differential geometry. It is easily seen from the definition that two functions $f, g \in C^\infty(\mathbb{R}^n)$ lie in the same coset in the quotient ring $\mathcal{F}_x(\mathbb{R}^n)$ if and only if there exists a neighborhood U of x such that $f(y) = g(y)$, $\forall y \in U$. The coset $[f]$ of f is called the germ of f. Using the germs of functions we can define the notions of differential forms and tangent vectors, which are the foundation of differential geometry.

Now, we can define several special types of rings.

Definition 2.2.19. Let R be a ring with identity. Denote the set of invertible elements in $R^* = \{0\}$ as U. Then U is a group with respect to the multiplication, called the **group of units** of R. An element in U is called a **unit** of R.

We leave the proof of the assertion that U is a group to readers. Notice that the notion of a unit is different from that of an identity.

Definition 2.2.20. A commutative ring with identity but without zero divisors is called a **domain**. A ring R with identity satisfying the condition that $R^* = U$, or in other words, that R^* is a group with respect to the multiplication, is called a **division ring** (or skew field, sfield). A commutative division ring is called a **field**.

It is obvious that a number field is a field. We now present an example of a field which is not a number field.

Example 2.2.21. In Example 2.2.9, consider the case when $m = p$ is a prime number. We know that

$$\mathbb{Z}_p = \{\bar{0}, \bar{1}, \bar{2}, \ldots, \overline{p-1}\}.$$

Now, we prove that $\mathbb{Z}_p$ is a field. It is obvious that $\mathbb{Z}_p$ is a commutative ring with identity. As p is prime, for any positive integer a, with $a < p$, there exist integers u, v such that $ua + vp = 1$. Thus, in the quotient ring $\mathbb{Z}_p$ we have

$$\overline{ua + vp} = \overline{ua} + \overline{vp} = \overline{ua} + \bar{0} = \bar{u}\bar{a} = \bar{1}.$$

Hence, $\bar{u}$ is an inverse of $\bar{a}$ with respect to the multiplication. Therefore, $\mathbb{Z}_p$ is a field. Notice that $\mathbb{Z}_p$ is a finite set; hence it cannot be a number field.

We have given several examples of fields. A natural problem is whether there is a non-commutative division ring. To construct such a ring, it is natural to consider some special set of matrices, since the multiplication of matrices is not commutative. In the next section, we will construct an explicit non-commutative division ring, called the ring of quaternions. The discovery of quaternions is a great event in the history of mathematics. Currently, quaternions play important roles in the representation theory of Lie groups and differential geometry.

Exercises

(1) In the following, determine whether the subset S of the ring R is a subring or an ideal:

    (i) $R = \mathbb{Q}$, $S = \{\frac{a}{b} | a, b \in \mathbb{Z}, 5 \nmid b\}$.

    (ii) $R = \mathbb{Q}$, $S = \{2^n \cdot m | m, n \in \mathbb{Z}\}$.

    (iii) $R = \mathbb{R}^{2 \times 2}$, $S = \left\{ \begin{pmatrix} a & b \\ -b & a \end{pmatrix} | a, b \in \mathbb{R} \right\}$.

    (iv) $R = \mathbb{R}^{2 \times 2}$, $S = \left\{ \begin{pmatrix} a & b \\ 0 & d \end{pmatrix} | a, b, d \in \mathbb{R} \right\}$.

    (v) $R = \left\{ \begin{pmatrix} a & b \\ 0 & 0 \end{pmatrix} | a, b \in \mathbb{R} \right\}$, $S = \left\{ \begin{pmatrix} a & 0 \\ 0 & 0 \end{pmatrix} | a \in \mathbb{R} \right\}$.

    (vi) $R = \mathbb{Z}_m$, $S = \{a + m\mathbb{Z} \in \mathbb{Z}_m \,|\, a = 0, \frac{m}{k}, 2\frac{m}{k}, \ldots, (k-1)\frac{m}{k}\}$, where m is a positive integer, and k is a factor of m.

(2) Let R be a ring, and $a \in R$. Show that the principal ideal generated by a is equal to

$$\left\{ \sum_{i=1}^{m} x_i a y_i + ra + as + na \,\middle|\, r, s \in R, x_i, y_i \in R, 1 \leq i \leq m, m > 0, n \in \mathbb{Z} \right\}.$$

(3) Let R be a ring without zero divisors, and I an ideal of R. Is the quotient ring R/I necessarily a ring without zero divisors?

(4) Let R be a domain, and I, J be two non-zero ideals of R. Show that $I \cap J \neq \{0\}$.

(5) Show that a finite domain must be a field.

(6) Let R be a ring with identity and without zero divisors, and it has only finite ideals (including left, right, and two-sided ideals). Show that R is a division ring.

(7) Determine the groups of units of the following rings:

(i) The ring $\mathbb{Z}[\sqrt{-1}]$; (ii) The quotient ring $\mathbb{Z}_7$;

(iii) The quotient ring $\mathbb{Z}_{24}$; (iv) The quotient ring $\mathbb{Z}_m$, $m > 0$.

(8) Let $\mathbb{F}$ be a number field. Show that the ring $\mathbb{F}^{n \times n}$ of $n \times n$ matrices over $\mathbb{F}$ has no non-trivial ideals.

(9) Let R be a finite ring without zero divisors. Show that R is a division ring.

(10) Suppose R is a ring with identity e, I is an ideal of R, and U is the group of units of R. Set

$$U_I = \{a \in U \mid a - e \in I\}.$$

Show that U_I is a normal subgroup of U.

Exercises (hard)

(11) Let R be a commutative ring, and I an ideal of R. Set

$$\sqrt{I} = \{a \in R \mid \text{there exists a positive integer } n \text{ such that } a^n \in I\}.$$

Show that $\sqrt{I}$ is also an ideal of R, called the **radical** of the ideal I. In particular, the set of nilpotent elements in R (namely, the radical of the ideal $\{0\}$) is an ideal of R, called the nilpotent radical of R, denoted as $\mathrm{Rad}\, R$. Show that for any ideal I of R, $\sqrt{\sqrt{I}} = \sqrt{I}$.

(12) Let R be a commutative ring, and $\mathrm{Rad}R$ the nilpotent radical of R. Show that the nilpotent radical of the quotient ring $R/\mathrm{Rad}R$ must be zero, hence there does not exist any non-zero nilpotent element in $R/\mathrm{Rad}R$.

(13) Let R be a commutative ring. Define a binary operation $\circ$ on R by

$$a \circ b = a + b - ab.$$

(i) Show that the associative law holds for $\circ$, and $0 \circ a = a$, $\forall a \in R$.

(ii) Show that R is a field if and only if $\{R - \{e\}; \circ\}$ is an abelian group, where e is the identity of R.

(14) (Kaplansky) Let R be a ring. An element a in R is called right quasi-regular if there exists $b \in R$ such that $a + b - ab = 0$. Show that if all but one elements in R are right quasi-regular, then R must be a division ring.

2.3 Quaternions

In this section, we will construct an example of a division ring which is not a field, called the ring of quaternions. This ring was discovered by the Irish mathematician Hamilton, and the discovery is a great breakthrough in the history of mathematics. The initial idea of Hamilton was to generalize the notion of numbers. It is well-known that the introduction of complex numbers led to revolutionary changes in the development of mathematics. Therefore, it is natural to consider whether there are other forms of *numbers*. If we view real numbers as univariate numbers, then complex numbers can be viewed as bivariate numbers. For many years, Hamilton tried to construct tri-variate numbers, but he failed. Finally, in 1843, he successfully constructed quaternions. Quaternions possess all properties of numbers except for the commutativity, and they play very important roles in the fields of algebra, representation theory of Lie groups, and differential geometry.

Now, we will construct the ring of quaternions. Recall that every complex number $a + b\sqrt{-1}$ can be viewed as a pair of real numbers (a, b). In this point of view, the addition and multiplication of complex numbers can be written as

$$(a_1, b_1) + (a_2, b_2) = (a_1 + a_2, b_1 + b_2);$$
$$(a_1, b_1)(a_2, b_2) = (a_1 a_2 - b_1 b_2, a_1 b_2 + a_2 b_1).$$

Following this idea, we can view quaternions as quadruples of real numbers, and then define the reasonable addition and multiplication among them. However, this process seems not to be natural. Now, we write complex numbers as real matrices, and then generalize this process to the complex case, which will also lead to quaternions. First notice that, if we write a complex number $a + b\sqrt{-1}$ as a 2×2 real matrix

$$\begin{pmatrix} a & b \\ -b & a \end{pmatrix},$$

then the set of complex numbers corresponds to a subset of $\mathbb{R}^{2\times2}$ as the following

$$\left\{ \begin{pmatrix} a & b \\ -b & a \end{pmatrix} \mid a, b \in \mathbb{R} \right\},$$

and in this correspondence, the addition and multiplication of complex numbers coincide with that of the matrices. This observation reminds us to consider the following subset of the set of 2×2 complex matrices $\mathbb{C}^{2\times2}$:

$$\mathbb{H} = \left\{ \begin{pmatrix} \alpha & \beta \\ -\bar{\beta} & \bar{\alpha} \end{pmatrix} \mid \alpha, \beta \in \mathbb{C} \right\}.$$

It is easy to check that the subset $\mathbb{H}$ is closed with respect to the addition and multiplication of matrices; hence it is a subring of $\mathbb{C}^{2\times2}$. We now explore the properties of $\mathbb{H}$. It is easily seen that:

(1) $\mathbb{H}$ contains the identity of $\mathbb{C}^{2\times2}$, namely, $\begin{pmatrix} 1 & 0 \\ 0 & 1 \end{pmatrix}$, hence it is a ring with identity;

(2) Let

$$\mathbf{i} = \begin{pmatrix} \sqrt{-1} & 0 \\ 0 & -\sqrt{-1} \end{pmatrix}, \quad \mathbf{j} = \begin{pmatrix} 0 & 1 \\ -1 & 0 \end{pmatrix}, \quad \mathbf{k} = \begin{pmatrix} 0 & \sqrt{-1} \\ \sqrt{-1} & 0 \end{pmatrix}.$$

Then we have

$$\mathbf{jk} = \mathbf{i}, \quad \mathbf{kj} = -\mathbf{i}.$$

Thus, $\mathbb{H}$ is not commutative.

(3) If

$$A = \begin{pmatrix} \alpha & \beta \\ -\bar{\beta} & \bar{\alpha} \end{pmatrix} \neq 0,$$

then A is invertible, and

$$A^{-1} = \frac{1}{|\alpha|^2 + |\beta|^2} \begin{pmatrix} \bar{\alpha} & -\beta \\ \bar{\beta} & \alpha \end{pmatrix}.$$

In summarizing, $\mathbb{H}$ is a non-commutative division ring. This is the famous ring of **quaternions**.

Now, let us go back to the above topic, namely, how to write a quaternion as a quadruple of real numbers, and define addition and multiplication among those quadruples. Using the special elements $\mathbf{i}, \mathbf{j}, \mathbf{k}$, one can write the matrix $\begin{pmatrix} \alpha & \beta \\ -\bar{\beta} & \bar{\alpha} \end{pmatrix}$, where $\alpha = a + b\sqrt{-1}$, $\beta = c + d\sqrt{-1}$, $a, b, c, d \in \mathbb{R}$, as

$$\begin{pmatrix} \alpha & \beta \\ -\bar{\beta} & \bar{\alpha} \end{pmatrix} = \begin{pmatrix} a + b\sqrt{-1} & c + d\sqrt{-1} \\ -c + d\sqrt{-1} & a - b\sqrt{-1} \end{pmatrix} = a\mathbf{1} + b\mathbf{i} + c\mathbf{j} + d\mathbf{k},$$

here $\mathbf{1}$ is the identity matrix. Then a quaternion can be viewed as a quadruple of real numbers. The addition of quadruples is just the addition of the corresponding components. To find the multiplication law, just notice that $\mathbf{1}$ is the identity, and the multiplication among $\mathbf{i}, \mathbf{j}, \mathbf{k}$ satisfies

$$\mathbf{i}^2 = \mathbf{j}^2 = \mathbf{k}^2 = \mathbf{ijk} = -\mathbf{1}.$$

These formulas can be linearly extended to any pair of quadruples of real numbers, which is the multiplication law of quaternions.

Just as complex numbers can be written as pairs of real numbers, quaternions can also be written as pairs of complex numbers. In fact, given $a, b, c, d \in \mathbb{R}$, we have

$$a\mathbf{1} + b\mathbf{i} + c\mathbf{j} + d\mathbf{k} = (a + b\mathbf{i}) + (c + d\mathbf{i})\mathbf{j}.$$

In this way, the quaternion $a\mathbf{1} + b\mathbf{i} + c\mathbf{j} + d\mathbf{k}$ corresponds to the pair of complex numbers $(a + b\sqrt{-1}, c + d\sqrt{-1})$. We leave as an exercise for readers to find the addition and multiplication of quaternions in this form.

Notice that both the field of complex numbers and the ring of quaternions can be viewed as vector spaces over the field of real numbers. Moreover, in these vector spaces, there is a multiplication, which is linear with respect to any factors. Furthermore, the multiplication is associative, and any non-zero element is invertible with respect to the multiplication. A vector space satisfying the above conditions is called a **division algebra** over the field of real numbers. It is a famous result in mathematics that, division algebras over the field of real numbers can only be one of $\mathbb{R}$, $\mathbb{C}$, or $\mathbb{H}$. Meanwhile, if we drop the requirement of associativity, then it can also be the set of **Octonions** $\mathbb{O}$. Octonions play very important roles in the theory of exceptional simple Lie groups and differential geometry. Readers interested in related topics are referred to Baez (2001).

Exercises

(1) Let R be a ring. Set $C(R) = \{a \in R \,|\, ab = ba, \forall b \in R\}$. Show that $C(R)$ is a subring of R, called the center of R. Construct an example to show that $C(R)$ is not necessarily an ideal of R.

(2) Show that the center of a division ring is a field. Find out the center $C(\mathbb{H})$ of the ring of quaternions $\mathbb{H}$.

(3) Let $x = a\mathbf{1} + b\mathbf{i} + c\mathbf{j} + d\mathbf{k} \in \mathbb{H}$. Define the conjugate element of x as $\bar{x} = a\mathbf{1} - b\mathbf{i} - c\mathbf{j} - d\mathbf{k}$. Show that for any $a, b \in \mathbb{R}$ and $x, y \in \mathbb{H}$, $\overline{ax + by} = a\bar{x} + b\bar{y}$, $\overline{xy} = \bar{x}\bar{y}$, $\bar{\bar{x}} = x$. Find the formula of conjugate elements when the quaternions are expressed in the matrix form.

(4) In the ring of quaternions $\mathbb{H}$, we define the norm of $x \in \mathbb{H}$ as $N(x) = x\bar{x}$. Prove the following assertions:

  (i) For any $x \in \mathbb{H}$, $N(x) \geq 0$, and equality holds if and only if $x = 0$.

  (ii) For any $x, y \in \mathbb{H}$, $\sqrt{N(x+y)} \leq \sqrt{N(x)} + \sqrt{N(y)}$, and the equality holds if and only if x, y are linearly dependent over the field of real numbers.

  (iii) $N(ax) = a^2 N(x)$, where $a \in \mathbb{R}$.

  (iv) $N(\bar{x}) = N(x)$.

  (v) $N(xy) = N(x)N(y)$.

(5) In the ring of quaternions $\mathbb{H}$, we define the trace of $x = a\mathbf{1} + b\mathbf{i} + c\mathbf{j} + d\mathbf{k}$ as $T(x) = 2a$. Show that for any $x \in \mathbb{H}$, the equality $x^2 - T(x)x + N(x) = 0$ holds.

(6) Let S be a subring of $\mathbb{H}$ and suppose S is a division ring. Show that if S satisfies the condition $yxy^{-1} \in S, \forall x \in S, y \in \mathbb{H}, y \neq 0$, then either $S = \mathbb{H}$ or $S \subset C(\mathbb{H})$.

(7) Show that the subset $\mathrm{Sp}(1) = \{x \in \mathbb{H} \,|\, N(x) = 1\}$ of $\mathbb{H}$ is a group with respect to the multiplication of quaternions.

(8) Show that the group Sp(1) is isomorphic to the following matrix group:

$$SU(2) = \left\{ A \in SL(2, \mathbb{C}) \,\middle|\, A\bar{A}' = I_2 \right\},$$

where I_2 is the 2×2 identity matrix.

Exercises (hard)

(9) Show that there does not exist a 3-dimensional real vector space which can be made into a division algebra by defining a multiplication on it.

(10) Let D be a division ring, and $C(D)$ the center of D.

(1) (Hua) If $a, b \in D$ and $ab \neq ba$, then

$$a = \left(b^{-1} - (a-1)^{-1} b^{-1} (a-1) \right) \cdot$$
$$\left(a^{-1} b^{-1} a - (a-1)^{-1} b^{-1} (a-1) \right)^{-1}.$$

(2) (Hua) Given $c \in D - C(D)$, let $L = \{ dcd^{-1} | d \in D, d \neq 0 \}$, and let D_1 be the minimal sub-division ring of D (that is, D_1 is a subring of D and is a division ring itself) containing L. Show that $ab = ba$, for any $a \in D - D_1$ and $b \in D_1$.

(3) (Cartan-Brauer-Hua) A subring S of D is called normal if $yxy^{-1} \in S$, $\forall x \in S$, $y \in D, y \neq 0$. Show that if S is normal and it is a division ring, then either $S = D$ or $S \subset C(D)$.

2.4 Homomorphisms of rings

In this section, we study homomorphisms between rings. One of the most important goals of ring theory is to give a classification of rings under isomorphisms. This is certainly unreachable if we do not restrict the problem to some special cases. Therefore, sometimes we will study certain special classes of rings, such as principal ideal domains and Euclidean domains that will be introduced in Section 2.7. The definition of homomorphisms between rings is similar to that of groups, namely, a homomorphism is just a map from a ring to another that keeps the operations.

Definition 2.4.1. Suppose R_1, R_2 are rings, and f is a map from R_1 to R_2. If for any $a, b \in R_1$, we have $f(a + b) = f(a) + f(b)$, and $f(ab) = f(a)f(b)$, then f is called a **homomorphism**. If the homomorphism is injective, then it is called an injective homomorphism. Similarly, we can define the notion of surjective homomorphisms. A homomorphism which is both injective and surjective is called an **isomorphism**. If there is an isomorphism between R_1 and R_2, we will say that R_1 is isomorphic to R_2, denoted as $R_1 \simeq R_2$.

Notice that a homomorphism from a ring R_1 to R_2 must be a group homomorphism between the additive groups $\{R_1; +\}$ and $\{R_2; +\}$. Thus, $f(0) = 0$, and $f(-a) = -f(a), \forall a \in R_1$.

Now, we will give several examples of homomorphisms of rings. First of all, as a trivial example, for any two rings R_1, R_2, we can define a map f from R_1 to R_2 such that $f(a) = 0$, $\forall a \in R_1$. Then f is a homomorphism, called the **null homomorphism**.

Example 2.4.2. Let V be an n-dimensional vector space over a number field $\mathbb{F}$, and $\mathrm{End}\, V$ the set of all linear transformations of V. It is easy to check that $\mathrm{End}\, V$ is a ring with respect to the addition and multiplication of linear transformations. Now, we will construct an isomorphism from this ring to a ring which we are familiar with. Fix a base $\varepsilon_1, \varepsilon_2, \ldots, \varepsilon_n$ of V. Define a map from $\mathrm{End}\, V$ to $\mathbb{P}^{n \times n}$ by

$$\phi(\mathcal{A}) = M(\mathcal{A}; \varepsilon_1, \varepsilon_2, \ldots, \varepsilon_n), \mathcal{A} \in \mathrm{End}\, V,$$

where $M(\mathcal{A}; \varepsilon_1, \varepsilon_2, \ldots, \varepsilon_n)$ is the matrix of $\mathcal{A}$ with respect to the base $\varepsilon_1, \varepsilon_2, \ldots, \varepsilon_n$. Then it follows from the results of linear algebra that ϕ is an isomorphism.

Example 2.4.3. Let I be an ideal of a ring R. Then the natural map from R to R/I:

$$\pi : R \to R/I, \ \pi(a) = a + I,$$

is a surjective homomorphism, called the **natural homomorphism**.

Example 2.4.4. Consider the ring $\mathrm{C}^{\infty}(\mathbb{R}^n)$ in Example 2.1.6. Given a point x_0 in $\mathbb{R}^n$, set

$$\varphi : \mathrm{C}^{\infty}(\mathbb{R}^n) \to \mathbb{R}, \ \varphi(f) = f(x_0).$$

Then φ is a homomorphism from the ring $\mathrm{C}^{\infty}(\mathbb{R}^n)$ to the field of real numbers.

It follows easily from the definition that, if f is a homomorphism from a ring R_1 to R_2, and g is a homomorphism from R_2 to R_3, then gf is a homomorphism from R_1 to R_3. Moreover, if f, g are injective, then gf is also an injective homomorphism; if f, g are surjective, then gf is a surjective homomorphism; if f, g are isomorphisms, then gf is also an isomorphism. When f is an isomorphism, f^{-1} is an isomorphism from R_2 to R_1. The proofs of the above assertions are left to readers. Next we study the properties of homomorphisms.

First, given a homomorphism $f : R_1 \to R_2$, we define $\operatorname{Ker} f = f^{-1}(0) = \{a \in R_1 \mid f(a) = 0\}$, called the **kernel** of f. We assert that $\operatorname{Ker} f$ is an ideal of R_1. In fact, since $f(0) = 0$, $\operatorname{Ker} f \neq \emptyset$. Moreover, for any $a, b \in \operatorname{Ker} f$, we have $f(a-b) = f(a) - f(b) = 0-0 = 0$, hence $a - b \in \operatorname{Ker} f$; Furthermore, for any $x \in R_1$, and $a \in \operatorname{Ker} f$, we have $f(ax) = f(a)f(x) = 0$, $f(xa) = f(x)f(a) = 0$, hence $ax, xa \in \operatorname{Ker} f$. By Proposition 2.2.5, $\operatorname{Ker} f$ is an ideal of R_1.

The following result is very important in ring theory, and is usually called the **Theorem of Homomorphisms of Rings**. In particular, the assertion (1) is particularly significant, and is referred to as the **Fundamental Theorem of Homomorphisms of rings** in the literature.

Theorem 2.4.5. *Let f be a homomorphism from a ring R_1 onto R_2. Denote $K = \operatorname{Ker} f$. Then the following assertions hold:*

(1) *Let π be the natural homomorphism from R_1 to R_1/K. Then there exists an isomorphism $\bar{f}$ from R_1/K to R_2 such that $f = \bar{f} \circ \pi$, namely, the following diagram is commutative:*

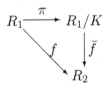

(2) *f establishes a bijection between the set of subrings of R_1 containing K and that of subrings of R_2, and this bijection sends ideals to ideals.*

(3) *If I is an ideal of R_1 containing K, then $R_1/I \simeq R_2/f(I)$.*

Proof. (1) First of all, as a homomorphism of rings, f must be a group homomorphism from the additive group $\{R_1; +\}$ onto $\{R_2; +\}$. Moreover, the natural homomorphism π is also a group homomorphism from the additive group R_1 to the quotient group $R_1/\mathrm{Ker}\, f$. By the Fundamental Theorem of Homomorphisms of groups, there is a group isomorphism $\bar{f}$ from $R_1/\mathrm{Ker}\, f$ to R_2 such that $f = \bar{f} \circ \pi$. Now, we prove $\bar{f}$ must be a homomorphism of ring. It is sufficient to prove that $\bar{f}$ keeps the multiplications. Given $a, b \in R_1$, we have

$$
\begin{aligned}
\bar{f}((a+K)(b+K)) &= \bar{f}(\pi(a)\pi(b)) = \bar{f}(\pi(ab)) \\
&= \bar{f} \circ \pi(ab) = f(ab) = f(a)f(b) \\
&= \bar{f}(\pi(a))\bar{f}(\pi(b)) = \bar{f}(a+K)\bar{f}(b+K).
\end{aligned}
$$

This proves the assertion. Thus, $\bar{f}$ is an isomorphism from $R_1/\mathrm{Ker}\, f$ to R_2, and the diagram is commutative.

(2) By the Fundamental Theorem of Homomorphisms of groups, f establishes a bijection between the set of the subgroups of $\{R_1; +\}$ containing K and that of subgroups of $\{R_2; +\}$, denoted as $\tilde{f}$. Recall that the bijection $\tilde{f}$ is defined as follows: given a subgroup H of $\{R_1; +\}$ containing K, we define $\tilde{f}(H) = f(H)$ (namely, the image of H under the map f), the later being a subgroup of $\{R_2; +\}$. We have shown that $\tilde{f}$ is a bijection. Moreover, given a subgroup H' of $\{R_2; +\}$, we have $\tilde{f}^{-1}(H') = f^{-1}(H')$. We now prove the following assertions:

(a) If H is a subring of R_1, and $H \supseteq K$, then $f(H)$ is a subring of R_2. To prove this, it is sufficient to prove that $f(H)$ is closed with respect to the multiplication of R_2. Given $a', b' \in f(H)$, there exist $a, b \in H$, such that $f(a) = a', f(b) = b'$. Thus, $a'b' = f(a)f(b) = f(ab) \in f(H)$. This proves the assertion.

(b) If H' is a subring of R_2, then $f^{-1}(H')$ is a subring of R_1 containing K. Obviously $f^{-1}(H')$ contains K. Hence, we only need to prove that $f^{-1}(H')$ is a subring of R_2. Similarly as before, we only need to prove that $f^{-1}(H')$ is closed with respect to the multiplication of R_1. Now, given $a, b \in f^{-1}(H')$, we have $f(ab) = f(a)f(b) \in H'$. Hence, $ab \in f^{-1}(H')$. Thus, the assertion holds.

(c) If $I \supseteq K$ and I is an ideal of R_1, then $f(I)$ is an ideal of R_2. By assertion (a), $f(I)$ is a subring of R_2. Moreover, given $a' = f(a) \in f(I)$ and $x' \in R_2$, by the assumption that f is surjective, there exists $x \in R_1$ such that $f(x) = x'$. Since I is an ideal, we have $a'x' = f(a)f(x) = f(ax) \in f(I)$, and $x'a' = f(x)f(a) = f(xa) \in f(I)$. Thus, $f(I)$ is an ideal of R_2.

(d) If I' is an ideal of R_2, then $f^{-1}(I')$ is an ideal of R_1 containing K. By assertion (b), $f^{-1}(I')$ is a subring of R_1 containing K. For any $a \in f^{-1}(I')$ and $x \in R_1$, by the assumption that I' is an ideal of R_2, we have $f(ax) = f(a)f(x) \in I'$, and $f(xa) = f(x)f(a) \in I'$. Therefore, $ax, xa \in f^{-1}(I')$. Hence, $f^{-1}(I')$ is an ideal of R_1.

Now, (2) follows from the above assertions.

(3) Suppose I is an ideal of R_1 containing K. Let π' be the natural map from R_2 to $R_2/f(I)$. Then $\pi' \circ f$ is a surjective homomorphism from R_1 to $R_2/f(I)$. We now prove that $\mathrm{Ker}\,(\pi' \circ f) = I$. If $x \in I$, then $f(x) \in f(I)$, hence $\pi'(f(x)) = 0$. Thus, $x \in \mathrm{Ker}\,(\pi' \circ f)$. On the other hand, If $x \in \mathrm{Ker}\,(\pi' \circ f)$, then by the definition of π' we have $f(x) \in f(I)$. Thus, $x \in f^{-1}(f(I))$. By assertion (2), f establishes a bijection $\tilde{f}$ between the set of ideals of R_1 containing K and that of ideals of R_2. Therefore, we have $\tilde{f}(f^{-1}(f(I))) = f(I) = \tilde{f}(I)$. Consequently, $f^{-1}(f(I)) = I$. This implies that $\mathrm{Ker}\,(\pi' \circ f) = I$. Now, it follows from (1) that $R_1/I \simeq R_2/f(I)$. $\qquad\square$

Corollary 2.4.6. *Let I_1, I_2 be ideals of a ring R, and $I_1 \subseteq I_2$. Then we have $R/I_2 \simeq (R/I_1)/(I_2/I_1)$.*

Proof. In Theorem 2.4.5, let $R_1 = R, R_2 = R/I_1$, and let f be the natural homomorphism from R_1 to R_2. Then the assertion follows. $\qquad\square$

Recall that the intersection of a family of (finitely or infinitely many) ideals of a ring is also an ideal. Now, for two ideals I, J of a ring R, we define

$$I + J = \{a + b \,|\, a \in I, b \in J\}.$$

Then it is easy to show that $I + J$ is also an ideal of R, called the sum of I, J. Similarly, we can define the sum of finitely many ideals, which is also an ideal of R.

Problem 2.4.7. Construct an example to show that the union of two ideals is not necessarily an ideal. Find out the condition for the union of two ideals to be an ideal.

Now, we can prove the following.

Corollary 2.4.8. *Let I, J be two ideals of a ring R. Then $(I+J)/I \simeq J/(I \cap J)$.*

Proof. We use the fundamental theorem of homomorphisms of rings. First, we define a map φ from $I + J$ to $J/(I \cap J)$ by

$$\varphi(a + b) = b + I \cap J, \quad a \in I, \ b \in J.$$

We show that φ is well-defined. If $a_1, a_2 \in I, b_1, b_2 \in J$ and $a_1 + b_1 = a_2 + b_2$, then we have $b_1 - b_2 = a_2 - a_1 \in I \cap J$. Thus, $b_1 + I \cap J = b_2 + I \cap J$. This proves our assertion. Now, it is easily seen that

$$\begin{aligned}
\operatorname{Ker} \varphi &= \{a + b \in I + J | \varphi(a + b) = 0 + I \cap J\} \\
&= \{a + b \in I + J | b + I \cap J = 0 + I \cap J\} \\
&= \{a + b \in I + J | b \in I \cap J\} = I.
\end{aligned}$$

Then by the fundamental theorem of homomorphism of rings, we have $(I + J)/I \simeq J/(I \cap J)$. $\qquad \square$

We now introduce some notions which are closely related to homomorphisms.

Definition 2.4.9. Let R_1, R_2 be rings. A map φ from R_1 to R_2 is called an **anti-homomorphism**, if

(1) φ is a homomorphism from the additive group $\{R_1; +\}$ to $\{R_2; +\}$;
(2) For any $a, b \in R_1$, we have $\varphi(ab) = \varphi(b)\varphi(a)$.

An anti-homomorphism which is also a bijection is called an **anti-isomorphism**. If there exists an anti-isomorphism between the rings R_1 and R_2, then we say that R_1 is anti-isomorphic to R_2.

Example 2.4.10. Let $\mathbb{F}$ be a number field. In the ring $\mathbb{F}^{n \times n}$ we define φ by $\varphi(A) = A'$. Then φ is an anti-isomorphism.

Example 2.4.11. In the ring of quaternions $\mathbb{H}$, we define a map $A : a\mathbf{1} + b\mathbf{i} + c\mathbf{j} + d\mathbf{k} \to a\mathbf{1} - b\mathbf{i} - c\mathbf{j} - d\mathbf{k}$. Then A is an anti-isomorphism. Readers are suggested to write down the above map in the matrix form.

Given a ring R, we can construct another ring R^{op}, such that R is anti-isomorphic to R^{op}. In fact, as a set we only need to set $R^{\text{op}} = R$. Then on R^{op} we define the same addition as R, and define the multiplication on R^{op} by $a \circ b = ba$, $a, b \in R$. It is easy to check that the identity map is an anti-isomorphism from R to R^{op}.

We now introduce a notion which is intermediate between homomorphisms and anti-homomorphisms.

Definition 2.4.12. Let R_1, R_2 be rings. A map φ from R_1 to R_2 is called a **semi-homomorphism**, if

(1) φ is a homomorphism from the additive group $\{R_1; +\}$ to $\{R_2; +\}$;
(2) For any $a, b \in R_1$, at least one of the equalities $\varphi(ab) = \varphi(a)\varphi(b)$ and $\varphi(ab) = \varphi(b)\varphi(a)$ holds.

It is an interesting problem whether there is a genius semi-homomorphism, or in other words, whether there exists a semi-homomorphism φ, such that φ is neither a homomorphism nor an anti-homomorphism. The following theorem of Loo-Keng Hua answers this problem.

Theorem 2.4.13. *Let φ be a semi-homomorphism from R_1 to R_2. Then φ is either a homomorphism or an anti-homomorphism.*

Proof. We apply the assertion in group theory that any group cannot be written as the union of two proper subgroups. Let φ be a semi-homomorphism from the ring R_1 to R_2. Given $a \in R_1$, we set

$$l_a = \{b \in R_1 \mid \varphi(ab) = \varphi(a)\varphi(b)\},$$
$$r_a = \{b \in R_1 \mid \varphi(ab) = \varphi(b)\varphi(a)\}.$$

We assert that l_a, r_a are both subgroups of the additive group $\{R_1; +\}$. We will only prove the assertion for l_a, the proof for r_a being similar. Since $0 \in l_a$, l_a is non-empty. Moreover, for $b_1, b_2 \in l_a$, we have

$$\varphi(a(b_1 - b_2)) = \varphi(ab_1 - ab_2) = \varphi(ab_1) - \varphi(ab_2)$$
$$= \varphi(a)\varphi(b_1) - \varphi(a)\varphi(b_2)$$
$$= \varphi(a)(\varphi(b_1) - \varphi(b_2))$$
$$= \varphi(a)\varphi(b_1 - b_2),$$

Thus, $b_1 - b_2 \in l_a$. This proves the assertion. By the assumption, we have $l_a \cup r_a = R_1$. Therefore, we have either $l_a = R_1$ or $r_a = R_1$. Now, we set

$$R_l = \{a \in R_1 \,|\, l_a = R_1\}, \quad R_r = \{a \in R_1 \,|\, r_a = R_1\}.$$

A similar argument as above shows that R_l, R_r are both subgroups of $\{R_1; +\}$. Then the above assertion implies that $R_l \cup R_r = R_1$. Consequently we have either $R_l = R_1$ or $R_r = R_1$, namely, φ is either a homomorphism or an anti-homomorphism. $\qquad\square$

Remark 2.1. Theorem 2.4.13 was proved by Loo-Keng Hua in 1949 Hua (1949). Hua applied this theorem to solve an important problem in affine geometry. The original proof of Hua is somehow complicated.

Exercises

(1) Let φ be a surjective homomorphism from a ring R_1 to R_2, where R_2 is not the null ring. Prove the following assertions:

    (i) If R_1 is commutative, then so is R_2;
    (ii) If R_1 is a ring with identity, then so is R_2;
    (iii) If R_1 is a division ring, then so is R_2;
    (iv) If R_1 is a field, then so is R_2;

    Moreover, suppose R_1 is a domain. Is R_2 necessarily a domain? Why?

(2) Let φ be a homomorphism from $\mathbb{Z}[x]$ to $\mathbb{R}$ defined by $\varphi(f(x)) = f(1 + \sqrt{2})$. Determine Ker φ.

(3) Show that the ring $\mathbb{Z}_p$, where p is a prime number, has only two endomorphisms, namely, the null homomorphism and the identity map. Is this assertion valid for a general division ring? Explain why.

(4) Let R be a division ring (maybe a field), and R^* the group of non-zero elements of R with respect to the multiplication. Show that the additive group $\{R; +\}$ is not isomorphic to the multiplicative group R^*.

(5) Let φ be a surjective homomorphism from $\mathbb{Z}$ to a ring R, and R_1 a subring of R. Show that R_1 is an ideal of R.

(6) Determine all the automorphisms of the number field $\mathbb{Q}[\sqrt{-1}] = \{a + b\sqrt{-1} \,|\, a, b \in \mathbb{Q}\}$.

(7) Show that the field of real numbers has only one automorphism.
(8) Show that the map $\varphi : a + b\sqrt{-1} \mapsto a - b\sqrt{-1}$ is an automorphism of $\mathbb{C}$.
(9) Let R be a commutative ring without zero divisors, and suppose $\mathrm{Ch}\,R = p$. Show that the map F from R to R defined by $F : a \mapsto a^p$ is a homomorphism of R, called the **Frobenius endomorphism**. Is F an injective homomorphism, or an isomorphism?
(10) Let R be a ring. On the set $\mathbb{Z} \times R$ we define addition and multiplication as

$$(m, a) + (n, b) = (m + n, a + b),$$
$$(m, a) \cdot (n, b) = (mn, na + mb + ab), \quad m, n \in \mathbb{Z}, a, b \in R.$$

Show that $\mathbb{Z} \times R$ is a ring with identity, and R is isomorphic to an ideal of $\mathbb{Z} \times R$.
(11) Let R be a ring without zero divisors, and $|R| = p$, where p is a prime number. Show that R is a field, and is isomorphic to $\mathbb{Z}_p$.
(12) Let R be a ring with identity. Show that the minimal subring of R containing the identity element e must be isomorphic to $\mathbb{Z}_m$ $(m > 0)$ or $\mathbb{Z}$.
(13) Let R_1, R_2 be two rings. On the set $R_1 \times R_2$ we define addition and multiplication by

$$(a_1, b_1) + (a_2, b_2) = (a_1 + a_2, b_1 + b_2),$$
$$(a_1, b_1)(a_2, b_2) = (a_1 a_2, b_1 b_2), \quad a_1, a_2 \in R_1, \; b_1, b_2 \in R_2.$$

Show that $R_1 \times R_2$ is a ring with respect to the above operations, called the direct sum of R_1 and R_2 (generally denoted as $R_1 \oplus R_2$). Determine which of the following maps are homomorphisms of rings:

  (i) $R_1 \to R_1 \times R_2 : a \mapsto (a, 0)$;
  (ii) $R_1 \to R_1 \times R_1 : a \mapsto (a, a)$;
  (iii) $R_1 \times R_1 \to R_1 : (a_1, b_1) \mapsto a_1 b_1$;
  (iv) $R_1 \times R_1 \to R_1 : (a_1, b_1) \mapsto a_1 + b_1$.

(14) Let R_1, R_2 be two rings with identity, and φ a surjective homomorphism from R_1 to R_2. Show that if u is a unit in R_1, then $\varphi(u)$ must be a unit of R_2. Is the map $u \mapsto \varphi(u)$ a surjective homomorphism from the group of units of R_1 to that of R_2?

(15) Let R be a division ring. Show that any non-null homomorphism from R to a ring must be injective.

(16) Let R be a commutative ring with identity 1. Define two binary operations $\oplus, \cdot$ on R as

$$a \oplus b = a + b + 1; \quad a \cdot b = a + b + ab, \quad a, b \in R.$$

Show that $\{R; \oplus, \cdot\}$ is a commutative ring isomorphic to R.

(17) Let m_1, m_2 be distinct positive integers. Show that, as abelian groups, $m_1\mathbb{Z}$ is isomorphic to $m_2\mathbb{Z}$. However, they are not isomorphic as rings.

(18) Let p, q be two positive integers which are relatively prime. Show that $\mathbb{Z}_{pq} \simeq \mathbb{Z}_p \oplus \mathbb{Z}_q$. Is the above assertion still true if p, q are not relatively prime? Explain why.

(19) In this exercise, we will show the existence and uniqueness of the field of fractions for a domain. Suppose R is a domain, and is a subring of a field F. If for any $a \in F$, there exist $b, c \in R$, $c \neq 0$, such that $a = bc^{-1}$, then F is called a **field of fractions** of R. We first prove the existence of a field of fractions by explicit construction. On the set $R \times R^*$ we define addition and multiplication as the following:

$$(a, b) + (c, d) = (ad + bc, bd),$$
$$(a, b)(c, d) = (ac, bd), \quad (a, b), (c, d) \in R \times R^*.$$

(i) Show that $R \times R^*$ is a commutative semigroup with identity $(0, 1)$ with respect to the addition and is a semigroup with identity $(1, 1)$ with respect to the multiplication.

(ii) On the set $R \times R^*$ we define a binary relation $\sim$ by $(a, b) \sim (c, d) \iff ad = bc$. Show that $\sim$ is an equivalent relation, and is a congruence relation with respect to the addition and multiplication.

(iii) Let $F = R \times R^*/ \sim$ be the quotient set, and denote by $\frac{a}{b}$ the equivalent class of (a, b). On the set F we define addition and multiplication as:

$$\frac{a}{b} + \frac{c}{d} = \frac{ad + bc}{bd},$$
$$\frac{a}{b}\frac{c}{d} = \frac{ac}{bd}$$

Show that the above operations make F into a field.

(iv) Show that the map from R to F defined by $\varphi : a \mapsto \frac{a}{1}$ is an injective homomorphism of rings, hence R can be viewed as a subring of F. Show that F is a field of fractions of R.

(v) Let K be a field which contains R as a subring. Show that there is a subfield F_2 of K such that $F_2 \supset R$, and F_2 is also a field of fractions of R. Hence, the field of fractions of a domain is unique.

Exercises (hard)

(20) Let R be a commutative ring. Show that for any ideals I, J of R, $\sqrt{I+J} = \sqrt{\sqrt{I}+\sqrt{J}}$.

(21) Let R', S be rings, and $R' \cap S = \emptyset$. Suppose S contains a subring R isomorphic to R'. Show that there exists a ring S', which is isomorphic to S, such that R' is a subring of S'.

(22) (Chinese Remainder Theorem). Let R be a ring with identity, and I_1, I_2 be ideals of R such that $R = I_1 + I_2$. Show that for any $a_1, a_2 \in R$, there exists $a \in R$ such that $a - a_1 \in I_1$ and $a - a_2 \in I_2$. Generalize the above assertion to the case of finitely many ideals, and deduce the classical Chinese Remainder Theorem for the ring of integers $\mathbb{Z}$.

(23) Let R, I_1, I_2 be the same as in the previous exercise. Show that $R/(I_1 \cap I_2) \simeq R/I_1 \oplus R/I_2$.

2.5   Factorizations in a domain

In this section, we study the factorial problem in a ring with respect to the multiplication. Recall that every positive integer not equal to 1 can be uniquely written as the product of finitely many prime numbers (without considering the order). This assertion can be generalized to the ring of integers, namely, every non-zero integer which is different from ± 1 can be factored into the product of finitely many integers of the form $\pm p$, where p is a prime number. Moreover, the factorization is unique up to order and the sign of the numbers. In linear algebra, we have studied the factorial theorem of the polynomial ring in one indeterminant over a number field. It is proved that every polynomial with degree greater than zero can be factored into the product of finitely many irreducible polynomials, and the factorization is unique up to order and non-zero constant multiples. These factorial theorems are vital in the theory of numbers or polynomials. For example, given two elements a, b, we can determine whether a is a divisor of b, or whether the equation $ax = b$ has a solution, provided that we know the explicit factorizations of a and b. On the other hand, the explicit factorization also gives us the greatest common divisor of any set of finitely many elements. The purpose of this section is to generalize the above results to rings. For the sake of simplicity, we will only consider domains.

We first define the notion of divisors.

Definition 2.5.1. Let R be a domain, and $a, b \in R$. If there exists $c \in R$ such that $a = bc$, then we say that b is a **factor** or **divisor** of a, and a is a **multiple** of b, denoted as $b \mid a$. If b is not a divisor of a, then we write $b \nmid a$.

It is easily seen from the definition that the divisibility is a binary relation in R, and the relation is reflexive and transitive, but not symmetric. Therefore, it is not an equivalence relation. Moreover, if u is a unit element, then $u \mid a$, $\forall a \in R$.

Now, we will generalize the notions of prime numbers and irreducible polynomials to a domain. Let us first recall their characterization. For convenience, in this section, we will use the term *prime number* to refer to an integer which is either a prime number in the strict sense (which must be a positive integer) or the negative of such a number. Then an integer p, $p \neq 0$, $p \neq \pm 1$, is prime if and only if the only divisors of p are ± 1 and $\pm p$; Meanwhile, it is easily seen that p is a prime number if and only if for any integers m, n, $p \mid mn$ implies that $p \mid m$ or $p \mid n$. In linear algebra, we have also proved that a polynomial $p(x)$ over a number field $\mathbb{F}$, which is non-zero and with degree > 0, is irreducible if and only if the divisors of $p(x)$ must be of the form c, or $cp(x)$, where c is a non-zero number in $\mathbb{F}$, and this is equivalent to the condition that for any two polynomials $f(x), g(x)$, $p(x) \mid f(x)g(x)$ implies that $p(x) \mid f(x)$ or $p(x) \mid g(x)$.

From the above observation, we see that the numbers ± 1 in the ring of integers and the polynomials c, $c \in \mathbb{F}^*$ in $\mathbb{F}[x]$ should not be considered when we deal with the factorial problem. Notice that, in the ring of integers, the group of units consists of ± 1, and that of $\mathbb{F}[x]$ is exactly the set $\mathbb{F}^*$. This leads to the following.

Definition 2.5.2. Let R be a domain, and $a, b \in R$. If there exists a unit u such that $a = ub$, then we say that a and b are **associates**, and denote as $a \sim b$.

The relation of associativity is obviously an equivalent relation as well as a congruence relation with respect to the multiplication. Moreover, it is easily seen that $u \in R^*$ is a unit if and only if $u \sim 1$.

Definition 2.5.3. Let R be a domain and $a \in R$. Then any unit and the associates of a are divisors of a, called the **trivial divisors** of a. If $b \mid a$, but $a \nmid b$, we say that b is a **proper factor** or **proper divisor** of a.

As we mentioned at the beginning of this section, there are two equivalent ways to define prime numbers or irreducible polynomials, one in terms of the set of divisors and the other in terms of the properties of divisibility. A natural idea is to generalize these conditions directly to a domain. However, this is not always reasonable. Let us first look at an example.

Example 2.5.4. Consider the set of numbers

$$\mathbb{Z}[\sqrt{5}] = \{a + b\sqrt{5} \mid a, b \in \mathbb{Z}\}.$$

It is obvious that $\mathbb{Z}[\sqrt{5}]$ is a domain with respect to the ordinary addition and multiplication of numbers. Now, we assert that

(1) The group of units of $\mathbb{Z}[\sqrt{5}]$ is

$$U = \{a + b\sqrt{5} \mid |a^2 - 5b^2| = 1\}.$$

In fact, if $|a^2 - 5b^2| = 1$, then

$$(a + b\sqrt{5})(a - b\sqrt{5}) = a^2 - 5b^2 = \pm 1.$$

Thus, $a + b\sqrt{5}$ is a unit. On the other hand, we define a function N on $\mathbb{Z}[\sqrt{5}]$ by $N(a + b\sqrt{5}) = |a^2 - 5b^2|$. Then it is easy to check that $N(\alpha\beta) = N(\alpha)N(\beta)$. If $a + b\sqrt{5}$ is a unit, then there exists $c + d\sqrt{5}$ such that $(a + b\sqrt{5})(c + d\sqrt{5}) = 1$. Thus, $N(a + b\sqrt{5})N(c + d\sqrt{5}) = 1$, hence $N(a + b\sqrt{5}) = |a^2 - 5b^2| = 1$.

(2) All the divisors of the element 2 are trivial. In fact, if $\alpha \mid 2$, then there exist $a, b \in \mathbb{Z}$, such that $2 = \alpha(a + b\sqrt{5})$. Then $N(a + b\sqrt{5})N(\alpha) = 4$. Since there is no $\beta \in \mathbb{Z}[\sqrt{5}]$ such that $N(\beta) = 2$ (see the exercise hereafter), We have either $N(a + b\sqrt{5}) = 1$ or $N(\alpha) = 1$. Thus, either $a + b\sqrt{5}$ or α is a unit, or in other words, α is a trivial divisor.

(3) It is easily seen that $2 \mid (1 + \sqrt{5})(1 - \sqrt{5})$, but $2 \nmid (1 + \sqrt{5})$, $2 \nmid (1 - \sqrt{5})$.

Problem 2.5.5. Show that there do not exist integers a, b such that $a^2 - 5b^2 = \pm 2$.

Example 2.5.4 shows that, the two equivalent conditions about the prime numbers in the ring of integers, or irreducible polynomials in the ring of polynomials over a number field, are no longer equivalent in a general domain. This leads to the following definitions.

Definition 2.5.6. Let R be a domain, and $a \in R^* - U$. If all the factors of a are trivial, then a is called an **irreducible element**. Otherwise, a is called a **reducible element**.

Definition 2.5.7. Let R be a domain, and $p \in R^* - U$. We say that a is a **prime element** if $p \mid ab$ implies $p \mid a$ or $p \mid b$.

From the above observation, we see that an element $a \in \mathbb{Z}$ is irreducible if and only if it is prime. This assertion also holds in the ring of polynomials over a number field. However, in a general domain, there may exist an irreducible element which is not prime. The following lemma gives the relationship between irreducible elements and prime elements.

Lemma 2.1. *Every prime element must be irreducible.*

Proof. Let R be a domain, and suppose $p \in R^* - U$ is a prime element in R. If a is a divisor of p, then there exists $b \in R^*$ such that $p = ab$, hence $p \mid ab$. Since p is prime, we have $p \mid a$ or $p \mid b$. If $p \mid a$, then $a \sim p$. If $p \mid b$, there exists $c \in R^*$ such that $b = pc$, whence $p = pac$. By the cancellation law of a domain, we have $ac = 1$. Thus, a is a unit. Therefore, a divisor of p must be either a unit or an associate of p. Hence, p is irreducible. $\qquad\square$

Now, let us turn to the factorial problem in a domain. We say that an element a in a domain can not be factored, if whenever a is written as the product of two elements $a = bc$, one of b, c must be a unit (and the other must be an associate of it). Given a domain R, our goal is to find some necessary and sufficient conditions such that every non-zero and non-unit element in R can be uniquely written as the product of elements which can not be factored. Furthermore, if the factorization exists, we need also consider the uniqueness.

We first consider the existence problem. To this end we introduce some definitions.

Definition 2.5.8. We say that a domain R satisfies the **finite factorization condition** if any non-zero and non-unit element a cannot be further factored after finite steps of factorizations in any manner.

Let us give some clarification about the above condition. Consider a non-zero and non-unit element a in R. If a is irreducible, it certainly can not be factored into the product of two proper divisors. If a is reducible, then there exist $b_1, b_2 \in R$ such that $a = b_1 b_2$, where b_1, b_2 are both proper divisors of a. Then we consider the factorization of b_1, b_2. If both b_1 and b_2 are irreducible, then in this way a cannot be further factored. If at least one of b_1, b_2 is reducible, then we consider the factorizations of b_1 and b_2, and so on. If in this process every factor becomes irreducible after finite steps, then we say that in this manner a can not be further factored after finite steps. Notice that in the first step, a might be factored as $a = c_1 c_2$, where c_1, c_2 are proper divisors of a, and none of c_1, c_2 is an associate of b_1 or b_2. We then consider factorization of c_1, c_2, and so on. The finite factorization condition means that, no matter in what manner, any non-zero and non-unit element can not be further factored after finite steps. It is easily seen that, in a domain satisfying the finite factorization condition, every non-zero and non-unit element can be written as the product of finitely many irreducible elements.

Definition 2.5.9. Let R be a domain. A sequence of (finitely or infinitely many) elements $a_1, a_2, \ldots$ in R is called a **proper divisor sequence** if for any $l \geq 1$, a_{l+1} is a proper divisor of a_l. If there does not exist any infinite proper divisor sequence in R, then R is said to satisfy the **divisor chain condition**.

We now prove the following.

Theorem 2.5.10. *A domain satisfies the finite factorization condition if and only if it satisfies the divisor chain condition.*

Proof.   Suppose the domain R satisfies the finite factorization condition, but there is an infinite proper division sequence $a_1, a_2, \ldots$. Then for any $l \geq 1$, a_{l+1} is a proper divisor of a_l, hence a_l can be factored as $a_l = a_{l+1} b_{l+1}$, where b_{l+1} is also a proper divisor of a_l (please explain why). This means that there exists a way to factor the element a endlessly, contradicting the assumption that R satisfies finite factorization condition. Therefore, R satisfies the divisor chain condition.

Conversely, suppose R satisfies the divisor chain condition, but does not satisfy the finite factorization condition. Then there exists a non-zero and non-unit element a and an endless process of factoring of a. Thus, there must be an infinite sequence b_l, $l = 1, 2, \ldots$, such that b_1 is a proper divisor of a, and b_{l+1} is a proper divisor of b_l, for any $l = 1, 2, \ldots$. But this is a contradiction with the divisor chain condition. This completes the proof of the theorem. $\square$

Now, we turn to the main topic of this section.

Definition 2.5.11. Let R be a domain. Suppose R satisfies the finite factorization condition and the factorization of any $a \in R^* - U$ is unique, in the sense that if there are two kinds of factorizations of a:

$$a = p_1 p_2 \ldots p_r = q_1 q_2 \ldots q_s,$$

where $p_1, p_2, \ldots, p_r, q_1, q_2, \ldots, q_s$ are irreducible, then $r = s$, and we can make $p_i \sim q_i$, $1 \leq i \leq r$, after some appropriate changes of the order. Then R is called a **unique factorization domain**. Sometimes we also say that R is **uniquely factorial**.

We stress here again that, since R satisfies the finite factorization condition, every non-zero and non-unit element can be factored into the product of finitely many irreducible elements; hence the uniqueness of the factorization makes sense. Moreover, as we mentioned above, the ring of integers and the ring of polynomials over a number field are both unique factorization domains, and there exist domains which are not uniquely factorial, for example, the ring $\mathbb{Z}[\sqrt{5}]$ or $\mathbb{Z}[\sqrt{-5}]$ (see the exercises of this section). The main goal of this section is to give several criteria for a domain to be uniquely factorial. For this we need some more definitions.

Definition 2.5.12. A domain R is said to satisfy the **primeness condition** if every irreducible element in R is prime.

Definition 2.5.13. Let R be a domain, and $a_1, a_2, \ldots, a_n \in R$. An element $d \in R$ is called a **common divisor** of $a_1, a_2, \ldots, a_n$ if $d \mid a_i$, for any $1 \leq i \leq n$. A common divisor d is called a **greatest common divisor** (abbreviated as g.c.d.) if for any common divisor c of $a_1, a_2, \ldots, a_n$, we have $c \mid d$. The domain R is said to satisfy the **greatest common divisor condition** (abbreviated as g.c.d. condition) if any two elements in R have a greatest common divisor.

We now give some properties of a domain satisfying the greatest common divisor condition. Let R be a domain satisfying the greatest common divisor condition. Then we have:

(1) The greatest common divisor of any two elements in R is unique up to a unit.

   The proof is easy and left to readers. In the following, we will denote the greatest common divisor of a and b as (a, b).

(2) Any finitely many elements of R have a greatest common divisor (unique up to a unit). Moreover, for any $a, b, c \in R$, $(a, (b, c))$ is a greatest common divisor of a, b, c. In particular, we have $(a, (b, c)) \sim ((a, b), c)$.

   The proof is easy and left to readers.

(3) For any $a, b, c \in R$, we have $c(a, b) \sim (ca, cb)$.

   Without loss of generality, we can assume that none of a, b, c is zero. Suppose $(a, b) = d_1$, $(ca, cb) = d_2$. Since $d_1 \mid a, d_1 \mid b$, we have $cd_1 \mid ca, cd_1 \mid cb$. Thus, $cd_1 \mid d_2$. Moreover, suppose $ca = u_1 d_2, d_2 = u_2 cd_1$. Then $ca = u_1 u_2 cd_1$, hence $a = u_1 u_2 d_1$. Thus, $u_2 d_1 \mid a$. Similarly, we have $u_2 d_1 \mid b$. Therefore, $u_2 d_1 \mid d_1$, which implies that $u_2 d_1 \sim d_1$ and u_2 is a unit. Therefore, $cd_1 = u_2^{-1} d_2 \sim d_2$.

(4) If $a, b, c \in R$ and $(a, b) \sim 1$, $(a, c) \sim 1$, then $(a, bc) \sim 1$.

   By 3), we have $(ac, bc) \sim c$, $(a, ac) \simeq a$. Thus, $(a, bc) \sim ((a, ac), bc) \sim (a, (ac, bc)) \sim (a, c) \sim 1$.

(5) Any irreducible element in R must be prime, or in other words, R satisfies the primeness condition.

If p is irreducible and $p \mid ab$, but $p \nmid a$, $p \nmid b$, then we have $(p, a) \sim 1$ and $(p, b) \sim 1$. By 4), we have $(p, ab) \sim 1$, which is a contradiction.

In the exercises of this section, we will present an example of a domain R in which there exist two elements a, b, such that the greatest common divisor of a, b does not exist.

Now, we can prove the main theorem of this section.

Theorem 2.5.14. *Let R be a domain. Then the following conditions are equivalent*:

(1) *R is a unique factorization domain*;
(2) *R satisfies the divisor chain condition and the primeness condition*;
(3) *R satisfies the divisor chain condition and the g.c.d. condition.*

Proof. (1) $\Rightarrow$ (2) Let R be a unique factorization domain. We first show that R satisfies the divisor chain condition. This follows directly from Theorem 2.5.10. However, for further applications, we would like to present another proof. Since every non-zero and non-unit element can be factored into the product of finitely many irreducible elements, and the uniqueness holds, we can define the length function l on R^* as follows:

$$l(a) = \begin{cases} 0, & \text{if } a \text{ is a unit;} \\ r, & \text{if } a = p_1 p_2 \ldots p_r, \text{ where the } p_i\text{'s are irreducible.} \end{cases}$$

It is easy to check that the function l is well-defined. In particular, it does not depend on the explicit factorization of the element. It is also easily seen that, if $a, b \in R^*$, then $l(ab) = l(a) + l(b)$. Moreover, $l(a) = 0$ if and only if a is a unit. It follows that, if $a \in R^* - U$ is a proper divisor of b, then $l(a) < l(b)$. This implies that any proper divisor chain in R is finite. Thus, R satisfies the divisor chain condition.

Next, we show that R satisfies the primeness condition. Let $p \in R^*$ be an irreducible element, and $p \mid ab$. Then there exists $c \in R$ such that $ab = pc$. We now prove that either $p \mid a$ or $p \mid b$. We need only to consider the following three cases:

(i) If at least one of a, b is equal to 0, then the assertion follows directly.

(ii) If one of a, b is a unit element, say b is a unit, then $a = abb^{-1} = pcb^{-1}$. Thus, $p \mid a$,

(iii) Suppose $a, b \in R^* - U$. We first show that $c \in R^* - U$. It is obvious that $c \neq 0$. If c is a unit element, then pc is an associate of p, hence it is irreducible. But it can be factored as $pc = ab$, and a, b are both non-unit elements. This is a contradiction. Thus, $c \in R^* - U$. By the assumption that R is a unique factorization domain, there exist irreducible elements p_i $(1 \leq i \leq t)$ such that

$$c = p_1 p_2 \ldots p_t.$$

Meanwhile, a, b can also be factored as

$$a = q_1 q_2 \ldots q_r; \quad b = q'_1 q'_2 \ldots q'_s,$$

where $q_1, q_2, \ldots, q_r, q'_1, q_2, \ldots, q'_s$ are irreducible elements. It follows from $pc = ab$ that

$$q_1 q_2 \ldots q_r q'_1 q'_2 \ldots q'_s = p p_1 p_2 \ldots p_t.$$

By the uniqueness, p must be an associate of certain q_i or certain q'_i. If $p \sim q_i$, then $p = q_i \varepsilon$, $\varepsilon \in U$. Thus,

$$a = q_1 \ldots q_i \varepsilon \varepsilon^{-1} q_{i+1} \ldots q_r = p(\varepsilon^{-1} q_1 \ldots q_{i-1} q_{i+1} \ldots q_r),$$

that is, $p \mid a$. Similarly, if p is an associate of one of q'_i, then $p \mid b$. Consequently, we have $p \mid q$ or $p \mid b$. Therefore, p is prime.

This completes the proof of $(1) \Rightarrow (2)$.

$(2) \Rightarrow (3)$ We only need to prove that R satisfies the g.c.d. condition. Given $a, b \in R$, we now show that a, b have a g.c.d. There are three cases as the following:

(i) If at least one of a, b is zero, say $a = 0$, then b is a g.c.d. of a, b.

(ii) If at least one of a, b is a unit, say a is a unit, then a is a g.c.d. of a, b.

(iii) Suppose $a, b \in R^* - U$. Since R satisfies the divisor chain condition, a, b can be factored as

$$a = q_1 q_2 \ldots q_r, \quad b = q'_1 q'_2 \ldots q'_s,$$

where $q_i, q'_j (i = 1, 2, \ldots, r; j = 1, 2, \ldots, s)$ are irreducible elements; hence they are prime elements by the primeness condition. Now, we can write the above equalities uniformly as:

$$a = \varepsilon_a p_1^{k_1} p_2^{k_2} \ldots p_n^{k_n},$$
$$b = \varepsilon_b p_1^{l_1} p_2^{l_2} \ldots p_n^{l_n},$$

where $\varepsilon_a, \varepsilon_b$ are unit elements, $k_i \geq 0, l_i \geq 0$, and $p_1, p_2, \ldots, p_n$ are irreducible and are not associative each other. Let $m_i = \min(k_i, l_i), i = 1, 2, \ldots, n$, and set

$$d = p_1^{m_1} p_2^{m_2} \ldots p_n^{m_n}.$$

Then it is obvious that $d \mid a, d \mid b$, hence, d is a common divisor of a, b. Now, suppose c is also a common divisor of a, b. Then $c \neq 0$. If c is a unit, then $c \mid d$. If c is not a unit, then it can be factored as

$$c = p_1' p_2' \ldots p_t',$$

where the p_i's are irreducible (hence prime). By the assumption $c \mid a$, we have $p_i' \mid a$, Then p_i' must be a divisor of certain p_j. As both p_i' and p_j are irreducible, they must be associative. Therefore, c can also be factored as

$$c = \varepsilon_c p_1^{h_1} p_2^{h_2} \ldots p_n^{h_n},$$

where ε_c is a unit. Since $c \mid a$, and $p_1, p_2, \ldots, p_n$ are not associative each other, we have $h_i \leq k_i$ (please explain why). Similarly, it follows from $c \mid b$ that $h_i \leq l_i$. Thus, $h_i \leq m_i$. Therefore, $c \mid d$. Thus, d is a g.c.d. of a, b.

This completes the proof of $(2) \Rightarrow (3)$.

$(3) \Rightarrow (2)$ In the above we have shown that a domain satisfying the g.c.d. condition must satisfy the primeness condition. This gives the proof of $(3) \Rightarrow (2)$.

$(2) \Rightarrow (1)$ Since R satisfies the divisor chain condition, by Theorem 2.5.10, it must satisfy the finite factorization condition. Thus, we only need to prove the uniqueness of the factorization. Suppose $a \in R^* - U$, and a have two kind of factorizations $a = p_1 p_2 \ldots p_r = q_1 q_2 \ldots q_s$, where p_i, q_j are irreducible. We now show that $r = s$ and we can make $p_i \sim q_i, 1 \leq i \leq r = s$ after appropriate changes of the order. We prove the assertion inductively on r. If $r = 1$, then

$$a = p_1 = q_1 q_2 \ldots q_s.$$

This implies that $s = 1$, and $p_1 = q_1$. Thus, the assertion holds for $r = 1$. Now, we assume that the assertion holds for $r = k - 1$. Then for $r = k$, we have $p_1 \mid q_1 q_2 \ldots q_s$. By the primeness condition, p_1 must be prime. Then p must be a divisor of one of $q_j, j = 1, 2, \ldots, s$. After some changes of the order, we can assume that $p_1 \mid q_1$. Since p_1, q_1 are irreducible, we have $p_1 \sim q_1$. Then we can write $p_1 = \varepsilon q_1$, where ε is a unit element. It follows that

$$(\varepsilon p_2) \ldots p_k = q_2 q_3 \ldots q_s.$$

By the inductive assumption, we get $k - 1 = s - 1 = r - 1$, and we can make $p_j \sim q_j$ after appropriate changes of the order. This means that the assertion holds for $r = k$.

The proof of the theorem is now completed. □

Exercises

(1) In the ring of Gauss integers $\mathbb{Z}[\sqrt{-1}]$, find all the associates of the element $a + b\sqrt{-1}$.

(2) Show that the elements 7 and 23 are irreducible in the ring of Gauss integers $\mathbb{Z}[\sqrt{-1}]$. Is 5 irreducible in this ring?

(3) Show that two integers $a, b \in \mathbb{Z}$ have the same greatest common divisor in the ring of Gauss integers $\mathbb{Z}[\sqrt{-1}]$ as they are in the ring $\mathbb{Z}$.

(4) Let $R = \{\frac{m}{2^n} \mid m, n \in \mathbb{Z}\}$. Show that R is a ring with respect to the addition and multiplication of numbers. Find all the irreducible elements, prime elements, and units in this ring.

(5) Is $x^2 + x + 1$ a divisor of $x^3 + 1$ in the ring $\mathbb{Z}_2[x]$? How about in the ring $\mathbb{Z}_3[x]$?

(6) Show that $\sqrt{-3}, 4 + \sqrt{-3}$ are prime elements in the ring $\mathbb{Z}[\sqrt{-3}] = \{a + b\sqrt{-3} \mid a, b \in \mathbb{Z}\}$.

(7) Show that in the ring $\mathbb{Z}[\sqrt{-5}] = \{a + b\sqrt{-5} \mid a, b \in \mathbb{Z}\}$, $3 \nmid 1 + 2\sqrt{-5}$, and 3 is irreducible but not prime, hence $\mathbb{Z}[\sqrt{-5}]$ is not a unique factorization domain.

(8) Let $R = \mathbb{Z}[\sqrt{-5}]$.

   (i) Show that if $a^2 + 5b^2 = 9$, then $a + b\sqrt{-5}$ is irreducible.

   (ii) Show that the elements $\alpha = 6 + 3\sqrt{-5}$ and $\beta = 9$ in R do not have a greatest common divisor.

(9) Show that the ring $\mathbb{Z}[\sqrt{-5}]$ satisfies the divisor chain condition.

(10) Let $R = \{\frac{a+b\sqrt{-3}}{2} | a, b \in \mathbb{Z}, a + b \text{ is even}\}$.

   (i) Show that R is a domain with respect to the addition and multiplication of numbers;

   (ii) Show that the set of unit elements of R consists of ± 1 and $\frac{\pm 1 \pm \sqrt{-3}}{2}$;

   (iii) Show that 2 is associative to $1 + \sqrt{-3}$ in R.

(11) In the domain $\mathbb{Z}[\sqrt{-3}]$, do $2(1 + \sqrt{-3})$ and 4 have a greatest common divisor?

(12) Show that $\mathbb{Z}[\sqrt{5}]$ is not a unique factorization domain.

(13) Construct an example to show that a subring of a unique factorization domain need not be a unique factorization domain.

(14) Let R be a unique factorization domain, and $a, b \in R^*$. Suppose $m \in R$ satisfies the conditions:

   (i) m is a common multiple of a, b, namely, $a \mid m, b \mid m$;

   (ii) for any common multiple n of a, b, $m \mid n$.

Then m is called a **least common multiple** of a, b. Prove the following assertions:

   (i) Let m be a minimal common multiple of a, b. Then an element m_1 is a minimal common multiple of a, b if and only if $m_1 \sim m$;

   (ii) Any two elements in R^* have a minimal common multiple;

   (iii) Denote by $[a, b]$ the minimal common multiple of a, b. Then

$$[a, b](a, b) \sim ab, \quad [a, (b, c)] \sim ([a, b], [a, c]).$$

(15) Let R be a unique factorization domain. If the g.c.d. of $a_1, a_2, \ldots, a_n \in R$ is a unit, then we say that $a_1, a_2, \ldots, a_n$ are **relatively prime**. Suppose $a_1 = db_1, a_2 = db_2$ and at least one of them is non-zero. Show that d is the g.c.d. of a_1, a_2 if and only if b_1, b_2 are relatively prime.

(16) Show that x, y are relatively prime in the domain $\mathbb{Q}[x, y]$. Do there exist polynomials $r(x, y), s(x, y) \in \mathbb{Q}[x, y]$ such that $r(x, y)x + s(x, y)y = 1$?

(17) Suppose R is a unique factorization domain, and there are only finite unit elements in R. Show that any non-zero element in R has only finitely many divisors.

Exercises (hard)

(18) Show that $\mathbb{Z}[x]$ is a unique factorization domain.
(19) The notions of divisors and associates defined for a domain can be naturally generalized to a commutative ring with identity. Show that if $a, b \in \mathbb{Z}_m$, and $a \mid b$, $b \mid a$, then a, b must be associative.
(20) Suppose R is a commutative ring with identity, and $a, b \in R$ satisfies the conditions $a \mid b$ and $b \mid a$. Are a, b necessarily associative?

2.6   Prime ideals and maximal ideals

In this section, we introduce two kind of special ideals–prime ideals and maximal ideals. Our motivation here is two-fold. First, the study of ideals is an important subject in ring theory; see for example the celebrated work of Jacobson. Therefore, it is helpful for us to become familiar with some special ideals. Second, the results of this section will be applied for several times in the following context of this book.

Recall that rings with *good properties* we have encountered in this chapter are generally domains or fields. In general, we will deal with many commutative rings with identity. If a commutative ring with identity is not a domain, then it contains zero divisors. Intuitively, this means that the ring is *too large*. A method to deal with this defect is to consider the quotient ring with respect to certain ideals, which can be a domain or a field. Let R be a commutative ring with identity and I an ideal of R. If R/I is a domain, then $(a + I)(b + I) = 0 + I$, $a, b \in R$ implies that $a + I = 0 + I$ or $b + I = 0 + I$. Equivalently, $ab \in I \Rightarrow a \in I$ or $b \in I$. This observation leads to the following.

Definition 2.6.1. Let I be an ideal of a ring R, and $I \neq R$. If $ab \in I$ implies $a \in I$ or $b \in I$, then I is called a **prime ideal** of R.

We first give some examples of prime ideals.

Example 2.6.2. Consider the domain $R = \mathbb{Z}$. We have shown that any ideal of R has the form $m\mathbb{Z}$, $m \geq 0$. It is trivial that the null ideal is prime in R. If $m \neq 0$, then $m\mathbb{Z}$ is prime if and only if $ab \in m\mathbb{Z} \Rightarrow a \in m\mathbb{Z}$ or $b \in m\mathbb{Z}$, that is, $m \mid ab \Rightarrow m \mid a$ or $m \mid b$. This is equivalent to the condition that m is a prime number. Consequently, $m\mathbb{Z}$ is a prime ideal if and only if $m = 0$ or m is prime.

Example 2.6.3. Consider the ring $R = \mathbb{Z}_4$. It is easily seen that there are three ideals of R: $\{\bar{0}\}$, $I = \{\bar{0}, \bar{2}\}$ and R itself. As R has a zero divisor $\bar{2}$, $\{\bar{0}\}$ is not a prime ideal of R. Now, consider the ideal I. If $a, b \in \{0, 1, 2, 3\}$, and $\bar{a}\bar{b} \in I$, then at least one of a, b is 0 or 2, hence $\bar{a} \in I$ or $\bar{b} \in I$. Therefore, I is a prime ideal of R.

Problem 2.6.4. Determine all the prime ideals of $\mathbb{Z}_m$.

Now, we study the fundamental properties of prime ideals. First, we have the following.

Theorem 2.6.5. *Let R be a commutative ring with identity, and I an ideal of R, $I \neq R$. Then I is a prime ideal if and only if R/I is a domain.*

Proof.   *The if part* have been proved before Definition 2.6.1. We now prove *the only if part*. Suppose I is a prime ideal. Since R is commutative with identity, R/I is a commutative ring with identity. If in the quotient ring R/I, $(a + I)(b + I) = 0 + I$, where $a, b \in R$, then we have $ab + I = 0 + I$. Hence, $ab \in I$. By the assumption that I is prime, we get $a \in I$ or $b \in I$. Thus, $a + I = 0 + I$ or $b + I = 0 + I$. This means that there is no zero divisors in R/I. Hence, R/I is a domain. □

Notice that a ring R is isomorphic to the quotient ring $R/\{0\}$. Therefore, as a corollary of the above theorem, we have the following.

Corollary 2.6.6. *Let R be a commutative ring with identity. Then R is a domain if and only if the ideal $\{0\}$ of R is prime.*

The following theorem gives the invariance of primeness of ideals under a homomorphism.

Theorem 2.6.7. *Let f be a surjective homomorphism from a commutative ring with identity R_1 onto R_2, and I a prime ideal of R_1 containing the kernel $K = \text{Ker} f$. Then $f(I)$ is a prime ideal of R_2.*

Proof. By the theorem of homomorphisms of rings, $f(I)$ is an ideal of R_2, and $R_1/I \simeq R_2/f(I)$. Since I is prime, by Theorem 2.6.5, R_1/I is a domain, hence $R_2/f(I)$ is also a domain. Then by the same theorem, $f(I)$ is a prime ideal of R_2. □

In the following we give another important feature of prime ideals in terms of set relation, which can also be viewed as the definition of prime ideals. Let R be a commutative ring with identity, and A, B non-empty subsets of R. Denote

$$AB = \left\{ \sum_{i=1}^{n} a_i b_i | a_i \in A, b_i \in B, n \in \mathbb{N}^* \right\}.$$

It is easily seen that if A, B are subrings (resp. ideals) of R, then AB is also a subring (resp. ideal) of R.

Theorem 2.6.8. *Let R be a commutative ring with identity, and I a proper ideal of R. Then I is a prime ideal of R if and only if for any ideals A, B, $AB \subseteq I$ implies $A \subseteq I$ or $B \subseteq I$.*

Proof. *The only if part* Let I be a prime ideal of R, and A, B ideals of R such that $AB \subseteq I$. Suppose conversely that neither $A \subseteq I$ nor $B \subseteq I$. Then there exist $a \in A$ and $b \in B$ such that $a \notin I$, $b \notin I$. Meanwhile, we also have $ab \in AB \subseteq I$, which is a contradiction with the primeness of I.

The if part Suppose conversely that there is an ideal I satisfying the condition of the theorem but it is not prime. Then there exist $a, b \in R$, such that $ab \in I$ but $a \notin I$, $b \notin I$. Now, we consider the ideals $A = \langle a \rangle$, $B = \langle b \rangle$. By the commutativity of R, it is easily seen that $AB = \langle ab \rangle \subseteq I$, which is a contradiction with the assumption. □

Next, we introduce the notion of maximal ideals. Similarly as in the above observation, if R is a commutative ring with identity but not a field, then it is *too large*, and we need to consider the quotient ring of R with respect to an ideal. Now, let us find out the condition for a quotient ring R/I to be a field. It is certainly necessary that $I \neq R$. Moreover, if R/I is a field, then for any $a \notin I$, $a + I$ is invertible in the quotient ring R/I, hence there is $b \in R$ such that $(a + I)(b + I) = 1 + I$, i.e., $1 - ab \in I$. In particular, if I_1 is an ideal of R such that $I \subset I_1$ and $I_1 \neq I$, then there exist $a \in I_1 - I$, $b \in R$ such that $1 - ab \in I$. Thus, for any $c \in R$, we have

$$c = (1 - ab + ab)c = (1 - ab)c + (ab)c = (1 - ab)c + a(bc).$$

Since I is an ideal and $1 - ab \in I$, we have $(1 - ab)c \in I \subseteq I_1$; Moreover, since I_1 is an ideal and $a \in I_1$, we have $a(bc) \in I_1$. Therefore, we have $c \in I_1$. Thus, $I_1 = R$. This leads to the following.

Definition 2.6.9. Let R be a commutative ring with identity. An ideal I of R is called a **maximal ideal** of R, if $I \neq R$, and the only ideals of R containing I are I and R.

Example 2.6.10. Consider the domain $R = \mathbb{Z}$ in Example 2.6.2. Let us determine all the maximal ideals of R. Firstly, it is easily seen that $\{0\}$ is not maximal. Secondly, if $m \geq 2$ is not a prime number, then there exist $n \geq 2$ such that $n|m$ and $n \neq m$. Then we have $m\mathbb{Z} \subset n\mathbb{Z}$ and $m\mathbb{Z} \neq n\mathbb{Z}$, $n\mathbb{Z} \neq R$. Thus, $m\mathbb{Z}$ is not a maximal ideal either. Now, we show that, if p is a prime number, then $\langle p \rangle$ is a maximal ideal of R. Suppose A is an ideal of R such that $\langle p \rangle \subset A$, $\langle p \rangle \neq A$. Then there exists $k \in A$ such that $k \notin \langle p \rangle$. Since p is prime, p, k must be relatively prime. Therefore, there exits integers a, b such that $ap + bk = 1$. This implies that $1 \in A$. Then for any integer l, we have $l = l \times 1 \in A$. Thus, $A = R$. Hence, $\langle p \rangle$ is a maximal ideal of R. Consequently, the maximal ideals of R are $p\mathbb{Z}$, where p is a prime number.

Problem 2.6.11. The above example shows that, in the ring of integers, a non-zero ideal is maximal if and only if it is prime. Construct an example to show that this is not true for a general domain.

Example 2.6.12. Let $\mathbb{F}$ be a field. We show that $\{0\}$ is a maximal ideal of $\mathbb{F}$. Given any non-zero ideal I of $\mathbb{F}$, there exists $a \in I$ such that $a \neq 0$. As any non-zero element in $\mathbb{F}$ is invertible, a has an inverse a^{-1}. Then we have $1 = aa^{-1} \in I$. Thus, for any $b \in \mathbb{F}$, $b = b \cdot 1 \in I$. Hence, $I = \mathbb{F}$. Therefore, $\{0\}$ is a maximal ideal.

Problem 2.6.13. Show that the converse statement of the assertion in Example 2.6.12 is also true, namely, if R is a commutative ring with identity and $\{0\}$ is a maximal ideal of R, then R is a field.

The next theorem gives a sufficient and necessary condition for an ideal to be maximal.

Theorem 2.6.14. *Let R be a commutative ring with identity, and M an ideal of R. Then M is a maximal ideal of R if and only if R/M is a field.*

Proof. Let M be a maximal ideal of R. To show that R/M is a field, it suffices to show that every non-zero element in R/M is invertible. If in the quotient ring $a + M \neq 0 + M$, then $a \notin M$. Consider the ideal $I = \langle M \cup \{a\} \rangle$. By the maximality of M, we have $I = R$. Thus, there exist $a_i \in M, r_i \in R, 1 \leq i \leq n$, and $r \in R$ such that

$$\sum_{i=1}^{n} a_i r_i + ar = 1.$$

Then in the quotient ring R/M we have $(a + M)(r + M) = 1 + M$, hence $a + M$ is invertible in R/M. Thus, R/M is a field.

Conversely, if R/M is a field, then $\{0 + M\}$ is a maximal ideal of R/M. Let A be an ideal of R containing M. Then A/M is an ideal of the quotient ring R/M. Thus, either $A/M = \{0 + M\}$ or $A/M = R/M$, that is, either $A = M$ or $A = R$. Therefore, M is a maximal ideal. $\square$

Exercises

(1) Determine the maximal ideals of the following rings:
   (i) $\mathbb{R} \times \mathbb{R}$; (ii) $\mathbb{R}[x]/\langle x^2 \rangle$;
   (iii) $\mathbb{R}[x]/\langle x^2 - 3x + 2 \rangle$; (iv) $\mathbb{R}[x]/\langle x^2 + x + 1 \rangle$.

(2) Show that $\langle x \rangle$ is a prime ideal of $\mathbb{Z}[x]$, but is not a maximal ideal.

(3) Let R be a domain, I an ideal of R, and π the natural homomorphism from R to R/I. Suppose M is an ideal of R containing I. Show that M is a maximal ideal of R if and only if $\pi(M)$ is a maximal ideal of R/I.

(4) Find out all the prime ideals and maximal ideals of $\mathbb{Z}_{36}$.

(5) Let $\mathbb{F}$ be a number field. Show that $\langle x \rangle$ is a maximal ideal of $\mathbb{F}[x]$.

(6) Let $R = \{2n | n \in \mathbb{Z}\}$. Show that $\langle 4 \rangle$ is a maximal ideal of R. Is the quotient ring $R/\langle 4 \rangle$ a field?

(7) Show that the quotient ring $\mathbb{Z}_2[x]/\langle x^3 + x + 1 \rangle$ is a field, but the quotient ring $\mathbb{Z}_3[x]/\langle x^3 + x + 1 \rangle$ is not a field.

(8) Let R be a ring, M an ideal of R. Suppose any $x \in R - M$ is a unit. Show that M is a maximal ideal of R and it is the only maximal ideal of R.

(9) Let A be an ideal of the ring R, and P a prime ideal of A. Show that P is an ideal of R.

(10) Suppose N is an ideal of R, and the quotient ring R/N is a divisible ring. Show that:

(i) N is a maximal ideal of R;

(ii) $\forall a \in R$, $a^2 \in N$ implies that $a \in N$.

(11) Let $m \in \mathbb{N}, m > 1$. Set

$$A = \{f(x) \quad |f(x) \in \mathbb{Z}[x], m|\, f(0)\}.$$

Show that A is an ideal of $\mathbb{Z}[x]$, and $\langle x \rangle \subsetneqq A$. Determine when A is a prime ideal.

(12) Suppose R is a ring, and for any $a \in R$, there exists $b \in R$ such that $a = b^2$. Show that any maximal ideal of R must be prime.

(13) Let P be an ideal of the ring R, and $Q = R - P$. Show that P is a prime ideal if and only if Q is a semi-group with respect to the multiplication of R.

(14) Let R be a unique factorization domain, and $p \in R$. Show that p is an irreducible element if and only if the principal ideal $\langle p \rangle$ is a prime ideal.

(15) Show that $\langle x, n \rangle$ is a maximal ideal in $\mathbb{Z}[x]$ if and only if n is a prime number.

Exercises (hard)

(16) Let R be a commutative ring with identity and I an ideal of R. Suppose J_i, $1 \le i \le m$ are prime ideals of R, and $I \subseteq \cup_{1 \le i \le m} J_i$. Show that there exists $1 \le l \le m$, such that $I \subseteq J_l$. Does the above assertion hold without the assumption that J_i are prime ideals?

(17) Let R be a commutative ring with identity, and $\mathcal{M}$ be the intersection of all maximal ideals of R. Show that

$$\mathcal{M} = \{x \in R \,|\, 1 - xy \in U, \quad \forall y \in R\},$$

where U is the group of units of R.

2.7   Principal ideal domains and Euclidean domains

Let us go back to the study of unique factorization domains. Recall that Theorem 2.5.14 gives some sufficient and necessary conditions for a domain to be uniquely factorial. Sometimes such conditions are not convenient. For example, in some special cases, we can use some sufficient (not necessary) conditions to assert that certain domains are uniquely factorial. We will give two classical examples of this type, namely principal ideal domains and Euclidean domains.

Definition 2.7.1. Let R be a commutative ring with identity. We say that R is a **principal ideal ring**, if every ideal of R is principal. If a principal ideal ring is a domain, then it is called a **principal ideal domain**.

We first give some examples.

Example 2.7.2. The ring of integers $\mathbb{Z}$ is a principal ideal domain. In fact, we have seen that any ideal of $\mathbb{Z}$ must be of the form $m\mathbb{Z} = \langle m \rangle$, where $m \geq 0$.

Example 2.7.3. Let us prove that $\mathbb{Z}[x]$ is not a principal ideal domain. Suppose conversely that this is not true. Then the ideal generated by the subset $S = \{3, x\}$ is a principal ideal. Thus, there exists $g(x) \in \mathbb{Z}[x]$ such that $\langle S \rangle = \langle g(x) \rangle$. On one hand, it follows from $3 \in \langle g(x) \rangle$ that $g(x) \mid 3$. Thus, $g(x)$ is equal to ± 1 or ± 3. On the other hand, it follows from $x \in \langle g(x) \rangle$ that there exists $u(x) \in \mathbb{Z}[x]$ such that $x = g(x)u(x)$. Therefore, we have $g(x) = \pm 1$. Thus, $\langle S \rangle = \mathbb{Z}[x]$, which is absurd, since $2 \notin \langle S \rangle$. Thus, $\mathbb{Z}[x]$ is not a principal ideal ring.

Problem 2.7.4. Let p be a prime number, and $u(x) \in \mathbb{Z}[x]$ a polynomial with degree > 0. Determine when the ideal $\langle p, u(x) \rangle$ is a principal ideal,

In Section 2.9, we will show that the polynomial ring over a unique factorization domain must be a unique factorization domain. Therefore, $\mathbb{Z}[x]$ is a unique factorization domain. This gives an example of a unique factorization domain which is not a principal ideal domain. However, in the following we will show that any principal ideal domain must be uniquely factorial.

We first give a lemma. As we mentioned above, the union of a family of ideals is generally not an ideal. But we have the following.

Lemma 2.7.5. *Suppose I_i, $i = 1, 2, \ldots$ is a sequence of ideals in a ring R, and $I_i \subset I_{i+1}$, $i = 1, 2, \ldots$. Then $I = \bigcup_i I_i$ is an ideal of R.*

Proof. We use the criterion for ideals directly. For any $x, y \in I$, and $r \in R$, there exist i, j such that $x \in I_i, y \in I_j$. Without loss of generality, we assume that $i \leq j$. Then by the assumption, we have $x, y \in I_j$. Since I_j is an ideal, we have $x - y \in I_j \subseteq I$ and $rx \in I_j \subseteq I$, $xr \in I_j \subseteq I$. Thus, I is an ideal of R. $\qquad\square$

Now, we can prove the following theorem.

Theorem 2.7.6. *A principal ideal domain must be a unique factorization domain.*

Proof. By the equivalent condition (2) of Theorem 2.5.14, we only need to prove that a principal ideal domain R satisfies the divisor chain condition and the primeness condition. The proof will be divided into the following two steps:

(1) R satisfies the divisor chain condition. Let

$$a_1, a_2, \ldots, a_n, \ldots$$

be a sequence of elements of R such that a_{k+1} is a proper divisor of a_k, $k = 1, 2, \ldots$. We now prove that the sequence must be finite. Consider the sequence of ideals $I_k = \langle a_k \rangle$, $k = 1, 2, \ldots$. Since a_{k+1} is a proper divisor of a_k, we have $I_k \subset I_{k+1}, k = 1, 2, \ldots$. Let I be the union of the above ideals. Then by Lemma 2.7.5, I is an ideal of R. Since R is a principal ideal domain, there exists $d \in R$ such that $I = \langle d \rangle$. As $d \in I$, there exists an index m such that $d \in \langle a_m \rangle$. We assert that a_m is the final element in the sequence. Otherwise, there exists an element a_{m+1}. Then we have $d \in \langle a_m \rangle, a_{m+1} \in \langle d \rangle$. It follows that $a_m \mid d, d \mid a_{m+1}$. Hence, $a_m \mid a_{m+1}$, but this is a contradiction with the assumption that a_{m+1} is a proper divisor of a_m. Therefore, R satisfies the divisor chain condition.

(2) R satisfies the primeness condition. Suppose p is an irreducible element in R. We first show that $\langle p \rangle$ is a maximal ideal in R. Suppose I is an ideal of R such that $\langle p \rangle \subseteq I$. Since R is a principal ideal domain, there exists $c \in R$ such that $I = \langle c \rangle$. Then we have $p \in \langle c \rangle$, hence there exists $r \in R$ such that $p = rc$. Since p is irreducible,

c is either a unit or an associate of p. If c is a unit, then we have $\langle c \rangle = R$, hence $I = R$; On the other hand, if c is an associate of p, then $I = \langle c \rangle = \langle p \rangle$. Therefore, $\langle p \rangle$ is a maximal ideal.

Suppose $p \mid ab$, $a, b \in R$. Then in the quotient $R/\langle p \rangle$ we have $(a + \langle p \rangle)(b + \langle p \rangle) = ab + \langle p \rangle = 0 + \langle p \rangle$. Since $\langle p \rangle$ is a maximal ideal, $R/\langle p \rangle$ is a field. Therefore, we have either $a + \langle p \rangle = 0 + \langle p \rangle$ or $b + \langle p \rangle = 0 + \langle p \rangle$, that is, either $p \mid a$ or $p \mid b$. Thus, p is a prime element. This shows that R satisfies the primeness condition. □

The converse statement of Theorem 2.7.6 is not true, as we have seen from Example 2.7.3, which indicates that $\mathbb{Z}[x]$ is uniquely factorial but it is not a principal ideal domain. We suggest readers to give a proof of the above assertion directly using the results about polynomials over the number field of rational numbers.

Next we give a special property of principal ideal domains.

Theorem 2.7.7. *Let R be a principal ideal domain, and d a greatest common divisor of $a, b \in R$. Then there exist $u, v \in R$ such that $d = ua + vb$.*

Proof. Since R is a principal ideal ring, there exists $d_1 \in R$ such that $\langle d_1 \rangle = \langle a, b \rangle$. We assert that d_1 is also a greatest common divisor of a, b. In fact, since $a, b \in \langle d_1 \rangle$, we have $d_1 \mid a, d_1 \mid b$. Hence, d_1 is a common divisor of a and b. Moreover, since $d_1 \in \langle a, b \rangle$, there exists $u_1, v_1 \in R$ such that $d_1 = u_1 a + v_1 b$. If c is a common divisor of a, b. Then we have $c \mid (u_1 a + v_1 b) = d_1$. Therefore, d_1 is a greatest common divisor of a, b. This then implies that $d_1 \mid d$ and $d \mid d_1$. Since R is a domain, d is an associate of d_1. Suppose $d = \epsilon d_1$, where ϵ is a unit. Then we have $d = \epsilon u_1 a + \epsilon v_1 b$. □

Problem 2.7.8. We have shown that $\mathbb{Z}[x]$ is not a principal ideal domain. Try to find two elements $f(x), g(x)$ in $\mathbb{Z}[x]$, such that there does not exist $u(x), v(x) \in \mathbb{Z}[x]$ satisfying $u(x)f(x) + v(x)g(x) = d(x)$, where $d(x)$ is a greatest common divisor of $f(x), g(x)$.

Now, we introduce another special type of principal ideal domains — Euclidean domains.

Definition 2.7.9. Let R be a domain. Suppose there exists a map δ from R^* to $\mathbb{N}$ such that for any $a, b \in R, b \neq 0$, there exists $q, r \in R$ such that

$$a = qb + r,$$

where $r = 0$ or $\delta(r) < \delta(b)$. Then (R, δ) is called a **Euclidean domain**. Sometimes we simply say that R is a Euclidean domain.

Vividly speaking, a Euclidean domain is a domain in which there is a division algorithm. Notice that in the literature there are several definitions of Euclidean domains, and some of them are different from the above one. But the most important feature of Euclidean domains is the division algorithm. We first give some examples.

Example 2.7.10. $\mathbb{Z}$ is a Euclidean domain.

This follows easily from the division algorithm of integers. In fact, we only need to define a map $\delta : \mathbb{Z}^* \to \mathbb{N}$ by $\delta(m) = |m|$. Then it is easy to check that $(\mathbb{Z}, \delta)$ is a Euclidean domain.

Example 2.7.11. Let $\mathbb{F}$ be a number field. Then the polynomial ring $\mathbb{F}[x]$ is a Euclidean domain.

Define a map δ by $\delta(f(x)) = \deg f(x)$, where $f(x) \neq 0$. Then it is easy to check that δ Satisfies the condition of Definition 2.7.9. Thus, $\mathbb{F}[x]$ is a Euclidean domain.

Now, we prove the following.

Theorem 2.7.12. *A Euclidean domain must be a principal ideal ring. In particular, it is a unique factorization domain.*

Proof. Let I be an ideal of a Euclidean domain (R, δ). We assume that $I \neq \{0\}$. Then the set of non-negative integers $\{\delta(x) | x \in I, x \neq 0\} \subseteq \mathbb{N} \cup \{0\}$ must have a minimum. Let $a \in R^* \cap I$ be an element such that $\delta(a)$ attains the minimum. Then for any $x \in I, x \neq 0$ we have $\delta(x) \geq \delta(a)$. By the definition, for any $b \in I$, there exist $q, r \in R$ such that $b = qa + r$, where $r = 0$ or $\delta(r) < \delta(a)$. Since $a, b \in I$, we have $r = b - qa \in I$. If $r \neq 0$, then we have $r \in I$ and $\delta(r) < \delta(a)$, which is a contradiction. Thus, $r = 0$, that is, $b = qa$. This implies that $I = \langle a \rangle$. Therefore, R is a principal ideal domain. $\square$

A natural problem is whether the converse statement of the above theorem holds, namely, whether there is a principal ideal domain which is not a Euclidean domain. This problem had been open for a long time in the history. However, it is now well-known that many principal ideal domains are not Euclidean. As an example, in the exercise of this section, we will show that $\mathbb{Z}[\frac{1}{2}(1+\sqrt{-19})]$ is a principal ideal domain but is not a Euclidean domain.

We mention here a very interesting historical event related to Euclidean domains. Let d be an integer such that $d \neq 0, 1$ and d has no non-trivial square factor. Set

$$\mathcal{O}_d = \begin{cases} \mathbb{Z}[\sqrt{d}], & d \equiv 2, 3 \pmod 4, \\ \mathbb{Z}\left[\frac{1+\sqrt{d}}{2}\right], & d \equiv 1 \pmod 4, \end{cases}$$

This is the **ring of algebraic integers over a quadratic field**. Gauss conjectured that if $d < 0$, then $\mathcal{O}_d$ is a principal ideal domain if and only if

$$d = -1, -2, -3, -7, -11, -19, -43, -67, -163.$$

This conjecture was proved to be true. In particular, when $d = -1, -2, -3, -7, -11$, $\mathcal{O}_d$ is a Euclidean domain. Therefore,

$$\mathcal{O}_{-19} = \mathbb{Z}\left[\frac{1}{2}(1 + \sqrt{-19})\right]$$

is the simplest principal ideal domain which is not a Euclidean domain. Furthermore, when

$$d = 2, 3, 5, 6, 7, 11, 13, 17, 19, 21, 29, 33, 37, 41, 57, 73,$$

$\mathcal{O}_d$ can be made into a Euclidean domain by defining appropriate map δ. However, up to now it is still an open problem to find out all the integers d such that $\mathcal{O}_d$ is a Euclidean domain.

Example 2.7.13. The ring of Gauss integers is a Euclidean ring. We define a map δ by $\delta(\beta) = |\beta|^2 = a^2 + b^2$, where $\beta = a + b\sqrt{-1} \in \mathbb{Z}[\sqrt{-1}]$. If $\beta = a_1 + b_1\sqrt{-1} \neq 0$, then for any $c_1 + d_1\sqrt{-1}$, we have

$$\frac{c_1 + d_1\sqrt{-1}}{a_1 + b_1\sqrt{-1}} = s + t\sqrt{-1},$$

where $s, t \in \mathbb{Q}$. Then there exist $c_2, d_2 \in \mathbb{Z}$ such that $|c_2 - s| \leq \frac{1}{2}$, $|d_2 - t| \leq \frac{1}{2}$. Consider $q = c_2 + d_2\sqrt{-1}$, $r = c_1 + d_1\sqrt{-1} - q\beta$. Then we have $c_1 + d_1\sqrt{-1} = q\beta + r$, and

$$\begin{aligned} \delta(r) &= \delta(\beta(s + t\sqrt{-1}) - q\beta) \\ &= \delta(\beta)\delta(s + t\sqrt{-1} - q) \\ &\leq \left(\frac{1}{4} + \frac{1}{4}\right)\delta(\beta) < \delta(\beta). \end{aligned}$$

Therefore, $\mathbb{Z}[\sqrt{-1}]$ is a Euclidean domain.

Exercises

(1) Let R be a domain, and $\langle a \rangle$, $\langle b \rangle$ two principal ideals of R. Show that $\langle a \rangle = \langle b \rangle$ if and only if a is an associate of b.

(2) Let R be a principal ideal domain. Show that for any $c \in R$, the set

$$S = \{d + \langle c \rangle \,|\, d, c \text{ are relatively prime}\}$$

is a group with respect to the multiplication of the quotient ring $R/\langle c \rangle$.

(3) Let R be a principal ideal domain and I a non-zero ideal of R. Prove the following assertions.

  (i) Every ideal of R/I is a principal ideal. Is R/I a principal ideal domain?

  (ii) There are only finitely many ideals in R/I.

(4) Find a polynomial $f(x)$ in $\mathbb{Q}[x]$ such that $\langle x^2 + 1, x^5 + x^3 + 1 \rangle = \langle f(x) \rangle$.

(5) Let $\mathbb{F}$ be a number field. Show that $\mathbb{F}[x, y]$ is not a principal ideal domain.

(6) Show that $\mathbb{Z}[\frac{1}{2}(1 + \sqrt{-3})]$ is a Euclidean domain.

(7) Show that the following domains with the map δ are Euclidean domains:

  (i) $\{\mathbb{Z}[\sqrt{-2}]; +, \cdot\}$, $\delta(a + b\sqrt{-2}) = a^2 + 2b^2$;

  (ii) $\{\mathbb{Z}[\sqrt{2}]; +, \cdot\}$, $\delta(a + b\sqrt{2}) = |a^2 - 2b^2|$;

  (iii) $\{R; +, \cdot\}$, where

$$R = \mathbb{Z}\left[\frac{1}{2}\left(1 - \sqrt{-7}\right)\right] = \left\{a + b\left(\frac{1 - \sqrt{-7}}{2}\right) \,|\, a, b \in \mathbb{Z}\right\},$$

$$\delta\left(a + b\left(\frac{1 - \sqrt{-7}}{2}\right)\right) = a^2 + ab + 2b^2.$$

(8) Construct an example in the ring $\mathbb{Z}[\sqrt{-1}]$ to show that, in the division algorithm of a Euclidean domain, the quotient and the remainder may be not unique.

(9) Show that any field is a Euclidean domain.

(10) Show that the domain $\mathbb{Z}[\sqrt{-6}]$ is not Euclidean.

Exercises (hard)

(11) Let (R, δ) be a Euclidean domain satisfying the condition $\delta(ab) = \delta(a)\delta(b)$, $\forall a, b \in R^*$. Show that a is a unit in $R \Leftrightarrow \delta(a) = \delta(1)$.

(12) In this exercise, we show that $R = \mathcal{O}_{-19} = \mathbb{Z}[\frac{1}{2}(1 + \sqrt{-19})]$ is a principal domain but not a Euclidean domain. Denote $\theta = \frac{1}{2}(1 + \sqrt{-19})$.

   (i) Show that the units of R are 1 and -1.

   (ii) Show that 2 and 3 are irreducible elements in R.

   (iii) Show that R is not a Euclidean domain. (*Hint*: Suppose conversely that (R, δ) is a Euclidean domain. Fix $m \in R$, $m \neq 0, \pm 1$, such that $\delta(m) = \min\{\delta(a) | a \neq 0, a \neq \pm 1\}$. Dividing 2 by m, one can show that $m = \pm 2$ or ± 3. Then divide θ by m to deduce a contradiction.)

   (iv) Let I be a non-zero ideal of R. Fix $b \in I$ such that $|b| \leq |c|$, $\forall c \in I^*$, where $|\cdot|$ denotes the length of a complex number. Show that if there exists $a \in I$ such that $\frac{a}{b} \notin R$, then there exists $a_1 \in I$, such that $\frac{a_1}{b} = x + \sqrt{-1}y \notin R$, where $-\frac{\sqrt{19}}{4} \leq y \leq \frac{\sqrt{19}}{4}$.

   (v) Show that if $\frac{a_1}{b} = x + y\sqrt{-1} \notin R$ and $-\frac{\sqrt{3}}{2} < y < \frac{\sqrt{3}}{2}$, then there exist an integer m_1, such that $|\frac{a_1}{b} - m_1| < 1$.

   (vi) Show that if $\frac{a_1}{b} = x + y\sqrt{-1} \notin R$ and $\frac{\sqrt{3}}{2} \leq y \leq \frac{\sqrt{19}}{4}$, then there exists an integer m_2, such that $|\frac{2a_1}{b} - \theta - m_2| < 1$.

   (vii) Show that R is a principal ideal domain.

2.8 Polynomials over a ring

From now on, we go to the second part of ring theory, which deals with polynomials over rings. The theory of polynomials is the foundation of commutative algebra and algebraic geometry, and is very useful in many branches of applied sciences.

Let us first introduce the notion of polynomials. We begin with polynomials in one indeterminate. Recall that, given a number field $\mathbb{F}$, the ring of polynomials in one indeterminate over $\mathbb{F}$, denoted by $\mathbb{F}[x]$, is a set consisting of elements of the form

$$a_n x^n + a_{n-1} x^{n-1} + \cdots + a_1 x + a_0, \ n \geq 0, \ a_i \in \mathbb{F}, \ i = 0, 1, \ldots, n.$$

Two elements $a_n x^n + a_{n-1} x^{n-1} + \cdots + a_1 x + a_0$ and $b_m x^m + b_{m-1} x^{m-1} + \cdots + b_1 x + b_0$, $m \geq n$, in $\mathbb{F}[x]$, are equal if and only if for any $0 \leq i \leq m$, $a_i = b_i$ (here we assume that $a_k = 0$, for $k > n$). The addition and multiplication in $\mathbb{F}[x]$ are defined in the usual way.

This definition can be easily generalized to a general field, or even a commutative ring with identity. But for a general ring R (which may have no identity), the generalization encounters some difficulty. For example, we even do not know what x stands for. Even if for a commutative ring with identity, this kind of generalization is not so intuitive. This leads to another way of thinking, namely, we write a polynomial $a_n x^n + a_{n-1} x^{n-1} + \cdots + a_1 x + a_0$ over $\mathbb{F}$ as an infinite sequence

$$\{a_0, a_1, \ldots, a_n, \ldots\},$$

where only finitely many of the a_is are non-zero. It is obvious that there is a bijection between the set $\mathbb{F}[x]$ and the set of infinite sequences of elements in $\mathbb{F}$ where only finitely many of the components are non-zero. This definition can be generalized to an arbitrary ring.

Let R be a ring. Denote by S the set of infinite sequences

$$f = \{a_0, a_1, a_2, \ldots\}, \quad a_i \in R,$$

where only finitely many of the a_is are non-zero. Define addition and multiplication on S as follows. Given

$$f_1 = \{b_0, b_1, b_2, \ldots\},$$
$$f_2 = \{c_0, c_1, c_2, \ldots\},$$

we set

$$f_1 + f_2 = \{b_0 + c_0, b_1 + c_1, b_2 + c_2, \ldots\},$$
$$f_1 f_2 = \{b_0 c_0, b_0 c_1 + b_1 c_0, b_0 c_2 + b_1 c_1 + b_2 c_0, \ldots\} = \{a_k\},$$

where

$$a_k = \sum_{i+j=k} b_i c_j, \quad k = 0, 1, 2, \ldots.$$

It is easy to check that S is a ring with respect to the above operations. If R is a ring with identity, then so is S. If R is commutative, then S is also commutative. The zero of the ring S is $\{0,0,0,\ldots\}$, and the negative of $f = \{a_0, a_1, a_2, \ldots\}$ is $-f = \{-a_0, -a_1, -a_2, \ldots\}$. If 1 is the identity of R, then $\{1, 0, 0, \ldots\}$ is the identity of S.

In the following, we usually refer to an element of the ring S as a **polynomial** over R, or a polynomial with coefficients in R. If $f = \{a_0, a_1, a_2, \ldots\}$ is a non-zero polynomial (i.e., at least one of the a_i is non-zero), and n is the maximal integer such that $a_n \neq 0$, then we say that the **degree** of f is n. The degree of f will be denoted by $\deg f$. We do not define the degree of the zero polynomial. If $\deg f = n$, then $a_0, a_1, \ldots, a_n$ are called the coefficients of f, and a_n is called the **leading coefficient** of f. A non-zero polynomial whose leading coefficient is equal to 1 is called **monic**. The proof of the following proposition is easy, and is left to readers.

Proposition 2.8.1. *Suppose $f, g \in S$ are both non-zero, and $\deg f = m$, $\deg g = n$. Then*

(1) $f + g = 0$ *or* $\deg(f + g) \leq \max(m, n)$;
(2) $fg = 0$ *or* $\deg(fg) \leq \deg f + \deg g$, *with equality holds if and only if the product of the leading coefficient of f and that of g is non-zero.*

In particular, if R is a domain, then so is S.

Problem 2.8.2. Construct an example to show that the polynomial ring of a principal ideal domain is not necessarily a principal ideal domain.

Define a map φ from R to S by $\varphi(a) = \{a, 0, 0, \ldots\}$. It is clear that φ is an injective homomorphism. Thus, R can be viewed as a subring of S.

Problem 2.8.3. Show that if R is a domain, then the set of units of S coincides with that of R.

Now, let R be a ring with identity 1. Let x be the polynomial

$$x = \{0, 1, 0, 0, \ldots\}.$$

Then it is easily seen that for any non-negative integer m and $a \in R$, we have $ax^m = \{c_0, c_1, \ldots, c_n, \ldots\}$, where $c_i = 0$, for $i \neq m$, and

$c_m = a$. Thus, a polynomial $f = \{a_0, a_1, \ldots\}$ with degree n can be uniquely written as

$$f = a_0 + a_1 x + a_2 x^2 + \cdots + a_n x^n.$$

In this way, we can express a polynomial as the familiar form in the indeterminate x. An element in S will be called a polynomial over R in one indeterminate, and S will be denoted as $S = R[x]$, called the **polynomial ring in one indeterminate** over the ring R.

The above definition of polynomials in one indeterminate can be naturally generalized to define polynomials in several indeterminates. Notice that an infinite sequence $(a_0, a_1, \ldots)$ is actually a map from the set of natural numbers $\mathbb{N}$ to the ring R defined by $i \to a_i$, where there are only finitely many integers whose image are non-zero. This method can be generalized to several indeterminates.

Let R be a ring. Denote by $(\mathbb{N})_n$ the set of all ordered n-tuples of natural numbers, namely,

$$(\mathbb{N})_n = \{(i_1, i_2, \ldots, i_n) | i_n \in \mathbb{N}\}.$$

For $(i) = (i_1, i_2, \ldots, i_n) \in (\mathbb{N})_n$ and $(j) = (j_1, j_2, \ldots, j_n) \in (\mathbb{N})_n$, we set $(i) + (j) = (i_1 + j_1, i_2 + j_2, \ldots, i_n + j_n)$.

Definition 2.8.4. Let R be a ring and n a positive integer. A **polynomial in n indeterminates** over R is a map f from $(\mathbb{N})_n$ to R such that there are only finitely many elements in $(\mathbb{N})_n$ whose image under f is non-zero. Given two polynomials in n indeterminates f and g, we define the addition and multiplication $h = f + g$ and $k = f \cdot g$ as

$$h(i) = f(i) + g(i), \quad k(i) = \sum_{(j)+(j')=(i)} f(j) \cdot g(j'), \quad (i) \in (\mathbb{N})_n.$$

It is clear that for $n = 1$ the above definition is just that of polynomials in one indeterminate. Denote by S the set of all polynomials in n indeterminates over R. Then it is easy to check that S becomes a ring with respect to the above addition and multiplication, called the **polynomial ring in n indeterminates** over R. Given $a \in R$, we define a polynomial in n indeterminates g_a by

$$g_a(i) = \begin{cases} a & \text{if } (i) = (0, 0, \ldots, 0); \\ 0 & \text{if } (i) \neq (0, 0, \ldots, 0). \end{cases}$$

Then g_0 is the zero of S. Moreover, if R is a ring with identity 1, then g_1 is the identity of S. It is also easy to check that $g_a + g_b = g_{a+b}$, $g_a \cdot g_b = g_{ab}$. Thus, the map $a \mapsto g_a$ is an injective homomorphism from the ring R to S. This fact makes it reasonable to identify a with g_a, and in this way R can be viewed as a subring of S.

If R has an identity, we can introduce another notation for polynomials in n indeterminates. Given an integer v such that $1 \leq v \leq n$, let $j^{(v)} \in (\mathbb{N})_n$ be the n-tuple of non-negative integers whose vth integer is 1, and all others are 0. Let $x_v \in S$ be the polynomial such that $x_v(j^{(v)}) = 1$, and $x_v(i) = 0$ for $(i) \neq j^{(v)}$. It is easily seen that, for $a \in R$ and non-negative integers $i_1, i_2, \ldots, i_n$, $a x_1^{i_1} x_2^{i_2} \ldots x_n^{i_n}$ is an element in S whose value at $(i_1, i_2, \ldots, i_n)$ is a, and the values at all other n-tuples are 0. It follows easily from this fact that every $f \in S$ can be uniquely expressed as a finite sum of the expressions of the form

$$a_{(i)} x_1^{i_1} x_2^{i_2} \ldots x_n^{i_n}.$$

A polynomial of the form $a x_1^{i_1} x_2^{i_2} \ldots x_n^{i_n}$ is called a monomial. From now on, for a unitary ring R, we denote the ring S as $R[x_1, x_2, \ldots, x_n]$.

Now, we define the degree of polynomials in n indeterminates. The degree of a monomial $a_{(i)} x_1^{i_1} x_2^{i_2} \ldots x_n^{i_n}$ $(a_{(i)} \neq 0)$ is defined to be $i_1 + i_2 + \cdots + i_n$. For $f \in R[x_1, x_2, \ldots, x_n]$, $f \neq 0$, the degree of f is the maximum of the degrees of all non-zero monomials in f. The degree of f will be denoted as $\deg f$. If all the monomials in f have the same degree, then f is called a **homogeneous polynomial**. It is clear that, if f, g are homogeneous, then fg is either equal to 0 or homogeneous.

To study polynomials, sometimes we need to view them as functions. We begin with polynomials in one indeterminate. Let R_1 be a commutative ring with identity, and R a subring containing the identity of R_1. Let $f(x) = a_0 + a_1 x + a_2 x^2 + \cdots + a_n x^n$ be a polynomial in x over R. Then for $a \in R_1$, $f(a) = a_0 + a_1 a + a_2 a^2 + \cdots + a_n a^n$ is said to be the value of $f(x)$ at a. In this way, $f(x)$ can be viewed as a map from R_1 to itself, or in other words, f can be viewed as a function on R_1 with values on R_1. It is easy to see that, for a fixed element a in R_1, the map $f(x) \to f(a)$ is a homomorphism from the ring $R[x]$ to R_1.

Problem 2.8.5. In linear algebra we have shown that, if two polynomials over a number field are equal as functions, then they must be equal as polynomials. Construct an example to show that, this assertion is not true for polynomials over a general ring.

Now, we prove the following.

Lemma 2.8.6. *Let R be a commutative ring with identity. Then there exists a ring $R_1 \supseteq R$, such that any two different polynomials in $R[x]$ are not equal to each other when viewed as functions on R_1.*

Proof. Set $R_1 = R[x]$. Then for a fixed polynomial $f(x) \in R_1$, the function defined by $f(x)$ is the map from R_1 to itself such that $g(x) \mapsto f(g(x))$. Given $f_1 \neq f_2$, consider $g(x) = x$. Then $f_1(g(x)) \neq f_2(g(x))$. This means that as functions on R_1, the values of f_1 and f_2 are different at the element $x \in R_1$. This completes the proof of the lemma. □

The value of the polynomial $f = a \in R$, when viewed as a function on R_1, is equal to a everywhere. Thus, elements in R are usually called **constants**.

Now, we introduce another definition of polynomial rings from the point of view of functions. We begin with polynomials in one indeterminate. Let R, R_1 be as above. Fix $a \in R_1$. Then a determines a homomorphism φ from the ring $R[x]$ to R_1, defined by $\varphi(f) = f(a)$. The image of $R[x]$ under this homomorphism is a subring of R_1, denoted as $R[a]$. It is easily seen that

$$R[a] = \{a_0 + a_1 a + a_2 a^2 + \cdots + a_m a^m \,|\, a_i \in R, m \geq 0\}.$$

$R[a]$ can also be defined as the minimal subring of R_1 containing R and a. This leads to the following definition.

Definition 2.8.7. An element $a \in R_1$ is called **algebraic** over R, if there exists a non-zero polynomial $f(x) \in R[x]$ such that $f(a) = 0$.

If a is not an algebraic element, then it is called **transcendental** over R.

Now, we can prove the following.

Proposition 2.8.8. *If a is transcendental over R, then the ring $R[a]$ is isomorphic to the polynomial ring $R[x]$ over R.*

Proof. Consider the homomorphism φ from $R[x]$ to R_1 defined by $\varphi(f) = f(a)$. Then φ is a surjective homomorphism from $R[x]$ onto $R[a]$. Since a is transcendental, the kernel of φ is equal to $\{0\}$. By the fundamental theorem of homomorphism of rings, we have $R[x] \simeq R[a]$. $\square$

The above proposition justifies the following (equivalent) definition of polynomial rings in one indeterminate.

Definition 2.8.9. Let R be a commutative ring with identity. A ring R_1 containing R is called a **polynomial ring** (in one indeterminate) over R, if there exists an element a in R_1 which is transcendental over R such that $R_1 = R[a]$. An elements satisfying the condition $R[a] = R_1$ is called a **generator** of R_1 over R.

Problem 2.8.10. Construct an example to show that, the generators of a polynomial ring over R may be not unique.

Now, we summarize the above conclusions as the following.

Theorem 2.8.11. *Let R be a commutative ring with identity. Then the polynomial ring over R exists. Moreover, any two polynomial rings over R must be isomorphic.*

The following theorem gives a more accurate statement of the assertion in Theorem 2.8.11.

Theorem 2.8.12. *Let S be a commutative ring with identity, and R a subring of S containing the identity. Suppose $a \in S$ is transcendental over R, S_1 is a commutative ring and $u \in S_1$. Then for any homomorphism η from R to S_1, there exists a unique homomorphism η_1 from $R[a]$ to S_1 such that $\eta_1(a) = u$ and $\eta_1|_R = \eta$.*

Proof. We first prove the existence of the extension η_1. For this we only need to define

$$\eta_1 \left(\sum a_i a^i \right) = \sum \eta(a_i)(u)^i = \sum \eta(a_i) u^i, \quad a_i \in R_1. \qquad (2.3)$$

Since a is transcendental over R, every element in $R[a]$ can be uniquely written as the sum

$$\sum a_i a^i.$$

Therefore, η_1 is a well-defined map from $R[a]$ to S_1. Obviously, $\eta_1(c) = \eta(c)$, $\forall c \in R$ and $\eta_1(a) = u$. It can be directly checked that η_1 is a homomorphism from $R[a]$ to S_1. This proves the existence of the extension.

Now, we prove the uniqueness. If η_2 is also an extension of η and $\eta_2(a) = u$, then by the definition of homomorphism it is easily deduced that (2.3) is also true for η_2. Since the expression of any element in $R[a]$ as the form $\sum a_i a^i$ is unique, we have $\eta_1(y) = \eta_2(y)$, $\forall y \in R[a]$. Therefore, the extension is unique. □

Theorem 2.8.11 has several corollaries as the following. The proofs are left to readers.

Corollary 2.8.13. *Let R_1, R_2 be commutative rings with identity. Suppose S_1 is a polynomial ring over R_1 and $a \in S_1$ is a generator. If S_2 is a ring with identity containing R_2 (as a subring), and $b \in S_2$, then any surjective homomorphism T_0 from R_1 to R_2 can be uniquely extended to a surjective homomorphism T from S_1 to $R_2[b]$ such that $T(a) = b$. Moreover, the homomorphism T is an isomorphism if and only if T_0 is an isomorphism and b is transcendental over R_2.*

Corollary 2.8.14. *Suppose R is a commutative ring with identity, and S_1, S_2 are polynomial rings over R, with generator a and b respectively, Then there exists a unique isomorphism T from S_1 to S_2 such that $T(a) = b$ and $T|_R$ is the identity map of R (T is called an R-isomorphism).*

Now, we generalize the above results to the case of several indeterminates. We first introduce the useful inductive method in dealing with polynomials in several indeterminates. Let R be a commutative ring with identity. Consider the set S_1 of polynomials in $S = R[x_1, x_2, \ldots, x_n]$ where the indeterminate x_n does not appear, that is,

$$S_1 = \{f \in S \mid f(i_1, i_2, \ldots, i_n) = 0, \forall (i_1, \ldots, i_n) \in (\mathbb{N})^n, i_n \neq 0\}.$$

It is obvious that there is a bijection between the set S_1 and the set of maps from $(\mathbb{N})^{n-1}$ to R. In fact, we only need to let $f \in S_1$ correspond to $f_1(x_1, x_2, \ldots, x_{n-1}) = f(x_1, \ldots, x_{n-1}, 0)$. This implies that we can identify S_1 with $R[x_1, \ldots, x_{n-1}]$. Moreover, the operations of the two rings coincide. Thus, S_1 can be viewed as a subring of S. On the other hand, we assert that S is a polynomial ring in one indeterminate over S_1. This follows from the facts that S is equal to the minimal subring of S containing both S_1 and x_n, and that x_n is transcendental over S_1.

The inductive method can be used to prove many important results about polynomials in several indeterminates.

Problem 2.8.15. Use the inductive method to show that, if R is a domain, then $R[x_1, x_2, \ldots, x_n]$ is also a domain.

Next we will understand polynomials in several indeterminates from the point view of functions. Suppose R is a commutative ring with identity, and is a subring of R_1 (which is also a commutative ring with identity, and the identities of R and R_1 coincide). Then a polynomial with n indeterminates over R

$$f(x_1, x_2, \ldots, x_n) = \sum a_{(i)} x_1^{i_1} x_2^{i_2} \ldots x_n^{i_n}$$

can be regarded as a function of n variables defined on R_1 with values in R_1, that is, for $a_1, a_2, \ldots, a_n \in R_1$,

$$f(a_1, a_2, \ldots, a_n) = \sum a_{(i)} a_1^{i_1} a_2^{i_2} \ldots a_n^{i_n}.$$

For fixed $a_1, a_2, \ldots, a_n$, the map $f \mapsto f(a_1, a_2, \ldots, a_n)$ is a homomorphism from the ring $R[x_1, x_2, \ldots, x_n]$ to R_1. We denote the image of this homomorphism as $R[a_1, a_2, \ldots, a_n]$, which is the minimal subring of R_1 containing R and $a_1, \ldots, a_n$.

Definition 2.8.16. The elements $a_1, a_2, \ldots, a_n$ is called **algebraically dependent** over R, if there exists a non-zero polynomial f in $R[x_1, x_2, \ldots, x_n]$ such that

$$f(a_1, a_2, \ldots, a_n) = 0.$$

Otherwise they are called **algebraically independent**.

Definition 2.8.17. Let R_1 be a ring with identity and R a subring of R_1. We say that R_1 is a polynomial ring over R if there exist elements $a_1, a_2, \ldots, a_n$ in R_1, which are algebraically independent over R, such that $R_1 = R[a_1, a_2, \ldots, a_n]$. Any such set $\{a_1, a_2, \ldots, a_n\}$ is called a **generating set**. Sometimes we also say that R_1 is a polynomial ring over R in $a_1, a_2, \ldots, a_n$.

By the definition, it is easily seen that R_1 is a polynomial ring in n-indeterminates over R if and only if there exists an R-isomorphism between $R[x_1, x_2, \ldots, x_n]$ and R_1. To prove the uniqueness of the polynomials ring, we need the following lemma.

Lemma 2.8.18. *Let R_2 be a ring with identity and R a subring (containing the identity of R_2) of R_2. Suppose $a_1, a_2, \ldots, a_n \in R_2$ and $n > 1$. Set $R_1 = R[a_1, a_2, \ldots, a_{n-1}]$. Then R_2 is a polynomial ring over R in $a_1, a_2, \ldots, a_n$ if and only if R_1 is a polynomial ring over R in $a_1, a_2, \ldots, a_{n-1}$, and R_2 is a polynomial ring over R_1 in a_n.*

This lemma can be proved inductively and will be left to readers. Now, we can prove.

Theorem 2.8.19. *Let R be a commutative ring with identity, and R_1 a polynomial ring over R in $a_1, a_2, \ldots, a_n$. Suppose $\bar{R}$ is a ring with identity and is a subring of the unitary ring $\bar{R}_1$, and $b_1, b_2, \ldots, b_n \in \bar{R}_1$. Then any surjective homomorphism T_0 from R to $\bar{R}$ can be uniquely extended to a surjective homomorphism T from R_1 to $\bar{R}[b_1, b_2, \ldots, b_n]$, such that $T(a_i) = b_i$. Moreover, T is an isomorphism if and only if T_0 is an isomorphism and $b_1, b_2, \ldots, b_n$ are algebraically independent over $\bar{R}$.*

This theorem can be proved inductively. It can also be proved similarly as in the one indeterminate case. We left the proofs of the theorem and the following two corollaries to readers.

Corollary 2.8.20. *Let R be a commutative ring with identity and R_1, R_2 polynomial rings over R, in $a_1, a_2, \ldots a_n$ and $b_1, b_2, \ldots, b_n$ respectively. Then there exists a unique isomorphism T from the ring R_1 to R_2 such that $T(a_i) = b_i$ and $T|_R = \mathrm{id}_R$.*

Corollary 2.8.21. *Suppose R_1 is a polynomial ring over a commutative unitary ring R in $a_1, a_2, \ldots, a_n$ and σ is an n-permutation. Then there exists a unique R-automorphism T (i.e., $T|_R = \mathrm{id}_R$) of R_1 such that $T(a_i) = a_{\sigma(i)}$.*

Exercises

(1) Determine the number of non-zero polynomials in $\mathbb{Z}_5[x]$ whose degree is less or equal to 3.

(2) In the polynomial ring $\mathbb{Z}_6[x]$, calculate the product of $f(x) = 2x - 5$ and $g(x) = 2x^4 - 3x + 3$.

(3) Let $f(x)$ be a polynomial over a unitary ring R. If $a \in R$ satisfies $f(a) = 0$, then a is called a root (or a zero point) of $f(x)$. Determine all the roots of the polynomial $x^3 + 2x + 2$ in $\mathbb{Z}_7$.

(4) Let $f(x) = a_n x^n + a_{n-1} x^{n-1} + \cdots + a_0$ be a polynomial over a commutative unitary ring R. The differential of f is defined to be

$$f'(x) = n a_n x^{n-1} + \cdots + a_1.$$

  (i) Show that for any polynomials $f(x)$ and $g(x)$, $(fg)' = f'g + fg'$.

  (ii) If $a \in R$ satisfies $f(a) = f'(a) = 0$, then it is called a **multiple root** of $f(x)$. Find all the multiple roots of the polynomials $x^{15} - x$ and $x^{15} - 2x^5 + 1$ over $\mathbb{Z}_5$.

(5) Prove Corollary 2.8.13 and 2.8.14.

(6) Prove Lemma 2.8.18.

(7) Prove Theorem 2.8.19, Corollary 2.8.20 and 2.8.21.

(8) Let R be a ring, an expression of the form

$$a_0 + a_1 t + a_2 t^2 + \cdots + a_n t^n + \cdots, \qquad a_i \in R,$$

is called a formal power series over R. Denote by $R[[t]]$ the set of all formal power series over R. Define addition and multiplication on $R[[t]]$ similarly as in the set of polynomials.

  (i) Show that $R[[t]]$ becomes a ring with respect to the addition and multiplication. Moreover, if R is commutative, then $R[[t]]$ is also commutative; if R is unitary, then $R[[t]]$ is also unitary.

  (ii) Suppose R is a domain. Is $R[[t]]$ necessarily a domain?

  (iii) Suppose R is a domain. Determine all the units in $R[[t]]$.

(9) Let p_1, p_2 be prime numbers, and $n_1, n_2 \in \mathbb{N}^*$. Show that $\sqrt[n_1]{p_1} + \sqrt[n_2]{p_2}$ is algebraic over $\mathbb{Q}$.

(10) Suppose R is a commutative ring with identity, and β_1, β_2 are both transcendental over R. Is $\beta_1 + \beta_2$ necessarily transcendental over R?

Exercises (hard)

(11) Let R be a commutative ring with identity, and I an ideal of R. Show that $I[x_1, \ldots, x_n]$ is an ideal of $R[x_1, \ldots, x_n]$, and the quotient ring $R[x_1, \ldots, x_n]/I[x_1, \ldots, x_n]$ is isomorphic to the polynomial ring $(R/I)[x_1, \ldots, x_n]$ over R/I.

(12) Let R_1 be a commutative ring with identity, and R a subring containing the identity of R_1. Show that if $\alpha \in R_1$ is algebraic over R, then $-\alpha$ is also algebraic over R. Moreover, if α, β are both algebraic over R, then $\alpha\beta$ is also algebraic over R.

2.9 Polynomial rings over a domain

In this section, we study polynomial rings over a domain. We will mainly focus on polynomial rings over a unique factorization domain.

We first consider the number of generators of a polynomial ring over a domain. Let R be a domain. If S is a polynomial ring in one indeterminate over R, and a is a generator, then any non-zero element y in S can be uniquely written as $y = f(a)$, where $f(x) \in R[x]$. We define the degree and the leading coefficient of y in a to be the degree and leading coefficient of the polynomial $f(x)$, respectively. Notice that the degree and leading coefficient of y depend not only on y but also on a. Now, we prove the following.

Theorem 2.9.1. *Let R be a domain, and S a polynomial ring in one indeterminate over R with generator a. Suppose the degree of $b \in S$ in a is n, and $f(x) \in R[x]$ is a polynomial with degree m. Then the degree of $f(b)$ in a is mn. Moreover, b is a generator of S over R if and only if it is linear in a (i.e., $n = 1$), and the leading coefficient of b in a is a unit in R. In this case, every non-zero element in S has the same degree in a or in b.*

Proof. Suppose $b = g(a)$ and the leading coefficients of $f(x)$ and $g(x)$ are c, d, respectively. Then the leading term of $f(b)$, when written as a polynomial in a, is $cd^m a^{mn}$. As R is a domain, the first assertion follows. If b is also a generator of S over R, then there exists a polynomial $h(x)$ such that $a = h(b)$. This implies that $n \deg h(x) = 1$, and the leading coefficient of $h(x)$ is a unit. Thus, $n = \deg h(x) = 1$. Therefore, b is linear in a, and the leading coefficient of b in a is a unit. Conversely, if b is linear in a and the leading coefficient is a unit, then $a \in R[b]$. Thus, $S = R[b]$, and for any non-zero polynomial $k(x)$ of degree m, the degree of $k(b)$ in a is also m. In particular, $k(b) \neq 0$. Therefore, b is transcendental over R. This completes the proof of the theorem. □

The proof of the following corollary will be left to readers.

Corollary 2.9.2. *Let φ be an R-automorphism (i.e., $\varphi|_R = \mathrm{id}$) of a polynomial ring $R[x]$ over R. Then $\varphi(x) = \varepsilon x + a$, where ε is a unit in R, and $a \in R$. Conversely, if $y = \varepsilon x + a$, and ε is a unit, then there exists a unique R-automorphism φ such that $\varphi(x) = y$.*

Next, we introduce a modified division algorithm in a polynomial over a domain. Recall that in linear algebra, we have a division algorithm in polynomial rings over a number field. If $f(x)$ is a polynomial over a number field $\mathbb{F}$, then for any non-zero $g(x) \neq 0$, there exist unique $q(x)$ and $r(x)$ such that

$$f(x) = g(x)q(x) + r(x), \tag{2.4}$$

where $r(x) = 0$ or $\deg r(x) < \deg g(x)$. The division algorithm is a key method in the theory of polynomials over a number field. Obviously, this algorithm can be easily generalized to polynomial rings over an arbitrary field, and from this it follows that the polynomial ring (in one indeterminate) over any field is a Euclidean ring. Unfortunately, this important technique can not be naturally generalized to polynomial rings over an arbitrary domain. In fact, we have shown in Section 2.7 that the polynomial ring $\mathbb{Z}[x]$ over the ring of integers $\mathbb{Z}$ is not a principal ideal domain, let alone a Euclidean domain. This implies that the division algorithm described as in (2.4) cannot be applied to $\mathbb{Z}[x]$.

However, there is a similar but relatively weaker algorithm in a polynomial ring over a domain, which can be viewed as a substitute of the division algorithm in (2.4). This technique can also be applied to deduce many important results.

Theorem 2.9.3. *Let R be a domain, and $R[x]$ the polynomial ring over R. Suppose the degrees of $f(x), g(x) \in R[x]$ are m and n, respectively. Denote $k = \max\{m-n+1, 0\}$. Let a be the leading coefficient of $g(x)$. Then there exist $q(x)$, $r(x)$ in $R[x]$ such that*

$$a^k f(x) = q(x)g(x) + r(x),$$

where $r(x) = 0$ or $\deg(r(x)) < n$. Moreover, the polynomials $q(x), r(x)$ satisfying the above conditions are unique.

Proof. If $m < n$, then $k = 0$. In this case, we only need to set $q(x) = 0$, and $r(x) = f(x)$. Now, we suppose $m \geq n - 1$. Then $k = m - n + 1$. We prove the assertion by induction on m. Notice that the assertion holds for $m = n - 1$, and when $m \geq n$, the degree of $af(x) - bx^{m-n}g(x)$ is at most $m-1$ (here b is the leading coefficient of $f(x)$). By induction there exist $q_1(x), r_1(x)$ such that

$$a^{(m-1)-n+1}(af(x) - bx^{m-n}g(x)) = q_1(x)g(x) + r_1(x),$$

where $r_1(x) = 0$ or $\deg(r_1(x)) < n$. Now, setting $q(x) = ba^{m-n}x^{m-n} + q_1(x), r(x) = r_1(x)$, we conclude by induction that the assertion holds for $m = n$. This proves the existence of $q(x)$ and $r(x)$.

Now, we prove the uniqueness. If there exists another pair of polynomials $q_2(x), r_2(x)$ such that $a^k f(x) = q_2(x)g(x) + r_2(x)$, where $r_2(x) = 0$ or $\deg(r_2(x)) < n$, then we have $(q(x) - q_2(x))g(x) = r_2(x) - r(x)$. Notice that R is a domain. If $q(x) - q_2(x) \neq 0$, then the degree of the polynomial $(q(x) - q_2(x))g(x)$ is at least n. This then implies that $r_2(x) - r(x)$ is non-zero; hence its degree is $< n$. This is a contradiction. Therefore, $q(x) - q_2(x) = 0, r(x) - r_2(x) = 0$.

The proof of the theorem is now completed. $\square$

Problem 2.9.4. Does the assertion of the above theorem hold for a general commutative ring with identity?

There are many applications of Theorem 2.9.3, e.g., it can be used to study the number of roots of a polynomial over a domain; see the exercises of this section. In the following, we give an important application of this theorem, namely, we prove that the polynomial ring over a unique factorization domain is also a unique factorization domain. We first give some definitions.

Definition 2.9.5. Let R be a unique factorization domain. A non-zero polynomial $f(x) \in R[x]$ is called **primitive** if the coefficients of $f(x)$ are relatively prime.

Given a non-zero polynomial $f(x)$ over R, let c be a greatest common divisor of the coefficients of $f(x)$. Then $f(x)$ can be written as $f(x) = cf_1(x)$, where $f_1(x)$ is primitive. This factorization is unique up to associates. We call c the **content** of $f(x)$, and denote it as $c(f)$. Then a polynomial is primitive if and only if its content is a unit. Now, we prove the following famous **Gauss's Lemma**.

Lemma 2.9.6. *Let R be a unique factorization domain. Then for any non-zero polynomials $f(x), g(x)$ over R, we have $c(fg) = c(f)c(g)$. In particular, the product of two primitive polynomials is primitive.*

Proof. Let $c = c(f), d = c(g)$. Then we have $f(x) = cf_1(x)$, $g(x) = dg_1(x)$, where $f_1(x)$, $g_1(x)$ are primitive polynomials. Since $f(x)g(x) = cdf_1(x)g_1(x)$, the first assertion follows from the second. Thus, we only need to prove that the product of two primitive polynomials is primitive. Suppose conversely that there are two primitive polynomials $f_1(x), g_1(x)$ such that $f_1(x)g_1(x)$ is not primitive. Then there is an irreducible element p in R such that p is a common divisor of all the coefficients of $f_1(x)g_1(x)$. Now, suppose $f_1(x) = \sum a_i x^i$, $g_1(x) = \sum b_j x^j$. Then by the primitiveness of the polynomials $f_1(x), g_1(x)$, at least one of the coefficients of $f_1(x)$ (as well as $g_1(x)$) can not be a multiple of p. Let a_s, b_t be the first coefficients of $f_1(x), g_1(x)$ which are not multiples of p, respectively. Now, consider the coefficient of x^{s+t} in the polynomial $f_1(x)g_1(x)$, which is

$$\ldots a_{s-1}b_{t+1} + a_s b_t + a_{s+1}b_{t-1} + \cdots .$$

Notice that p is a divisor of all the terms of the above summation except for $a_s b_t$. Thus, $p \mid a_s b_t$. Since a unique factorization domain satisfies the primeness condition, the irreducible element p is prime. Thus, $p \mid a_s$ or $p \mid b_t$, which is a contradiction with the assumption on a_s and b_t. $\square$

Lemma 2.9.7. *Suppose $f(x), g(x)$ are non-zero polynomials over a unique factorization domain R, and $g(x)$ is primitive. If $b \in R^*$, and $g(x) \mid bf(x)$, then $g(x) \mid f(x)$.*

Proof. Suppose $bf(x) = g(x)h(x)$, where $h(x) \in R[x]$. Then by Guass's Lemma, we have $c(gh) = c(h) = c(bf) = bc(f)$. This means that $b \mid c(h)$. In particular, $b \mid h(x)$. Thus, $g(x) \mid f(x)$. □

Now, we can prove the main result of this section.

Theorem 2.9.8. *Let R be a unique factorization domain. Then the polynomial ring $R[x]$ over R is also a unique factorization domain.*

Proof. We will apply condition (2) in Theorem 2.5.14, namely, we prove that $R[x]$ satisfies the divisor chain condition and the primeness condition.

We first show that $R[x]$ satisfies the divisor chain condition. Suppose

$$f_1(x), f_2(x), \dots, f_n(x), \dots$$

is a sequence of elements in $R[x]$, such that for any $i \geq 1$, $f_{i+1}(x)$ is a proper divisor of $f_i(x)$. Write $f_i(x)$ as $c_i g_i(x)$, where c_i is the content of $f_i(x)$, and $g_i(x)$ is primitive. It is easily seen that $c_i \mid c_{i+1}$ and $g_i(x) \mid g_{i+1}(x)$. If c_{i+1} is not a proper divisor of c_i, then c_i is associative to c_{i+1}. In this case, $g_{i+1}(x)$ must be a proper divisor of $g_i(x)$. By the primitiveness of the polynomials $g_{i+1}(x)$ and $g_i(x)$, we conclude that the degree of $g_i(x)$ must be greater than that of $g_{i+1}(x)$. Consequently, for any i, either c_{i+1} is a proper divisor of c_i, or the degree of $g_i(x)$ is greater than that of $g_{i+1}(x)$. Since R is a unique factorization domain, it satisfies the divisor chain condition. This implies that the sequence $f_1(x), f_2(x), \dots, f_n(x), \dots$ must be finite.

Next we show that $R[x]$ satisfies the primeness condition. Suppose $p(x)$ is an irreducible element in $R[x]$ and $p(x) \mid f(x)g(x)$. If the degree of $p(x)$ is 0, then $p(x) = p$ itself is an irreducible element in R (and by the primeness condition of R, p is also prime). Then p is a divisor of $c(fg) = c(f)c(g)$. Thus, $p \mid c(f)$ or $p \mid c(g)$. Therefore, $p \mid f(x)$ or $p \mid g(x)$. Now, we suppose the degree of $p(x)$ is positive and $p(x) \nmid f(x)$. We proceed to prove that $p(x) \mid g(x)$. Consider the following subset of $R[x]$:

$$G = \{u(x)p(x) + v(x)f(x)|u(x), v(x) \in R[x]\}.$$

Let $k(x)$ be a non-zero polynomial in G whose degree is minimal among all non-zero polynomials of G, and suppose its leading coefficient is a. By Theorem 2.9.3, there exist a non-negative integer m and polynomials $q(x)$, $r(x)$ such that $a^m f(x) = k(x)q(x) + r(x)$, where $r(x) = 0$ or $\deg r(x) < \deg k(x)$. Since $k(x) \in G$, we have $r(x) = a^m f(x) - k(x)q(x) \in G$. If $r(x) \neq 0$, then $r(x)$ is a non-zero polynomial in G whose degree is less than $\deg k(x)$, which is a contradiction with the assumption on $k(x)$. Therefore, we have $r(x) = 0$, hence $a^m f(x) = k(x)q(x)$. Now, we factor $k(x)$ as $k(x) = ck_1(x)$, where $c = c(k)$ is the content of $k(x)$, and $k_1(x)$ is primitive. Then by Lemma 2.9.7, $k_1(x) \mid f(x)$. Similarly we can prove that $k_1(x) \mid p(x)$. Since $p(x)$ is irreducible and is not a divisor of $f(x)$, $k_1(x)$ must be a unit in $R[x]$, hence it is a unit in R. Therefore, $k(x) = k \in R$. Suppose $k = u(x)p(x) + v(x)f(x)$. Then we have $kg(x) = u(x)p(x)g(x) + v(x)f(x)g(x)$. This implies that $p(x) \mid kg(x)$. Since the degree of $p(x)$ is positive and it is irreducible, it must be a primitive polynomial. Then by Lemma 2.9.7, we have $p(x) \mid g(x)$. Therefore, $R[x]$ satisfies the primeness condition. $\square$

Exercises

(1) Prove Corollary 2.9.2.

(2) Let R be a domain. Show that the set of all the R-automorphisms of the polynomial ring $R[x]$ is a group, and it is isomorphism to the semi-direct product of the group U and the abelian group $\{R; +\}$, where U is the group of units in R.

(3) Let R be a domain and $f(x) \in R[x]$. Show that $a \in R$ is a root of $f(x)$ if and only if $x - a \mid f(x)$.

(4) Let R be a domain, and $f(x) \in R[x]$. Show that if $a_1, a_2, \ldots, a_s$ are distinct roots of $f(x)$, then $(x - a_1)(x - a_2)\ldots(x - a_s)$ is a divisor of $f(x)$. Deduce from this result that a polynomial over a domain with degree m has at most m distinct roots.

(5) Determine all the roots of the polynomial $x^5 - 10x^4 + 35x^3 - 50x^2 + 24x$ over the ring $\mathbb{Z}_{16}$.

(6) Let p be a prime number. Show that in the polynomial ring $\mathbb{Z}_p[x]$ we have a factorization:

$$(x^p - x) = x(x - 1)(x - 2) \cdots (x - (p - 1)).$$

(7) Prove the Wilson Theorem: if p is a prime number, then $p \mid (p - 1)! + 1$.

(8) Let p be a prime number, and $f(x) \in \mathbb{Z}_p[x]$, $f(x) \neq 0$. Show that if there exists $g(x) \neq 0$, such that $g^2(x) \mid f(x)$, then the quotient ring $\mathbb{Z}_p[x]/\langle f(x) \rangle$ has a non-zero nilpotent element.

(9) Let R_1 be a commutative ring with identity. Suppose R is a subring of R_1 which contains the identity of R_1. Show that if $a_1, a_2, \ldots, a_n \in R_1$ are algebraically independent over R, then any of the elements $a_1, a_2, \ldots, a_n$ is transcendental over R.

(10) Suppose R is a domain, and $f(x_1, \ldots, x_n)$ is a non-zero polynomial in n indeterminates over R. Show that if S is an infinite subset of R, then there exist $a_1, \ldots, a_n \in S$ such that $f(a_1, \ldots, a_n) \neq 0$.

(11) Show that the ring of formal power series over the field of real numbers $\mathbb{R}$, $\mathbb{R}[[t]]$, is a unique factorization domain.

Exercises (hard)

(12) Suppose R is a domain but not a field. Show that $R[x]$ is not a principal ideal domain.

(13) Suppose (R, δ) is a Euclidean domain and satisfies the conditions $\delta(ab) = \delta(a)\delta(b)$, $\delta(a + b) \leq \max(\delta(a), \delta(b))$, $\forall a, b \in R^*$. Show that R is either a field or a polynomial ring in one indeterminate over a field.

(14) (Eisenstein's Criterion). Suppose R is a unique factorization domain, $\mathbb{F}$ is the field of fractions of R, and $f(x) = a_n x^n + a_{n-1} x^{n-1} + \cdots + a_1 x + a_0 \in R[x]$. Show that if there exists an irreducible element $p \in R$, such that $p \mid a_i, 0 \leq i \leq n - 1$, but $p \nmid a_n$, $p^2 \nmid a_0$, then $f(x)$ is an irreducible element in $\mathbb{F}[x]$.

2.10   Symmetric polynomials

In this section, we study symmetric polynomials over a commutative ring with identity. Let R be a commutative ring with identity, and

$R[x_1, x_2, \ldots, x_n]$ the polynomial ring in n indeterminates over R. Given an n-permutation $\sigma \in S_n$, there exists a unique automorphism T of $R[x_1, x_2, \ldots, x_n]$ such that $T(x_i) = x_{\sigma(i)}$ and $T|_R = \mathrm{id}_R$. We denote this automorphism by σ^*. It is obvious that $(\sigma\tau)^* = \sigma^*\tau^*$, $\forall \sigma, \tau \in S_n$.

Definition 2.10.1. A polynomial $f(x_1, x_2, \ldots, x_n) \in R[x_1, \ldots, x_n]$ is said to be **symmetric** if for any $\sigma \in S_n$ we have $\sigma^*(f) = f$.

It is easily seen that, if f, g are symmetric, then so are $f + g$ and fg. Thus, the set of all the symmetric polynomials in $R[x_1, \ldots, x_n]$ is a subring of $R[x_1, x_2, \ldots, x_n]$, denoted as $\mathcal{S}[x_1, x_2, \ldots, x_n]$. Obviously, $R \subseteq \mathcal{S}[x_1, x_2, \ldots, x_n]$. Our main goal of this section is to show that $\mathcal{S}[x_1, x_2, \ldots, x_n]$ is a polynomial ring in n indeterminates over R, and then find out a generating set of this polynomial ring.

Example 2.10.2. Given a commutative ring R with identity 1, set

$$p_1 = x_1 + x_2 + \cdots + x_n;$$
$$p_2 = x_1 x_2 + x_1 x_3 + \cdots + x_1 x_n + x_2 x_3 + \cdots = \sum_{i<j} x_i x_j;$$
$$p_3 = \sum_{i<j<k} x_i x_j x_k;$$
$$\cdots\cdots\cdots$$
$$p_n = x_1 x_2 \ldots x_n.$$

Then $p_1, p_2, \ldots, p_n$ are symmetric polynomials. These polynomials are called **elementary symmetric polynomials**.

To state and prove the main result of this section, we need the notion of lexicographic ordering for polynomials in several indeterminates: given two non-zero monomials $f = a x_1^{i_1} x_2^{i_2} \ldots x_n^{i_n}$, $g = b x_1^{j_1} x_2^{j_2} \ldots x_n^{j_n}$ over R, we say that f is higher than g, denoted as $f \succ g$, if there exists an index $k \geq 1$, such that for any $l < k$, $i_l = j_l$, but $i_k > j_k$. For example, $x_1^2 x_2 x_3 \succ x_1 x_2^5 x_3^4$, $x_1 x_2 x_3^3 x_4 \succ x_1 x_2 x_3^2 x_4^9$, etc.

The following theorem is very important, and is called the **Fundamental Theorem of symmetric polynomials** in the literature.

Theorem 2.10.3. *Let R be a commutative ring with identity. Then the elementary symmetric polynomials $p_1, p_2, \ldots, p_n$ are algebraically* *independent over R, and $S[x_1, x_2, \ldots, x_n] = R[p_1, p_2, \ldots, p_n]$. Equivalently, every symmetric polynomial can be uniquely expressed as a polynomial of elementary symmetric polynomials.*

Proof. We first show that every symmetric polynomial can be expressed as a polynomial of elementary symmetric polynomials. Let f be a non-zero symmetric polynomial, and suppose

$$f = f_1 + f_2 + \cdots + f_m$$

is the decomposition of f into the sum of homogeneous polynomials. Notice that for any $\sigma \in S_n$, σ^* sends a homogeneous polynomial of degree k to a homogeneous polynomial of degree k. This fact implies that if f is symmetric, then so is f_j, for any $j = 1, 2, \ldots, m$. Therefore, to prove our assertion, we can assume, without loss of generality, that f itself is a non-zero homogeneous symmetric polynomial. Suppose the degree of f is m, and the highest term in f (lexicographic ordering) is

$$a x_1^{k_1} x_2^{k_2} \ldots x_n^{k_n}, \quad a \neq 0.$$

We assert that $k_1 \geq k_2 \geq \cdots \geq k_n$. Otherwise, there exists $1 \leq i \leq n-1$ such that $k_i < k_{i+1}$. Consider $\sigma = (i, i+1) \in S_n$. It is easily seen that the monomial $a x_1^{k_1} \ldots x_{i-1}^{k_{i-1}} x_i^{k_{i+1}} x_{i+1}^{k_i} \ldots x_n^{k_n}$ appears in the expansion of $f = \sigma^*(f)$. But

$$a x_1^{k_1} \ldots x_{i-1}^{k_{i-1}} x_i^{k_{i+1}} x_{i+1}^{k_i} \ldots x_n^{k_n} \succ a x_1^{k_1} x_2^{k_2} \ldots x_n^{k_n},$$

which is a contradiction with the assumption that $a x_1^{k_1} x_2^{k_2} \ldots x_n^{k_n}$ is the highest term in f. This proves our assertion. Now, set $d_1 = k_1 - k_2, d_2 = k_2 - k_3, \ldots, d_{n-1} = k_{n-1} - k_n, d_n = k_n$. Then it is easy to see that the highest term of $p_1^{d_1} p_2^{d_2} \ldots p_n^{d_n}$ is

$$x_1^{k_1} x_2^{k_2} \ldots x_n^{k_n}.$$

Therefore, either the homogeneous symmetric polynomial

$$g_1 = f - a p_1^{d_1} \ldots p_n^{d_n}$$

is equal to 0, or its degree is m but the highest term $b x_1^{l_1} \ldots x_n^{l_n}$ satisfies $a x_1^{k_1} \ldots x_n^{k_n} \succ b x_1^{l_1} \ldots x_n^{l_n}$. As there are only finitely many monomials of the same degree, this procedure applied repeatedly will reduce the polynomial to zero after finite steps. Then by reversing the procedure, we can express f as a polynomial of $p_1, p_2, \ldots, p_n$.

Now, we prove that $p_1, p_2, \ldots, p_n$ are algebraically independent over R. Notice that for $d_1, d_2, \ldots, d_n \geq 0$, the highest term in the polynomial $p_1^{d_1} p_2^{d_2} \ldots p_n^{d_n}$ is

$$x_1^{d_1+d_2\cdots+d_n} x_2^{d_2+\cdots+d_n} \ldots x_n^{d_n}.$$

This implies that $p_1^{c_1} p_2^{c_2} \ldots p_n^{c_n} = p_1^{d_1} p_2^{d_2} \ldots p_n^{d_n}$ if and only if they have the same highest term as polynomials in $x_1, x_2, \ldots, x_n$. Suppose $F(x_1, x_2, \ldots, x_n) \in R[x_1, x_2, \ldots, x_n]$ is a non-zero polynomial. We select a fixed non-zero monomial of F

$$a x_1^{d_1} x_2^{d_2} \ldots x_n^{d_n}, \quad a \neq 0,$$

satisfying the following two conditions:

(1) Among all the non-zero monomials of $F(x_1, x_2, \ldots, x_n)$, the value of $d_1 + 2d_2 + \ldots + nd_n$ attains the maximum;
(2) Among all the monomials satisfying condition (1), the monomial defined by $x_1^{k_1} x_2^{k_2} \ldots x_n^{k_n}$, where $k_i = \sum_{j=i}^{n} d_j$, is the highest according to the lexicographic ordering.

Obviously the monomial satisfying conditions (1) and (2) exists and is unique. Now, we consider $F(p_1, p_2, \ldots, p_n)$. It is easily seen that the monomial

$$a x_1^{k_1} x_2^{k_2} \ldots x_n^{k_n}, \quad a \neq 0.$$

appears in the expansion of F in $x_1, \ldots, x_n$. However, other non-zero monomials in the expansion either have less degree or is not as high as the above monomial according to the lexicographic ordering. Therefore, we have $F(p_1, p_2, \ldots, p_n) \neq 0$. Thus, $p_1, p_2, \ldots, p_n$ are algebraically independent over R.

The proof of the theorem is now completed. $\square$

We now give an example to illustrate the proof of the above theorem.

Example 2.10.4. Express the symmetric polynomial $x_1^3 + x_2^3 + x_3^3$ over $\mathbb{Q}$ as a polynomial of the elementary symmetric polynomials.

Solution. Since the highest monomial of $x_1^3 + x_2^3 + x_3^3$ is x_1^3, we first set

$$g_1 = (x_1^3 + x_2^3 + x_3^3) - p_1^3 = -3(x_1^2 x_2 + x_2^2 x_1 + \cdots) - 6x_1 x_2 x_3.$$

Now, the highest monomial of g_1 is $-3x_1^2 x_2$. Then we set

$$g_2 = g_1 - (-3p_1 p_2) = 3x_1 x_2 x_3 = 3p_3.$$

Therefore,

$$x_1^3 + x_2^3 + x_3^3 = p_1^3 - 3p_1 p_2 + 3p_3.$$

Finally, we introduce a method of undermined coefficients which is frequently used in expressing a symmetric polynomial as a polynomial of the elementary symmetric polynomials. We will describe this method based on Example 2.10.4. Notice that the degree of $x_1^3 + x_2^3 + x_3^3$ is 3, and the highest term is x_1^3, which corresponds to the triple of non-negative integers $(3, 0, 0)$. To express the above polynomial as a polynomial of the elementary symmetric polynomials in x_1, x_2, x_3, we need to carry out the following three steps:

Step 1: Determine all the triples of non-negative integers (i, j, k) satisfying the following conditions: (1) $i + j + k = 3$, and $i \geq j \geq k$; (2) (i, j, k) is lower than or equal to $(3, 0, 0)$ in the lexicographic ordering. It is easily seen that these triples are $(3, 0, 0), (2, 1, 0), (1, 1, 1)$.

Step 2: Write down all the monomials $p_1^{i-j} p_2^{j-k} p_3^k$, where the triples (i, j, k) are the ones in Step 1, namely, p_1^3, $p_1 p_2$ and p_3.

Step 3: It is easy to prove that there exist constants $a, b, c \in \mathbb{Q}$ such that

$$x_1^3 + x_2^3 + x_3^3 = ap_1^3 + bp_1 p_2 + cp_3.$$

Then we can determine the constants a, b, c by selecting some special values of x_1, x_2, x_3.

In this example, we can proceed as follows. Letting $x_1 = 1, x_2 = x_3 = 0$, we get $a = 1$. Then letting $x_1 = 1, x_2 = 0, x_3 = -1$, we get $b = -3$. Finally, letting $x_1 = x_2 = x_3 = 1$ we get $c = 3$.

We remark here that in this example it seems that the method of undetermined coefficients does not make the calculation much easier.

But this method is very effective in dealing with some complicated polynomials; see the exercises of this section. On the other hand, we must mention that the symmetric polynomial in Example 2.10.4 is homogeneous. When dealing with a symmetric polynomial which is not homogeneous using the method of undetermined coefficients, we need first write the polynomial as the sum of homogeneous symmetric polynomials, and then deal with the homogeneous components one by one.

Exercises

(1) Suppose $\Delta_n = \Pi_{1 \le i < j \le n}(x_i - x_j)$. Show that Δ_n^2 is a symmetric polynomial, and express Δ_3^2 as a polynomial of the elementary symmetric polynomials.

(2) Use the method of undetermined coefficients to express the symmetric polynomial $f(x_1, x_2, x_3) = (x_1^3 + x_2^3)(x_2^3 + x_3^3)(x_1^3 + x_3^3)$ as a polynomial of the elementary symmetric polynomials.

(3) Express the symmetric polynomial $(x_1^2 + x_2 x_3)(x_2^2 + x_1 x_3)(x_3^2 + x_1 x_2)$ as a polynomial of the elementary symmetric polynomials.

(4) Let R be a commutative ring with identity, $x_1, x_2, \ldots, x_n \in R$, and $f(x) = (x - x_1)(x - x_2) \ldots (x - x_n) = x^n - p_1 x^{n-1} + \cdots + (-1)^n p_n$. Set $s_0 = n$, $s_k = x_1^k + x_2^k + \cdots + x_n^k$, $k = 1, 2, \ldots$. Show that

$$x^{k+1} f'(x) = (s_0 x^k + s_1 x^{k-1} + \cdots + s_k) f(x) + g(x),$$

where $f'(x)$ is the differential of $f(x)$, and $\deg g(x) < n$.

(5) Let R be a commutative ring with identity. Express the symmetric polynomial $x_1^2 x_2 x_3 + x_1 x_2^2 x_3 + x_1 x_2 x_3^2$ in $R[x_1, x_2, x_3]$ as a polynomial of s_1, s_2, s_3.

(6) Let R be a commutative ring with identity, and $a \in R$. Express the symmetric polynomial $\Pi_{i=1}^n(x_1 + \cdots + x_{i-1} + a x_i + x_{i+1} + \cdots + x_n)$ as a polynomial of the elementary symmetric polynomials over R.

(7) Let a, b, c be complex numbers. Show that the three roots of the polynomial $x^3 + ax^2 + bx + c$ form an arithmetic sequence if and only if $2a^3 - 9ab + 27c = 0$.

(8) Let x_1, x_2, x_3, x_4, x_5 be the roots of the complex polynomial $x^5 - 3x^3 - 5x + 1$. Calculate $\sum_{i=1}^5 x_i^4$.

Exercises (hard)

(9) Prove the Newton's formulas:

$$s_k - s_{k-1}p_1 + \cdots + (-1)^{k-1}s_1p_{k-1} + (-1)^k k p_k = 0, \quad k \leq n;$$
$$s_k - s_{k-1}p_1 + \cdots + (-1)^n s_{k-n}p_n = 0, \quad k > n.$$

(10) Show that $\{s_1, s_2, \ldots, s_n\}$ is a generating set of the polynomial ring $\mathcal{S}[x_1, x_2, \ldots, x_n]$ over R.

2.11 Summary of the chapter

In this chapter, we have introduced the fundamental notions and properties of rings. We can see that, since a ring has two binary operations, the content of ring theory is much richer compared to group theory, and there are much more techniques involved in ring theory. For example, there is no counterpart of polynomial theory in group theory. The applications of ring theory are rather extensive due to its universality.

Meanwhile, we can also see that, many definitions and theorems in ring theory are very similar to those in group theory, including the content or the proofs of the theorems. For example, the fundamental theorem of homomorphisms, subsystems, quotient systems, etc. Nevertheless, we must pay more attention to the differences between group theory and ring theory. For example, many notions in ring theory cannot be defined for a group, e.g., rings without zero divisors, characteristic, as well as domains, division rings, and fields. Of course, the most important point is that the polynomial theory in ring theory, which even cannot be defined for a group, has become the foundation of many branches in modern algebra, such as commutative algebra and algebraic geometry.

There are many topics in ring theory which have not been dealt with in this book, e.g., theory of ideals, Noether rings, Artin rings, etc. Readers who are interested in these topics can consult the references of this book.

Chapter 3

Modules

In this chapter, we introduce the notion of a module and study the fundamental properties. The precise definition of a module was first given by the German mathematician Noether. It was Noether who found for the first time that the relationship between the matrix representation theory of finite groups and the theory of algebraic structures can be clarified by the notion of a module, and then carried out a series of studies. Due to this, the study of modules has always been related to the matrix representation theory and algebraic theory.

Modules are generalizations of vector spaces over a number field in linear algebra. On the one hand, every vector space over a number field can be viewed as a module over the base field. On the other hand, it is well known that the theory of linear transformations is the most important subject in linear algebra. But we will soon show that, given a linear transformation of a vector space V over a number field $\mathbb{F}$, V can be viewed as a module over the polynomial ring $\mathbb{F}[\lambda]$. This point of view will lead to the theory of canonical forms of linear transformations. As another important application, the module theory can be used to classify finitely generalized abelian groups. In this chapter, we will give a detailed introduction to the structure theory of modules over a principal ideal domain, and the applications of this theory to canonical forms of linear transformations, as well as the classification of finitely generated abelian groups.

3.1   Fundamental concepts

In this section, we will introduce fundamental concepts of modules. The notion of a module is very similar to that of a vector space over a number field. The precise definition is the following:

Definition 3.1.1. Let R be a ring with identity, and $\{M; +\}$ an abelian group. If there is a map from $R \times M$ to M: $(a, x) \mapsto ax$ satisfying the following conditions:

(1) $(a + b)x = ax + bx$, $\forall a, b \in R$, $x \in M$;
(2) $a(x + y) = ax + ay$, $\forall a \in R$, $x, y \in M$;
(3) $1x = x$, $\forall x \in M$, where 1 is the identity of R;
(4) $a(bx) = (ab)x$, $\forall a, b \in R$, $x \in M$,

then M is called a **left R-module**.

Similarly, if M is an abelian group, and there is a map from $M \times R$ to M: $(x, a) \mapsto xa$ satisfying analogous conditions as the above (1)–(4), then M is called a **right R-module**. If M is simultaneously a left R-module and a **right R-module**, and satisfies the condition $a(xb) = (ax)b$, $\forall a, b \in R, x \in M$, then M is called a **bi-module** over R. In this book, unless otherwise stated, an R-module will refer to a **left R-module**. In general, assertions that are valid for left modules will also be valid for right modules, and readers can judge whether the assertion holds for a bi-module.

In the special case that $R = \mathbb{F}$ is a field, we will say that the module M is a vector space over $\mathbb{F}$. Thus, the notion of a module is a natural generalization of that of a vector space. We stress here that all the fundamental properties of a vector space over a number field in linear algebra are also valid for a vector space over a general field. In particular, we can define and study the dimension of a general vector space as well as the linear transformations on a vector space.

Next, we give some explicit examples of modules. These examples show that the notion of modules is abstracted from various mathematical objects, hence the theory of modules can be applied very broadly.

Example 3.1.2. Let G be an abelian group. Define a map from $\mathbb{Z} \times G$ to G by: $(m, x) \mapsto mx$. Then it is easy to check that G is a $\mathbb{Z}$-module. This will lead to a classification of finitely generated abelian groups.

Example 3.1.3. Any unitary ring R can be viewed as an R-module. In fact, R is an abelian group with respect to the addition, and the multiplication gives a map from $R \times R$ to R which satisfies the conditions (1)–(4) in Definition 3.1.1.

Example 3.1.4. Recall that a vector space V over a field $\mathbb{F}$ is an $\mathbb{F}$-module. Now, fix a linear transformation $\mathcal{A}$ on V. Then V can be made into a $\mathbb{F}[\lambda]$-module, where $\mathbb{F}[\lambda]$ is the polynomial ring in one indeterminate over $\mathbb{F}$. In fact, we only need to define a map from $\mathbb{F}[\lambda] \times V$ to V as follows: given $f(\lambda)$ and $v \in V$, set

$$f(\lambda)v = f(\mathcal{A})(v).$$

Then it is easy to check that the conditions (1)–(4) in Definition 3.1.1 are satisfied. Therefore, V becomes an $\mathbb{F}[\lambda]$ module. This point of view will lead to the theory of canonical forms of linear transformations on vector spaces.

As an exercise, we suggest readers provide all the details in the above examples.

Now, let us give some simple properties of a module. First notice that, for an R-module M, we will denote the zero element of the abelian group M as well as the zero element of the additive group R as 0, provided that this will cause no confusion. Then it is easy to show that, for any $a \in R$, we have $a0 = 0$; meanwhile, for any $x \in M$, we have $0x = 0$. From this we deduce that $a(-x) = (-a)x = -(ax)$, $\forall a \in R, x \in M$. Moreover, by the definition of a module, it is easily seen that for any $a_1, a_2, \ldots, a_n \in R$, and $x_1, x_2, \ldots, x_m \in M$, we have

$$\left(\sum_{i=1}^{n} a_i \right) \left(\sum_{j=1}^{m} x_j \right) = \sum_{i=1}^{n} \sum_{j=1}^{m} a_i x_j.$$

Now, we introduce the notions of submodules, quotient modules, and homomorphism of modules, and study the fundamental properties.

Definition 3.1.5. Let R be a unitary ring and M, an R-module. A non-empty subset N of M is called a **submodule** of M if N is an R-module with respect to the addition of M and the map from $R \times N$ to N is defined by the restriction of the map from $R \times M$ to M.

Problem 3.1.6. Show that a non-empty subset N of a module M is a submodule if and only if: (1) N is a subgroup of the additive group M, (2) for any $a \in R$, $x \in N$, $ax \in N$.

The above criterion applied to Examples 3.1.2–3.1.4 gives a series of examples of submodules:

(1) A non-empty subset W of a vector space V over a field $\mathbb{F}$ is a submodule of V if and only if it is a linear subspace of V.
(2) A non-empty subset H of an abelian group G is a submodule of G (as a $\mathbb{Z}$-module) if and only if it is a subgroup of G.
(3) If we regard a unitary ring R as an R-module, then a non-empty subset R_1 is a submodule if and only if R_1 is a left ideal of R.
(4) Given a linear transformation $\mathcal{A}$ of a vector space V over a field $\mathbb{F}$, a non-empty subset W of V is a submodule of V (viewed as an $\mathbb{F}[\lambda]$-module) if and only if W is an invariant subspace of $\mathcal{A}$.

Given two submodules M_1, M_2 of an R-module M, it is easy to check that the intersection $M_1 \cap M_2$ is also a submodule of M. Moreover, define

$$M_1 + M_2 = \{x_1 + x_2 | x_1 \in M_1, x_2 \in M_2\}.$$

Then $M_1 + M_2$ is also a submodule of M, called the **sum of the submodules** M_1 and M_2. Notice that the intersection of a family of (finitely or infinitely many) submodules must be a submodule. However, the notion of the sum of submodules can only be generalized to the case of finitely many submodules.

Now, we introduce a method to construct a submodule from an arbitrary non-empty subset of a module. Let S be a non-empty subset of a module M. Then the intersection of all the submodules containing S (such submodules exist, e.g., M) is a submodule of M, called the submodule of M generated by S, denoted as $[S]$. If $[S] = M$, then we say that S is a generating subset of M. If there exists $y \in M$ such that $[y] = M$, then M is called a **cyclic module**. A module M is called finitely generated if there exists a finite subset S of M such that $[S] = M$. It is obvious that an abelian group G viewed as a $\mathbb{Z}$-module is a cyclic module if and only if it is a cyclic group, and G is finitely generated as a $\mathbb{Z}$-module if and only if it is a finitely generated group.

Problem 3.1.7. Show that $[S] = \{\sum_{i=1}^{m} a_i y_i \,|\, m \in \mathbb{N}, a_i \in R, y_i \in S\}$.

Next we introduce the notion of **direct sum of submodules**.

Definition 3.1.8. If the submodules $M_1, M_2, \ldots, M_s$ of M satisfy the condition $M = M_1 + M_2 + \cdots + M_s$ and the expression of any element u in M as $u = a_1 + a_2 + \cdots + a_s$, $a_i \in M_i$, is unique, then we say that M is the (internal) direct sum of $M_1, M_2, \ldots, M_s$, denoted as $M = M_1 \oplus M_2 \oplus \cdots \oplus M_s$.

Theorem 3.1.9. *Suppose M_i, $1 \leq i \leq s$ are submodules of M, and $M = M_1 + M_2 + \cdots + M_s$. Then the following three conditions are equivalent:*

(1) $M = M_1 \oplus M_2 \oplus \cdots \oplus M_s$;
(2) *If $a_i \in M_i, 1 \leq i \leq s$ and $a_1 + a_2 + \cdots + a_s = 0$, then $a_1 = a_2 = \ldots a_s = 0$, in other words, the expression of 0 is unique.*
(3) *For any i, we have*

$$M_i \cap \left(\sum_{j \neq i} M_j \right) = \{0\}. \tag{3.1}$$

The proof of this theorem is similar to the case of the direct sum of vector spaces in linear algebra, and is omitted.

Definition 3.1.10. Suppose $N_1, N_2, \ldots, N_n$ are R-modules. Define addition on $N = N_1 \times N_2 \times \cdots \times N_n$ and a map from $R \times N$ to N as follows:

$$(x_1, \ldots, x_n) + (y_1, \ldots, y_n) = (x_1 + y_1, \ldots, x_n + y_n),$$

$$a(x_1, \ldots, x_n) = (ax_1, \ldots, ax_n), \; x_i, y_i \in N_i, 1 \leq i \leq n, a \in R.$$

Then N becomes an R-module, called the **external direct sum** of the modules $N_1, N_2, \ldots, N_n$, denoted as

$$N = N_1 \otimes N_{\otimes} \cdots \otimes N_n.$$

The following proposition shows that there is no essential difference between internal direct sum and external direct sum.

Proposition 3.1.11. *Let* $N = N_1 \otimes N_2 \otimes \cdots \otimes N_n$ *be the external direct sum of the* R-*modules* $N_1, N_2, \ldots, N_n$. *Set*

$$N_i' = \{(0, 0, \ldots, 0, x_i, 0, \ldots, 0) \in N \mid x_i \in N_i\}.$$

Then the N_i'*'s are submodules of* N, *and* N *is the internal direct sum of* $N_1', N_2' \ldots, N_n'$.

Proof.   It is obvious that for any $\alpha_1, \alpha_2 \in N_i'$, we have $\alpha_1 - \alpha_2 \in N_i'$. Thus, the N_i''s are subgroups of N with respect to the addition. Meanwhile, for any $a \in R, x_i \in N_i$, we have $a(0, \ldots, 0, x_i, 0, \ldots, 0) = (0, \ldots, 0, ax_i, 0, \ldots, 0) \in N_i'$. Thus, the N_i''s are submodules of N. Now, for any $x_i \in N_i, 1 \leq i \leq n$,

$$(x_1, \ldots, x_n) = (x_1, 0, \ldots, 0) + \cdots + (0, \ldots, 0, x_n).$$

Hence, $N = N_1' + N_2' + \cdots + N_n'$. Moreover, the condition (3.1) is obviously satisfied. Therefore, $N = N_1' \oplus N_2' \oplus \cdots \oplus N_n'$. □

Finally, we introduce quotient modules and the Fundamental Theorem of Homomorphisms of modules.

Theorem 3.1.12. *Let* N *be a submodule of an* R-*module* M. *On the quotient group* M/N *of the abelian group* M, *we define a map from* $R \times M/N$ *to* M/N *as follows:*

$$a(x + N) = ax + N, \quad a \in R, x \in M.$$

Then M/N *becomes an* R-*module with respect to the addition of the quotient group and the above map, called the* **quotient module** *of* M *with respect to* N.

Proof.   We first check that the map described above is well-defined. If $x_1, x_2 \in M$ satisfy $x_1 + N = x_2 + N$, then $x_1 - x_2 \in N$. Since N is a submodule of M, we have $a(x_1 - x_2) = ax_1 - ax_2 \in N, \forall a \in R$. Thus, $ax_1 + N = ax_2 + N$. Therefore, the map is well-defined. Now, it is easy to check that the conditions (1)–(4) in Definition 3.1.1 are satisfied. Therefore, M/N is an R-module. □

Definition 3.1.13. Let M_1, M_2 be two R-modules. A map ϕ from M_1 to M_2 is called a **homomorphism** if the following conditions are satisfied:

(1) $\phi(x + y) = \phi(x) + \phi(y)$, $\forall x, y \in M_1$;
(2) $\phi(ax) = a\phi(x)$, for any $a \in R$ and $x \in M_1$.

Condition (1) means that ϕ is a homomorphism between the abelian groups M_1 and M_2. We can also define the notions of surjective homomorphisms and injective homomorphisms of modules. A homomorphism is called an isomorphism if it is both surjective and injective. It is obvious that the composition of two homomorphisms is also a homomorphism, and the inverse map of an isomorphism is an isomorphism. If there exists an isomorphism between the modules M_1 and M_2, then we say that the modules M_1 and M_2 are isomorphic, denoted as $M_1 \simeq M_2$. It is also easy to see that the binary relation *isomorphic* is an equivalence relation in the set of all R-modules.

Example 3.1.14. In Proposition 3.1.11, define a map φ_i from N_i to N_i' as follows:

$$\varphi_i(x_i) = (0, \ldots, 0, x_i, 0 \ldots, 0), \ x_i \in N_i,$$

where the ith component of the element of the right-hand side is x_i, and all the other components are 0. Then φ_i is an isomorphism. Thus, the module N_i is isomorphic to N_i'.

Proposition 3.1.15. *Let ϕ be a homomorphism from the module M_1 to M_2. Set*

$$\operatorname{Ker} \phi = \{x \in M_1 \mid \phi(x) = 0\}, \quad \operatorname{Im} \phi = \{\phi(x) \mid x \in M_1\}.$$

Then $\operatorname{Ker} \phi$ is a submodule of M_1, and $\operatorname{Im} \phi$ is a submodule of M_2, called the kernel and image of ϕ, respectively.

Proof. We only prove for $\operatorname{Ker} \phi$. The proof for $\operatorname{Im} \phi$ will be left to readers. First, since ϕ is a homomorphism of abelian groups, $\operatorname{Ker} \phi$ is not empty, and is a subgroup of the abelian group M_1. Now for any $a \in R, x \in \operatorname{Ker} \phi$, we have $\phi(ax) = a\phi(x) = a0 = 0$. Thus, $ax \in \operatorname{Ker} \phi$. Therefore, $\operatorname{Ker} \phi$ is a submodule of M_1. $\quad\square$

Sometimes we will also denote the image $\operatorname{Im} \phi$ directly as $\phi(M_1)$, and this notation will be convenient in dealing with some problems related to submodules.

Now, we state the **Fundamental Theorem of Homomorphisms of modules**. Since the proof is similar to that for rings, we will leave it to the readers.

Theorem 3.1.16. *Let ϕ be a surjective homomorphism from the module M_1 to M_2. Then we have the following*:

(1) *The quotient module $M_1/\operatorname{Ker}\phi$ is isomorphic to M_2.*
(2) *There exists a bijection between the set of submodules containing $\operatorname{Ker}\phi$ and the set of submodules of M_2.*
(3) *If N_1 is a submodule of M_1 containing $\operatorname{Ker}\phi$, then $M_1/N_1 \simeq M_2/\phi(N_1)$.*

Finally, we give an interesting example.

Example 3.1.17. From the content of this section it is easily seen that the same abelian group may be viewed as a module over different rings, and we will see in the following sections that this point of view will be useful. We have also seen that any abelian group can be viewed as a $\mathbb{Z}$-module, and this will lead to a classification of finitely generated abelian groups. However, the abelian group itself has no direct relationship with the ring of integers. Now, we present another point of view, namely, given an abelian group G, we first construct a ring uniquely determined by G, and then view G as a module over this ring. Denote by $\operatorname{End}(G)$ the set of all endomorphisms of G into itself. Then define addition and multiplication on $\operatorname{End}(G)$ as

$$(\xi + \eta)(x) = \xi(x) + \eta(x);$$
$$(\xi\eta)(x) = \xi(\eta(x)), \ \xi, \eta \in \operatorname{End}(G), x \in G.$$

It is easy to check that $\operatorname{End}(G)$ becomes a ring with respect to the above addition and multiplication, called the ring of endomorphisms of G. Now, we define a map from $\operatorname{End}(G) \times G$ to G by $(\xi, x) \mapsto \xi(x)$. Then it is easy to check that G becomes an $\operatorname{End}(G)$-module.

Problem 3.1.18. Let p be a prime number. Determine the ring of endomorphisms of $\mathbb{Z}_p$, and describe the structure of $\mathbb{Z}_p$ as an $\operatorname{End}(\mathbb{Z}_p)$-module.

Exercises

(1) Prove Theorem 3.1.9.

(2) Let G be an abelian group. Suppose there exists a map ϕ from $\mathbb{Q} \times G$ to G which makes G a $\mathbb{Q}$-module. Show that such a map ϕ is unique.

(3) Show that a finite abelian group with more than one element cannot be made into a $\mathbb{Q}$-module.

(4) Let I be an ideal of a commutative unitary ring R, and M an R-module. Show that $IM = \{\sum_{i=1}^{m} a_i x_i |, m \in \mathbb{N}, a_i \in I, x_i \in M\}$ is a submodule of M.

(5) Let M be an R-module, and N_1, N_2 be submodules of M. Show that

$$I = \{a \in R | ax \in N_2, \forall x \in N_1\}$$

is an ideal of R.

(6) A complex number is called an **algebraic integer** if it is a root of a non-zero monic polynomial with integer coefficients. Let α be an algebraic integer, and $M = \mathbb{Z}[\alpha]$. Show that for any positive integer m, $mM = \{m\gamma | \gamma \in M\}$ is a submodule of M, and the quotient module R/mR is a finite set.

(7) An R-module M is called **simple** if M has only two submodules: M and $\{0\}$. Prove **Schur's lemma**: if φ is a homomorphism from a simple module M to a simple module M', then either $\varphi = 0$ or φ is an isomorphism.

(8) Let M be an R-module. Define the **annihilator** of M as

$$\text{ann}(M) = \{a \in R \,|\, ax = 0, \forall x \in M\}.$$

Show that $\text{ann}(M)$ is an ideal of R, and determine the annihilator of the $\mathbb{Z}$-module $\mathbb{Z}_3 \otimes \mathbb{Z}_6 \otimes \mathbb{Z}_8$.

(9) Let M be an R-module, and φ a homomorphism from a ring R_1 to R. Define a map $R_1 \times M \to M$ by $(a, x) = \varphi(a)x$, $a \in R_1, x \in M$. Show that this map makes M into an R_1-module.

(10) Prove Theorem 3.1.16.

(11) Let M be a module, and M_1, M_2 two submodules of M. Show that

   (i) If $M_1 \subseteq M_2$, then $M/M_2 \simeq (M/M_1)/(M_2/M_1)$;

   (ii) If $M = M_1 \oplus M_2$, then $M_1 \simeq (M_1 + M_2)/M_2$, $M_2 \simeq (M_1 + M_2)/M_1$.

(12) Let M be a module, and M_1, M_2 submodules of M. Show that $(M_1 + M_2)/M_2 \simeq M_1/(M_1 \cap M_2)$.

(13) Let M be a module, and $M_1, M_2, \ldots, M_s$ be submodules of M. Show that $M = M_1 \oplus M_2 \oplus \cdots \oplus M_s$ if and only if

   (i) $M = M_1 + M_2 + \cdots + M_s$;

   (ii) for any $2 \leq j \leq s$, $M_j \cap \sum_{i=1}^{j-1} M_i = \{0\}$.

(14) Does the conclusion of Exercise 3.1 hold if we change the condition (2) to $M_i \cap M_j = \{0\}, \forall i \neq j$?

(15) Let M be a module, and $M_1, M_2, \ldots, M_s$ submodules of M such that $M = M_1 \oplus M_2 \oplus \cdots \oplus M_s$. Suppose N is a submodule of M, and $N = N_1 \oplus N_2 \oplus \cdots \oplus N_s$, where $N_i \subset M_i$, $i = 1, 2, \ldots, s$. Show that $M/N \simeq M_1/N_1 \oplus M_2/N_2 \oplus \cdots \oplus M_s/N_s$.

(16) Let M, N be R-modules, and $f : M \to N$, $g : N \to M$ be homomorphisms such that $f(g(y)) = y, \forall y \in N$. Show that $M = \operatorname{Ker} f \oplus \operatorname{Im} g$.

Exercises (hard)

(17) A module is called **indecomposable** if it cannot be written as the direct sum of two non-zero submodules. Show that $\mathbb{Z}$ is indecomposable as a $\mathbb{Z}$-module, and $\mathbb{Z}_m, m > 1$ is indecomposable as a $\mathbb{Z}$-module if and only if $m = p^e$, where p is a prime number and $e > 0$.

(18) Let R be a domain. Show that R is indecomposable as an R-module.

(19) Let R be a domain, and $\mathbb{F}$ be the fractional field of R. Show that the additive group $\mathbb{F}$ is indecomposable as an R-module. In particular, the additive group $\{\mathbb{Q}; +\}$ of rational numbers is indecomposable as a $\mathbb{Z}$-module.

3.2   Matrices over a ring and the endomorphism rings of modules

In this section, we will first study the properties of matrices over a ring, and then apply the related results to deal with homomorphisms of modules.

Let R be a ring with identity 1. An $m \times n$ matrix over R is a rectangle consisting of m rows and n columns with an entry (element, coefficient, or coordinate) in R on each of the intersectional points of rows and columns. A generic matrix can be written as

$$(a_{ij})_{m \times n} = \begin{pmatrix} a_{11} & a_{12} & \cdots & a_{1n} \\ a_{21} & a_{22} & \cdots & a_{2n} \\ \vdots & \vdots & \vdots & \vdots \\ a_{m1} & a_{m2} & \cdots & a_{mn} \end{pmatrix},$$

where $a_{ij} \in R$ is called the entry in the ith row and jth column. Similarly as in linear algebra, we can define the identity matrix, triangular matrices, and diagonal matrices.

Denote by $R^{m \times n}$ the set of all $m \times n$ matrices with entries in R. Similarly as in the case of matrices over a number field, we can define addition on $R^{m \times n}$ and a map from $R \times R^{m \times n}$ to $R^{m \times n}$. Then it is easy to check that $R^{m \times n}$ becomes an R-module. In the special case of $n = 1$, we usually denote $R^{m \times 1}$ as R^m. It is easy to see that, as an R-module, R^m is equal to the direct sum of m copies of R.

The module R^m has some special properties, which will be studied in detail in the next section. We now consider the set of homomorphisms from R^m to R^n. In general, given two R-modules M, N, we denote the set of homomorphisms from M to N as $\mathrm{Hom}_R(M, N)$ (sometimes as $\mathrm{Hom}(M, N)$, if this will not cause any confusion). From now on, we assume that R is a commutative ring with identity. Then we define addition on $\mathrm{Hom}(M, N)$, and a map from $R \times \mathrm{Hom}(M, N)$ to $\mathrm{Hom}(M, N)$, as follows:

$$(\eta + \xi)(x) = \eta(x) + \xi(x),$$
$$(a\eta)(x) = a\eta(x), \quad \eta, \xi \in \mathrm{Hom}(M, N), x \in M, a \in R.$$

It is easy to check that $\mathrm{Hom}(M, N)$ is an R-module with respect to the above operations.

Theorem 3.2.1. *Let R be a commutative ring with identity. Then* $\mathrm{Hom}(R^m, R^n)$ *is isomorphic to* $R^{m \times n}$ *as an R-module.*

Proof.   Consider the following elements in R^m:

$$e_1 = (1, 0, 0, \ldots, 0),$$
$$e_2 = (0, 1, 0, \ldots, 0),$$
$$\ldots \quad \ldots \quad \ldots \quad \ldots \quad \ldots$$
$$e_m = (0, 0, \ldots, 0, 1).$$

It is easily seen that every element $(x_1, x_2, \ldots, x_m)$ in R^m can be uniquely expressed as

$$(x_1, x_2, \ldots, x_m) = \sum_{i=1}^{m} x_i e_i.$$

Similarly, we can define n elements $f_j \in R^n$, $j = 1, 2, \ldots, n$ such that the jth entry of f_j is 1, and all the other elements are 0. Then the element $(y_1, y_2, \ldots, y_n)$ of R^n can be uniquely expressed as $(y_1, y_2, \ldots, y_n) = \sum_{j=1}^{n} y_j f_j$.

Suppose $\eta \in \mathrm{Hom}(R^m, R^n)$. Then for any $1 \leq i \leq m$, there exists a unique system of elements a_{ij}, $j = 1, 2, \ldots, n$, in R such that $\eta(e_i) = \sum_{j=1}^{n} a_{ij} f_j$. Setting $M(\eta) = (a_{ij})_{m \times n}$, we get a map φ from $\mathrm{Hom}(R^m, R^n)$ to $R^{m \times n}$ defined by $\varphi(\eta) = M(\eta)$. It is obvious that φ is a homomorphism of modules. Moreover, φ is injective. Now, given $B = (b_{ij})_{m \times n} \in R^{m \times n}$, consider the map $\xi : R^m \to R^n$ such that

$$\xi(x_1, x_2, \ldots, x_m) = \sum_{i=1}^{m} \sum_{j=1}^{n} x_i b_{ij} f_j.$$

Then it is easy to check that ξ is a homomorphism from R^m to R^n, and $\varphi(\xi) = B = (b_{ij})_{n \times n}$. Thus, φ is also surjective. Therefore, $\mathrm{Hom}(R^m, R^n)$ is isomorphic to $R^{m \times n}$.   □

Similarly, as matrices over a field, we can define multiplication among matrices over a commutative unitary ring R. Note that matrix multiplication can be defined only when the number of columns of the first matrix is equal to the number of rows of the second matrix. Moreover, although R is commutative, the multiplication between two matrices in $R^{n \times n}$ is in general not commutative. It is also easily seen that the identity matrix is an identity element in $R^{n \times n}$.

Now, we consider the set $\text{Hom}(M, M)$ of endomorphisms of an R-module M. Besides addition, we can also define multiplication (composition) in $\text{Hom}(M, M)$. Then it is easy to check that $\text{Hom}(M, M)$ becomes a ring, denoted as $\text{End}_R(M)$ (or simply $\text{End}(M)$, if R is unambiguous), called the **ring of endomorphisms** of M. Similarly as Theorem 3.2.1, we can prove

Theorem 3.2.2. *If R is a commutative unitary ring, then the ring $\text{End}(R^n)$ is isomorphic to the ring $R^{n \times n}$.*

Problem 3.2.3. Construct an example to show that if R is not commutative, then the assertion of Theorem 3.2.2 need not be true.

An $n \times n$ matrix A over R is called invertible if there exists $B \in R^{n \times n}$ such that $AB = BA = I_n$. In general, it is hard to determine whether a matrix is invertible. However, when the ring R is commutative and unitary, we can apply the properties of determinant to accomplish this.

Now, we define the determinant of an $n \times n$ matrix over a commutative unitary ring R. Similarly as in linear algebra, for $A = (a_{ij})_{n \times n} \in R^{n \times n}$, we define

$$
\det(A) = \begin{vmatrix} a_{11} & a_{12} & \cdots & a_{1n} \\ a_{21} & a_{22} & \cdots & a_{2n} \\ \vdots & \vdots & \vdots & \vdots \\ a_{n1} & a_{n2} & \cdots & a_{nn} \end{vmatrix} = \sum_{\sigma \in S_n} (-1)^{\text{sg}(\sigma)} a_{1i_1} a_{2i_2} \cdots a_{ni_n},
$$

where

$$
\sigma = \begin{pmatrix} 1 & 2 & \cdots & n \\ i_1 & i_2 & \cdots & i_n \end{pmatrix} \in S_n,
$$

and $\text{sg}(\sigma) \in R$ is the signature of σ, that is, if σ is an odd permutation, then $(-1)^{\text{sg}(\sigma)} = -1$; and if σ is even, $(-1)^{\text{sg}(\sigma)} = 1$. It is easy to check that all the properties of the determinant of matrices over a number field are valid for determinants of matrices over a commutative unitary ring. In particular, we can define the notions of cofactor minors and algebraic cofactor minors. Given a matrix A, denote the algebraic cofactor minor of a_{ij} as A_{ij}. Then for any $1 \leq i, k \leq n$, we have

$$\sum_{j=1}^{n} a_{ij} A_{kj} = \delta_{ik} \det A, \qquad (3.2)$$

$$\sum_{j=1}^{n} a_{ji} A_{jk} = \delta_{ik} \det A. \qquad (3.3)$$

The adjoint matrix of A is defined to be

$$A^* = \begin{pmatrix} A_{11} & A_{21} & \cdots & A_{n1} \\ A_{12} & A_{22} & \cdots & A_{n2} \\ \cdots & \cdots & \cdots & \cdots \\ A_{1n} & A_{2n} & \cdots & A_{nn} \end{pmatrix}.$$

Then it follows from (3.2), (3.3) that $AA^* = A^*A = (\det A)I_n$. From this we conclude the following:

Proposition 3.2.4. *Let R be a commutative ring with identity. Then a matrix $A \in R^{n \times n}$ is invertible if and only if $\det A$ is a unit in R. Moreover, if $A, B \in R^{n \times n}$ satisfy $AB = I_n$, then $B = A^{-1}$, hence $BA = I_n$.*

Finally, we study the normal form of matrices over a principal ideal domain. In the following sections, we will apply the related results to study the structure of finitely generated modules over a principal ideal domain.

Let D be a principal ideal domain. Two $m \times n$ matrices A, B over D are said to be **equivalent** if there exists an invertible matrix $P \in D^{m \times m}$ and an invertible matrix $Q \in D^{n \times n}$ such that $A = PBQ$. In linear algebra, we have shown that every $m \times n$ matrix A over a number field $\mathbb{F}$ is equivalent to a matrix of the form

$$\begin{pmatrix} I_r & 0 \\ 0 & 0 \end{pmatrix} = \begin{pmatrix} 1 & & & & & \\ & 1 & & & 0 & \\ & & \ddots & & & \\ & & & 1 & & \\ & 0 & & & 0 & \\ & & & & & \ddots \end{pmatrix},$$

where $r \geq 0$ is uniquely determined by A, called the rank of A. The matrix

$$\begin{pmatrix} I_r & 0 \\ 0 & 0 \end{pmatrix}$$

is called the **normal form** of A. Obviously, this assertion is not valid for matrices over a general principal ideal domain (Readers are suggested to construct an example to show this). Our problem is to find a special form of simple matrices over a principal ideal domain, such that every matrix is equivalent to such a form.

To solve the above problem, we need abundant invertible matrices. In linear algebra, we have introduced elementary matrices, and these matrices can be similarly defined for a principal ideal domain. Now, we list all the elementary matrices, and indicate the change of a matrix when it is multiplied by an elementary matrix from left or right.

(1) Denote by P_{ij} the matrix obtained by interchanging the ith row and the jth row of the identity matrix I_n, and leaving other rows unchanged. P_{ij} is invertible, since $P_{ij}^2 = I_n$. Left multiplication of A by P_{ij} amounts to interchanging the ith and the jth row of A, and leaving other rows unchanged. Similarly, right multiplication of A by P_{ij} amounts to interchanging the ith and the jth column of A, and leaving other columns unchanged.

(2) Suppose $i \neq j$, $a \in D$. Denote by $T_{ij}(a) = I_n + ae_{ij}$, where e_{ij} is the matrix with 1 in the (i, j)-entry, and 0s elsewhere. $T_{ij}(a)$ is invertible, since $T_{ij}(a)T_{ij}(-a) = I_n$. Left multiplication of A by $T_{ij}(a)$ gives a matrix whose ith row is obtained by multiplying the jth row by a and adding it to the ith row of A, and whose remaining rows are the same as A. Similarly, right multiplication of A by $T_{ij}(a)$ gives a matrix whose ith column is obtained by multiplying the jth column by a and adding it to the ith column of A, and whose remaining columns are the same as A.

(3) Let u be a unit element in D. Denote by $D_i(u)$ the matrix obtained by multiplying the ith row of the identity matrix by u, and leaving other rows unchanged. Then $D_i(u)$ is invertible, since $D_i(u)D_i(u^{-1}) = I_n$. Left multiplication of A by $D_i(u)$ yields a matrix whose ith row is equal to multiplication of the ith row of A by u, and whose other rows are the same as A; Similarly,

right multiplication of A by $D_i(u)$ gives a matrix whose ith column is obtained by multiplying the ith column of A by u, and whose other columns are the same as A.

Now, we can give a solution to the above-mentioned problem. In the proof of the following theorem, we will only deal with the case that D is a Euclidean domain. In this case, the division algorithm can be applied.

Theorem 3.2.5. *Let D be a principal ideal domain. Then every (non-zero) $m \times n$ matrix over D is equivalent to a matrix of the following form*:

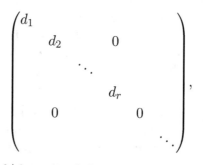

where $d_i \neq 0$ and $d_i | d_{i+1}$, $i = 1, 2, \ldots, r-1$.

Proof. We first deal with the case that D is a Euclidean domain. Let δ be the map from D^* to $\mathbb{N}$, and $A \in D^{m \times n}$, $A \neq 0$. We carry out the following three types of operations to A:

Type I: By interchanging rows or columns of A, we get a matrix B, such that B is equivalent to A, $b_{11} \neq 0$, and among all the non-zero entries of B, the value of $\delta(b_{11})$ attains the minimum.

Type II: Suppose there exists an entry $b_{1j}, j > 1$ in the matrix B, such that $b_{11} \nmid b_{1j}$. By the division algorithm, we can assume that $b_{1j} = q_{11}b_{11} + b'_{1j}$, where $b'_{1j} \neq 0$, and $\delta(b'_{1j}) < \delta(b_{11})$. Multiplying B by $T_{1j}(-q_{11})$ from the right, then applying operations of Type I, we get a matrix $B_1 = (b_{ij}^{(1)})$, such that $b_{11}^{(1)} \neq 0$, $\delta(b_{11}^{(1)}) \leq \delta(b_{ij}^{(1)})$, $\forall b_{ij}^{(1)} \neq 0$, and $\delta(b_{11}^{(1)}) < \delta(b_{11})$. Since the values of δ are non-negative integers, the process will terminate after finitely many steps. Then we get a matrix $B_2 = (b_{ij}^{(2)}) \sim A$, such that $b_{11}^{(2)}$ is non-zero, and $\delta(b_{11}^{(2)})$ attains the minimum among all the integers $\delta(b_{ij}^{(2)})$, $b_{ij}^{(2)} \neq 0$.

Moreover, we also have $b_{11}^{(2)}|b_{1j}^{(2)}$, $j = 2, \ldots, n$. Applying appropriate operations to B_2, we get a matrix $B_3 = (B_{ij}^{(3)})$, such that $b_{11}^{(3)} \neq 0$ and $b_{1j}^{(3)} = 0$, $j = 2, \ldots, n$. The same reasoning applied to the first column shows that A is equivalent to a matrix C, such that $c_{11} \neq 0$, and $c_{i1} = c_{1j} = 0$, $i, j \geq 2$.

Type III: Suppose there exists c_{ij}, $i, j \geq 2$ in the matrix C such that $c_{11} \nmid c_{ij}$. Multiplying C by $T_{1i}(1)$, then applying operations of type II, we get a matrix C_1, which is equivalent to A, such that $c_{11}^{(1)} \neq 0$, $c_{i1}^{(1)} = c_{1j}^{(1)} = 0$, $i, j \geq 2$, and $\delta(c_{11}^{(1)}) < \delta(c_{11})$. Since the values of δ are non-negative integers, using finitely many operations as above, we will get a matrix M, which is also equivalent to A, such that $m_{11} \neq 0$, and for any $i, j \geq 2$, $m_{i1} = m_{1j} = 0$, $m_{11}|m_{ij}$.

Now, we write the above matrix M in block form as

$$M = \begin{pmatrix} m_{11} & 0 \\ 0 & M_1 \end{pmatrix}, \quad M_1 \in D^{(m-1)\times(n-1)}.$$

If $M_1 = 0$, then A has been transformed to a normal form. If $M_1 \neq 0$, then we apply the three types of operations to M_1 to transform M_1 into a normal form. Notice that the three types of operations to M_1 keep the properties that any of the entries of the matrix is a multiple of m_{11}. Thus, the proof of the theorem can be completed by induction.

In the general case, the map δ is not available. But we have the **length function** l on D^* defined as follows: for a unit u in D, $l(u) = 0$; if a is non-unit and non-zero, we define $l(a)$ to be the numbers of irreducible factors in the factorizations of a (counting the multiples). This length function can be applied to prove the theorem for the general case. We leave the details to readers as several exercises in this section. $\qquad\square$

Now, we consider the uniqueness of the normal form in Theorem 3.2.5. For this we need the notions of minors and invariant factors of a matrix. Similarly as in the case of number fields, given a matrix $A \in D^{m\times n}$, $k \leq \min(m, n)$, fix k rows of A: $i_1, i_2, \ldots, i_k$, and k columns of A: $j_1, j_2, \ldots, j_k$. Then we get a $k \times k$ matrix consisting of the entries of A at the intersectional points of the above rows and columns, denoted as

$$A \begin{pmatrix} i_1 i_2 \ldots i_k \\ j_1 j_2 \ldots j_k \end{pmatrix}.$$

The determinant of this $k \times k$ matrix is called a k-minor of A. If $A \neq 0$, and r is a positive integer such that all the $(r+1)$-minors of A are zero, but there exists at least one non-zero r-minor of A, then we say that the rank of A is r (the rank of the matrix 0 is defined to be 0). Now, given a matrix A of rank r, and $i \leq r$, the ith **determinant factor** of A is defined to be the greatest common divisor of all the i-minors of A (which is unique up to a unit), and is denoted as Δ_i (or $\Delta_i(A)$). By the properties of determinant, it is easily seen that every $(i+1)$-minor of A is the D-linear combination of some i-minors. Thus, $\Delta_i \mid \Delta_{i+1}$, $i = 1, 2, \ldots, r-1$.

Theorem 3.2.6. *Let D be a principal ideal domain, and $A \in D^{m \times n}$, $A \neq 0$. If*

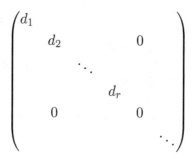

is a normal form of A, then r is equal to the rank of A, and up to unit multipliers we have $d_i = \Delta_i/\Delta_{i-1}$, $1 \leq i \leq r$, here $\Delta_0 = 1$.

Proof. We first prove that if two matrices A, B are equivalent, then they have the same determinant factors in all orders (up to unit multipliers). Let P be an invertible $m \times m$ matrix, and $A_1 = PA$. Then every row of A_1 is a D-linear combination of the rows of A. By the properties of determinants, it is easily seen that any k-minor of A_1, with $k \leq \min(m, n)$, is a D-linear combination of some k-minors of A. It follows that every determinant factor $\Delta_k(A_1)$ of A_1 is a divisor of the determinant factor $\Delta_k(A)$ of A. On the other hand, we can rewrite the above identity as $A = P^{-1}A_1$. Then the same reasoning as above implies that $\Delta_k(A)$ is also a divisor of $\Delta_k(A_1)$.

Thus, $\Delta_k(A) \sim \Delta_k(A_1)$; hence, A, A_1 has the same determinant factors in all orders. Moreover, if Q is an invertible $n \times n$ matrix, and $A_2 = AQ$, then we have $A_2' = Q'A'$. Notice that for any matrix M, the matrices M' and M have the same set of minors of any order (see Exercise (14) of this section). Thus, M' and M have the same determinant factors. The above argument then shows that for any $k \leq \min(m, n)$, $\Delta_k(A_2) \sim \Delta_k(A)$. Therefore, for any invertible matrices P, Q, A and PAQ have the same determinant factors in all orders.

Now, suppose

$$N = \begin{pmatrix} d_1 & & & & & \\ & d_2 & & & 0 & \\ & & \ddots & & & \\ & & & d_r & & \\ & 0 & & & 0 & \\ & & & & & \ddots \end{pmatrix}$$

is a normal form of A, where $d_i \neq 0$, $d_i | d_{i+1}$, $i = 1, 2, \ldots, r - 1$. Then the above argument shows that the matrices N and A have the same determinant factors in all orders. But a direct calculation shows that the determinant factors of N are $\Delta_k(N) = d_1 d_2 \ldots d_k, 1 \leq k \leq r$; $\Delta_k(N) = 0, k > r$. Therefore, r is equal to the rank of A, and $d_i = \Delta_i(A)/\Delta_{i-1}(A)$, $i = 1, 2, \ldots, r$. This completes the proof of the theorem. □

Problem 3.2.7. On the surface, it seems that the assumption that D is a principal ideal domain is not used in the proof of the above theorem. Is Theorem 3.2.6 really true without the assumption that D is a principal ideal domain?

Exercises

(1) Prove Theorem 3.2.2.
(2) Let R be a commutative ring with identity, and M a simple R-module. Show that the ring of endomorphisms $\mathrm{End}(M)$ is a field.

(3) Show that if R is a domain, then for any $n \times n$ matrices over R, $(AB)^* = B^* A^*$.

(4) Let D be a principal ideal domain, and l, the length function of D. Given $a, b \in D^*$, show that

   (i) If $a \sim b$, then $l(a) = l(b)$;
   (ii) If $a \mid b$, but $b \nmid a$, then $l(a) < l(b)$;
   (iii) If $a \mid b$, then $l(a) = l(b) \iff a \sim b$.

(5) Let D be a principal ideal domain. Suppose $A = (a_{ij})_{m \times n} \in D^{m \times n}$, where $a_{11} \neq 0, a_{1j} \neq 0$, and $j > 1$. Let $d = (a_{11}, a_{1j})$. Fix $u, v \in D$ such that $u a_{11} + v a_{1j} = d$, and suppose $a_{11} = b_{11}d, a_{1j} = b_{1j}d$. Consider the matrix $Q = (q_{ij}) \in D^{n \times n}$, where

$$q_{11} = u, \quad q_{1j} = -b_{1j}, \quad q_{1k} = 0, \quad k \neq 1, j;$$

$$q_{j1} = v, \quad q_{jj} = b_{11}, \quad q_{jk} = 0, \quad k \neq 1, j;$$

$$q_{ik} = \delta_{ik}, \quad i \neq 1, j.$$

Show that Q is invertible, and compute AQ.

(6) Apply the conclusions of Exercises (4) and (5) of this section to complete the proof of Theorem 3.2.5.

(7) Determine the determinant factors and the invariant factors of the matrix

$$\begin{pmatrix} 1 & 1 & 0 & 0 \\ 2 & 1 & 0 & 0 \\ 3 & 2 & 2 & 2 \\ 4 & 3 & 3 & 3 \end{pmatrix}.$$

(8) Determine the normal form of the following matrix over the ring $\mathbb{Z}[\lambda]$:

$$\begin{pmatrix} \lambda & \lambda & 1 & 0 \\ \lambda^2 & \lambda & 0 & 1 \\ 3\lambda & 2\lambda & \lambda^2 & \lambda - 2 \\ 4\lambda & 3 - \lambda & 2 & 1 \end{pmatrix}.$$

(9) Let S be a non-empty subset of a commutative unitary ring R. Set $C(S) = \{a \in R \mid ab = ba, \forall b \in S\}$. Show that $C(S)$ is a subring of R. In particular, the subring $C(R)$ is called the center

of R. Now, suppose R_1 is a commutative ring with identity, and $R = R_1^{n \times n}$. Let $S \subset R$ be the subset of R consisting of the matrix

$$\begin{pmatrix} 0 & 1 & 0 & \cdots & 0 & 0 \\ 0 & 0 & 1 & \cdots & 0 & 0 \\ \vdots & \vdots & \vdots & \vdots & \cdots & \vdots \\ 0 & 0 & 0 & \cdots & 0 & 1 \\ 0 & 0 & 0 & \cdots & 0 & 0 \end{pmatrix}.$$

Determine $C(S)$.

(10) Let R be a commutative unitary ring. Determine the center of $R^{n \times n}$.

(11) Let $\mathbb{F}$ be a field. Show that a non-zero matrix A in $\mathbb{F}^{n \times n}$ is a zero divisor if and only if A is not invertible. Is this assertion true for a general commutative ring with identity?

(12) Let D be a principal ideal domain, and $c_1, c_2, \ldots, c_m \in D$. Suppose d is the greatest common divisor of $c_1, c_2, \ldots, c_m$. Show that there exists an invertible matrix $A \in D^{m \times m}$ such that $(c_1, c_2, \ldots, c_m)A = (d, 0, \ldots, 0)$. Is the above assertion true for a general unique factorization domain? Explain why.

(13) Suppose D is a principal ideal domain, and $a_1, a_2, \ldots, a_n \in D$ are relatively prime. Show that there exists an invertible matrix A in $D^{n \times n}$ such that the first row of A is equal to $(a_1, a_2, \ldots, a_n)$.

(14) Let A be an $n \times n$ matrix over a principal ideal domain D. Show that A is equivalent to A'.

Exercises (hard)

(15) Let A be an $m \times n$ matrix over a commutative unitary ring R. Define a homomorphism of modules φ from R^n to R^m by $\varphi(\alpha) = A\alpha$, $\alpha \in R^n$. Show that the following conditions are equivalent:

(i) φ is surjective;
(ii) The ideal of R generated by all the m-minors of A is equal to R;
(iii) There exists a matrix $B \in R^{n \times m}$ such that $AB = I_m$.

(16) Let $\{v_1, v_2, \ldots, v_m\}$ be a generating set of a module M over a commutative unitary ring R, and J, an ideal of R. Let JM be the set of all the finite sums of the elements of the form $av, a \in J, v \in M$.

   (i) Show that if $JM = M$, then there exists $A \in J^{m \times m}$ such that
   $$(v_1, v_2, \ldots, v_m)(I_m - A) = 0.$$

   (ii) Under the same assumption of (1), show that $\det(I_m - A) = 1 + \alpha$, where $\alpha \in J$, and $\det(I_m - A) \in \mathrm{ann}(M)$.

   (iii) An R-module M is called faithful if $\mathrm{ann}(M) = 0$. Prove **Nakayama's lemma**: if M is a finitely generated faithful R-module, and $JM = M$, then $J = R$.

(17) Let R be a commutative unitary ring, and M, a finitely generated R-module. Show that if for any maximal ideal J of R, $JM = M$, then $M = \{0\}$.

(18) Let R be a commutative unitary ring. Show that for any $n \times n$ matrices A, B over R, $(AB)^* = B^* A^*$.

3.3   Free modules

In this section, we will introduce the modules with the nicest structure, namely, free modules. This type of modules is a natural generalization of finite dimensional vector spaces over a field. The theory of free modules is also the foundation of module theory.

Definition 3.3.1. Let R be a ring with identity, and M, an R-module. If there exist n elements $u_1, u_2, \ldots, u_n$ in M, such that for any R-module N and any n elements $v_1, v_2, \ldots, v_n$ in N, there exists a unique homomorphism ϕ from M to N such that $\phi(u_i) = v_i$, $i = 1, 2, \ldots, n$, then M is called a **free module** of rank n over R, and the n-tuple $(u_1, u_2, \ldots, u_n)$ is called a base of M.

For convenience, we agree that the null module is a free module with rank 0.

We stress here that, the uniqueness of the homomorphism in the above definition is very important, and this is used in the study of free modules very frequently. Moreover, by the definition we immediately have

Proposition 3.3.2. *Let M_1, M_2 be R-modules. Suppose M_1 is a free module and $(u_1, u_2, \ldots, u_n)$ is a base of M. If there exists an isomorphism $\phi : M_1 \to M_2$, then M_2 is also a free module and*

$$(\phi(u_1), \phi(u_2), \ldots, \phi(u_n))$$

is a base of M_2.

We first give several important examples of free modules.

Example 3.3.3. Let V be an n-dimensional vector space over a field $\mathbb{F}$. Suppose $(\alpha_1, \alpha_2, \ldots, \alpha_n)$ is a base of V. Then V is a free $\mathbb{F}$-module of rank n, and $(\alpha_1, \alpha_2, \ldots, \alpha_n)$ is a base of V as a free $\mathbb{F}$-module.

Example 3.3.4. Let R be a ring with identity. Then R can be viewed as a R-module. Let R^m be the external direct sum of m copies of R. Consider the following m elements in R^m:

$$e_1 = (1, 0, 0, \ldots, 0),$$

$$e_2 = (0, 1, 0, \ldots, 0),$$

$$\cdots \cdots \cdots \cdots \cdots$$

$$e_m = (0, 0, \ldots, 0, 1),$$

where 1 is the identity of R. Then it is easy to see that every element $(x_1, x_2, \ldots, x_m)$ in R^m can be uniquely expressed as $(x_1, x_2, \ldots, x_m) = \sum_{i=1}^{m} x_i e_i$. Now given any R-module M, and any n elements $v_1, v_2, \ldots, v_m$ in M, define

$$\phi \left(\sum_{i=1}^{m} x_i e_i \right) = \sum_{i=1}^{m} x_i v_i.$$

Then ϕ is a homomorphism from R^m to M, and $\phi(e_i) = v_i$. Moreover, if $\varphi : R^m \to M$ is a homomorphism of modules, and $\varphi(e_i) = v_i$, then

$$\varphi \left(\sum_{i=1}^{m} x_i e_i \right) = \sum_{i=1}^{m} x_i \varphi(e_i) = \sum_{i=1}^{m} x_i v_i.$$

Thus, $\varphi = \phi$. Therefore, the homomorphism satisfying the condition $\phi(e_i) = v_i$ is unique. Thus, R^m is a free module over R of rank m, and $(e_1, e_2, \ldots, e_m)$ is a base of R^m.

It follows from the above examples and Proposition 3.3.2 that, if an R-module M is isomorphic to R^m, then M must be a free module of rank m. In the following, we will show that the converse statement of this assertion is also true. To this end, we first notice that the existence of a base is an important feature of a free module. Thus, it is crucial to find the sufficient and necessary condition for a subset of a free module to be a base. We now give the following:

Definition 3.3.5. Let $u_1, u_2, \ldots, u_m$ be m elements in an R-module M. If the following conditions are satisfied:

$$\sum a_i u_i = 0 \iff a_i = 0, \quad a_i \in R,$$

then we say that the elements $u_1, u_2, \ldots, u_m$ are **linearly independent**.

Now, we can prove

Theorem 3.3.6. *Let M be a module over a unitary ring R, and $u_1, u_2, \ldots, u_m$ be m elements in M. Then M is a free module, and $(u_1, u_2, \ldots, u_m)$ is a base of M if and only if*

(1) *$\{u_1, u_2, \ldots, u_m\}$ is a set of generators of M;*
(2) *the elements $u_1, u_2, \ldots, u_m$ are linearly independent.*

Proof.   *The only if part.*   Let $(u_1, u_2, \ldots, u_m)$ be a base of the free module M. We first show that $\{u_1, u_2, \ldots, u_m\}$ is a set of generators of M. Denote by N the submodule of M generated by $u_1, u_2, \ldots, u_m$. Since M is a free module, and $(u_1, u_2, \ldots, u_m)$ is a base, there exists a unique homomorphism ξ from M to N such that $\xi(u_i) = u_i$. Meanwhile, denote by η the embedding of N into M, that is, $\eta(x) = x$, $\forall x \in N$. Then η is a homomorphism of modules, and $\eta(u_i) = u_i$. Thus, the composition $\eta\xi$ is a homomorphism from M into M and $(\eta\xi)(u_i) = u_i$. On the other hand, the identity map id_M of M into itself is also a homomorphism satisfying the condition $\mathrm{id}_M(u_i) = u_i$. By the uniqueness, we have $\eta\xi = \mathrm{id}_M$. Thus, η is surjective. In particular, $N = \eta(N) = M$. Hence, $\{u_1, u_2, \ldots, u_m\}$ is a set of generators of M.

We now show that $u_1, u_2, \ldots, u_m$ are linearly independent. By the assumption, there exists a homomorphism φ from M to R^m, such that $\varphi(u_i) = e_i$, $i = 1, 2, \ldots, m$. Since R^m is a free module, and

$(e_1, \ldots, e_m)$ is a base, there exists a homomorphism ϕ from R^m to M such that $\phi(e_i) = u_i$, $i = 1, 2, \ldots, m$. Thus, $\varphi\phi(e_i) = e_i$. This implies that $\varphi\phi = \mathrm{id}_{R^m}$. Moreover, since $\{u_1, u_2, \ldots, u_m\}$ is a generating set of M, we also have $\phi\varphi = \mathrm{id}_M$. Thus, φ is an isomorphism of modules. In particular, $\mathrm{Ker}\,\varphi = \{0\}$, which implies that $u_1, u_2, \ldots, u_m$ are linearly independent.

The if part. Suppose the elements $u_1, u_2, \ldots, u_m$ in an R-module M satisfy the conditions (1) and (2), N is an R-module, and $v_1, v_2, \ldots, v_m$ are m elements in N. Then every element of M can be uniquely expressed as an R-linear combination of $u_1, u_2, \ldots, u_m$. Now, we define a map ψ from M to N such that

$$\psi\left(\sum_{i=1}^m a_i u_i\right) = \sum_{i=1}^m a_i v_i, \quad a_i \in R.$$

Then it is easily seen that ψ is a homomorphism from M to N, and $\psi(u_i) = v_i$, $i = 1, 2, \ldots, m$. The homomorphism satisfying the above condition is obviously unique. Therefore, M is a free module and $(u_1, u_2, \ldots, u_m)$ is a base of M. □

Combining Example 3.3.4 and the proof of the above theorem, we get the following:

Corollary 3.3.7. *Let M be an R-module. Then M is a free module with rank m if and only if M is isomorphic to R^m.*

In linear algebra, we have shown that the dimension of a finite dimensional vector space over a number field is uniquely determined by the vector space. In particular, if $(\alpha_1, \alpha_2, \ldots, \alpha_n)$ and $(\beta_1, \beta_2, \ldots, \beta_m)$ are two bases of a vector space V over a number field, then $m = n$. This assertion can be easily generalized to a vector space over a general field. A natural problem is whether the assertion is true for a free module. The answer is negative. In fact, there exist a unitary ring R and two positive integers $m \neq n$, such that $R^m \simeq R^n$ (see Artin, 1991, p. 490). Thus, the assertion is not true for a general unitary ring. But we will show immediately that if R is a commutative ring with identity, then the **rank** of a free module over R is unique.

Theorem 3.3.8. *Let R be a commutative ring with identity. If M is a free module over R, and $(u_1, u_2, \ldots, u_n)$, $(u'_1, u'_2, \ldots, u'_m)$ are two bases of M, then $n = m$.*

Proof. Since $(u_1, u_2, \ldots, u_n)$ is a base of M, each of u'_i, $i = 1, 2, \ldots, m$ can be uniquely expressed as an R-linear combination of $u_1, \ldots, u_n$:

$$u'_i = \sum_{j=1}^{n} a_{ij} u_j, \quad a_{ij} \in R. \tag{3.4}$$

Similarly, each of u_k, $k = 1, 2, \ldots, n$ can also be uniquely expressed as an R-linear combination of $u'_1, u'_2, \ldots, u'_m$:

$$u_k = \sum_{l=1}^{m} b_{kl} u'_l, \quad b_{kl} \in R. \tag{3.5}$$

Substituting (3.4) into (3.5), and taking into account the fact that $u_1, u_2, \ldots, u_n$ are linearly independent, we get

$$\sum_{s=1}^{m} b_{ks} a_{sl} = \delta_{kl}, \quad 1 \le k, l \le n. \tag{3.6}$$

Similarly, we have

$$\sum_{l=1}^{n} a_{il} b_{lj} = \delta_{ij}, \quad 1 \le i, j \le m. \tag{3.7}$$

If $n < m$, we consider the following two $m \times m$ matrices:

$$A = \begin{pmatrix} a_{11} & \cdots & a_{1n} & 0 & \cdots & 0 \\ a_{21} & \cdots & a_{2n} & 0 & \cdots & 0 \\ \vdots & \vdots & \vdots & \vdots & \vdots & \vdots \\ a_{m1} & \cdots & a_{mn} & 0 & \cdots & 0 \end{pmatrix},$$

$$B = \begin{pmatrix} b_{11} & b_{12} & \cdots & b_{1m} \\ \vdots & \vdots & \vdots & \vdots \\ b_{n1} & b_{n2} & \cdots & b_{nm} \\ 0 & 0 & \cdots & 0 \\ \vdots & \vdots & \vdots & \vdots \\ 0 & 0 & \cdots & 0 \end{pmatrix}.$$

Then it follows from (3.7) that $AB = I_m$, where I_m is the $m \times m$ identity matrix. Since R is commutative and unitary, this implies that $C = (c_{ij})_{m \times m} = BA = I_m$. But a direct calculation shows that for any $i > n$, we have $c_{ij} = 0$, $j = 1, 2 \dots, m$. This is a contradiction. Therefore, $n \geq m$. Similarly, we can prove that $m \geq n$. Thus, $n = m$. $\square$

Problem 3.3.9. Construct an example to show that there exists a non-commutative unitary ring R, and two $n \times n$ matrices A, B over R, such that $AB = I_n$, but $BA \neq I_n$.

Corollary 3.3.10. *If R is a commutative ring with identity, then $R^m \simeq R^n \iff n = m$.*

Exercises

(1) Let M be a free module over a domain R. Show that if $ax = 0$, where $a \in R$, $x \in M$, then either $a = 0$ or $x = 0$. If R is not a domain, is the above assertion true?

(2) Let R be a commutative ring with identity. Suppose any submodule of any free R-module is a free module. Show that R must be a principal ideal domain.

(3) Suppose $R = \mathbb{C}[x, y]$, and M is the ideal of R generated by x, y. Is M a free R-module?

(4) Let R be a commutative ring with identity. Suppose I is an ideal of R and R/I is a free R-module. Show that $I = \{0\}$.

(5) Let I be an ideal of a commutative unitary ring R. Show that I is a free R-module if and only if I is a principal ideal generated by an element which is not a zero divisor.

(6) Suppose R is a commutative ring with identity, and any finitely generated R-module is free. Show that R is a field.

(7) Suppose R is a commutative ring with identity, and φ is an endomorphism of the free module R^m. Show that if φ is surjective, then φ must be an isomorphism.

(8) Suppose $\mathbb{F}$ is a field and φ is an injective endomorphism of $\mathbb{F}^n$. Show that φ is an isomorphism. Is this assertion true for a general commutative ring with identity?

(9) Let M be a module over a commutative unitary ring R. Show that if M is the direct sum of two submodules $M = M_1 \oplus M_2$, and M_1, M_2 are free, then M is a free module and $r(M) = r(M_1) + r(M_2)$.

Exercises (hard)

(10) Suppose φ is a homomorphism from the free $\mathbb{Z}$-module $\mathbb{Z}^n$ to $\mathbb{Z}^m$, and A is the matrix of φ with respect to the standard bases of $\mathbb{Z}^n$ and $\mathbb{Z}^m$. Show that:
 (i) φ is injective if and only if the rank of A is n.
 (ii) φ is surjective if and only if the m-minors of A are equal to 1.
(11) Suppose R is a commutative ring with identity, M is a free R-module, and $f \in \mathrm{End}_R(M, M)$. Show that f is injective if and only if f is not a left zero divisor in $\mathrm{End}_R(M, M)$.

3.4 Finitely generated modules over a principal ideal domain

In this section, we study finitely generated modules over a principal ideal domain. The main results of this section are some theorems which give a complete description of the structure of such modules, which will be very useful in various branches of mathematics. In the next two sections, we will apply the results of this section to classify finitely generated abelian groups, and study the canonical forms of linear transformations of a vector space over a field.

Let D be a principal ideal domain, and M a finitely generated module over D. Suppose $\{x_1, x_2, \ldots, x_n\}$ is a set of generators of M. Then for any base $(\alpha_1, \alpha_2, \ldots, \alpha_n)$ of the free D-module D^n, there exists a surjective homomorphism φ from D^n to M, such that $\varphi(\alpha_i) = x_i$, $i = 1, 2, \ldots, n$. By the Fundamental Theorem of Homomorphisms of modules, we have $M \simeq D^n/K$, where $K = \mathrm{Ker}\,\varphi$ is a submodule of D^n. This observation gives us a clue that, to understand the structure of finitely generated modules over D, it is important to study the properties of submodules of D^n. The following theorem is a key for us.

Theorem 3.4.1. *Any submodule of a free module over a principal ideal domain must be a free module, and the rank of the submodule does not exceed the rank of the module itself.*

Proof. Let K be a submodule of D^n. We will prove inductively on n that K is a free module. When $n = 1$, K is a submodule of

$D^1 = D$ if and only if K is an ideal of D. Since D is a principal ideal ring, there exists $a \in K$ such that $K = \langle a \rangle$. If $a = 0$, the assertion is certainly true. Suppose $a \neq 0$. Then a is a generator of K. Since D has no zero divisors, the expression of any element $b \in K$ as a multiple of a is unique. This implies that the element a forms a base of K. Therefore, the assertion holds for $n = 1$.

Now, we suppose $n > 1$ and the assertion holds for $n - 1$. Fix a base $(e_1, e_2, \ldots, e_n)$ of D^n, and set

$$D_1 = [e_2, \ldots, e_n].$$

Then D_1 is a free module of rank $n - 1$. If $K \subseteq D_1$, then by the inductive assumption, K is a free module and the rank of K satisfies $r(K) \leq n - 1 < n$. If $K \not\subseteq D_1$, then there exists $f = a_1 e_1 + a_2 e_2 + \cdots + a_n e_n \in K$ such that $f \notin D_1$, where $a_1 \neq 0$. Now, we set

$$I = \left\{ b_1 \in D \mid \text{there exists } b_2, \ldots, b_n \in D \text{ such that } \sum_{i=1}^{n} b_i e_i \in K \right\}.$$

We assert that I is an ideal of D. In fact, I is obviously non-empty. Since K is a submodule, I is closed with respect to subtraction. Moreover, for any $b \in I$ and $r \in D$, we have $rb \in I$. This proves our assertion. Since $a_1 \in I$ and $a_1 \neq 0$, $I \neq \{0\}$. Thus, there exists $d \neq 0$ such that $I = \langle d \rangle$. In particular, there exist $d_2, d_3, \ldots, d_n \in D$ such that $g = de_1 + d_2 e_2 + \cdots + d_n e_n \in K$. Now, given $h \in K$, write

$$h = c_1 e_1 + c_2 e_2 + \cdots + c_n e_n,$$

where $c_i \in D$, $i = 1, 2, \ldots, n$. Then $c_1 \in I$. Thus, there exists $c_1' \in D$ such that $c_1 = c_1' d$. Then we have $h - c_1' g \in K \cap D_1$.

Now, we consider the following two submodules of K: $K_1 = [g]$, $K_2 = K \cap D_1$. The above argument shows that $K = K_1 + K_2$. Moreover, it is obvious that $K_1 \cap K_2 = \{0\}$. Thus, $K = K_1 \oplus K_2$. On the other hand, since $K_2 \subseteq D_1$, by the inductive assumption, K_2 is a free module, and $r(K_2) \leq n - 1$. It is easily seen that K_1 is a free module and $r(K_1) = 1$. Then by Exercise (9) of Section 3.3, K is a free module, and $r(K) = r(K_1) + r(K_2) \leq 1 + (n - 1) = n$. This means that the assertion also holds for n, and the proof of the theorem is completed. $\qquad \square$

Problem 3.4.2. In linear algebra, it has been shown that if a subspace W of a vector space V over a number field $\mathbb{F}$ has the same dimension as V, then $W = V$. Show that this is also true for vector spaces over a general field. Is this true for the rank of free modules over a principal ideal domain?

By the above observation and Theorem 3.4.1, any finitely generated D module is isomorphic to the quotient module D^n/K of D^n with respect to a submodule K, and the submodule K is also free. Therefore, our next goal is to find a base $(u_1, u_2, \ldots, u_n)$ of D^n, and a base of $(v_1, v_2, \ldots, v_m)$ of K, where $m \leq n$, such that the relationship between $u_1, u_2, \ldots, u_n$ and $v_1, v_2, \ldots, v_m$ is as simple as possible. Then the two bases can be used to give an accurate description of the structure of D^n/K. Using the theorem about the normal forms of matrices over a principal ideal domain in Section 3.2, we now prove the following:

Theorem 3.4.3. *Let K be a (non-zero) submodule of the free module D^n over a principal ideal domain D. Then there exists a base $(v_1, v_2, \ldots, v_m)$ of K and a base $(u_1, u_2, \ldots, u_n)$ of D^n, such that $v_i = d_i u_i$, $i = 1, 2, \ldots, m$, where $d_i \neq 0$, and $d_i | d_{i+1}$, $i = 1, 2, \ldots, m-1$.*

Proof. We first fix an arbitrary base $(w_1, w_2, \ldots, w_n)$ of D^n and an arbitrary base $(x_1, x_2, \ldots, x_m)$, $m \leq n$, of K. Then each of x_i can be expressed as a linear combination of $w_1, w_2, \ldots, w_n$:

$$x_i = \sum_{j=1}^{n} a_{ij} w_j, \quad 1 \leq i \leq m,$$

where $a_{ij} \in D$. These equations can be written in the form of matrices as

$$\begin{pmatrix} x_1 \\ \vdots \\ x_m \end{pmatrix} = A \begin{pmatrix} w_1 \\ \vdots \\ w_n \end{pmatrix},$$

where $A = (a_{ij})_{m \times n}$ is an $m \times n$ matrix over D. Now, for any base $(w_1', v_2', \ldots, w_n')$ of D^n, and any base $(x_1', x_2', \ldots, x_m')$ of K, there exist invertible matrices $P \in D^{m \times m}$ and $Q \in D^{n \times n}$ such that

$$\begin{pmatrix} w'_1 \\ \vdots \\ w'_n \end{pmatrix} = Q \begin{pmatrix} w_1 \\ \vdots \\ w_n \end{pmatrix}, \quad \begin{pmatrix} x'_1 \\ \vdots \\ x'_m \end{pmatrix} = P \begin{pmatrix} x_1 \\ \vdots \\ x_m \end{pmatrix}.$$

Then we have

$$\begin{pmatrix} x'_1 \\ \vdots \\ x'_m \end{pmatrix} = PAQ^{-1} \begin{pmatrix} w'_1 \\ \vdots \\ w'_m \end{pmatrix}.$$

By Theorem 3.2.5, there exist invertible matrices P_1 and Q_1 such that

$$P_1 A Q_1^{-1} = \begin{pmatrix} d_1 & & & & & \\ & d_2 & & & 0 & \\ & & \ddots & & & \\ & & & d_r & & \\ & 0 & & & 0 & \\ & & & & & \ddots \end{pmatrix},$$

where r is the rank of A, $d_i \neq 0$, $i = 1, 2, \ldots, r$, and $d_i | d_{i+1}$, $i = 1, 2, \ldots, r - 1$. Therefore, if the bases $(w'_1 w'_2, \ldots, w'_n)$ and $(x'_1, x'_2, \ldots, x'_m)$ are chosen such that the transitive matrices P, Q above are exactly P_1 and Q_1, respectively (this is possible, please explain why), then we have $x'_i = d_i w'_i$, $i = 1, 2, \ldots, r$. Moreover, it is obvious that $r \leq m$. Since m is equal to the rank of K, we have $r = m$. This completes the proof of the theorem. $\qquad \square$

This theorem will lead to the structure theorem of finitely generated modules over a principal ideal domain. To state the result, we introduce the notion of the annihilator of an element in a module.

Definition 3.4.4. Let M be a module over a commutative unitary ring R, and $x \in M$. Then the subset $\text{ann}(x) = \{a \in R | ax = 0\}$ of R is an ideal of the ring R, called the **annihilator** of x. Suppose $x \in M$, and $x \neq 0$. If $\text{ann}(x) = \{0\}$, then x is called a **free element**, otherwise it is called a **torsion element**. The module M is called

torsion-free if every non-zero element in M is free; M is called a **torsion module** if every non-zero element in M is a torsion element.

In the exercises of this section, we will list some properties of the annihilator of an element. Now, we can state the first result about the structure of finitely generated modules over a principal ideal domain.

Theorem 3.4.5. *Let M be a finitely generated module over a principal ideal domain D. Then M is the direct sum of some cyclic submodules*:

$$M = Dy_1 \oplus Dy_2 \oplus \cdots \oplus Dy_s,$$

where $\operatorname{ann}(y_1) \supseteq \operatorname{ann}(y_2) \supseteq \cdots \supseteq \operatorname{ann}(y_s)$.

Proof.   By the above argument, we only need to prove the assertion for the quotient module D^n/K. Without loss of generality, we can assume that $K \neq 0$. By Theorem 3.4.3, there exists a base $(w_1, w_2, \ldots, w_n)$ of D^n, and $d_j \in D$, $1 \leq j \leq m$, satisfying $d_j | d_{j+1}$, $1 \leq j \leq m - 1$, where $m = r(K) \leq n$, such that $(d_1 w_1, d_2 w_2, \ldots, d_m w_m)$ is a base of K. Then we have

$$D^n = Dw_1 \oplus Dw_2 \oplus \cdots \oplus Dw_n,$$

$$K = D(d_1 w_1) \oplus D(d_2 w_2) \oplus \cdots \oplus D(d_m w_m).$$

Notice that $D(d_j w_j)$ is a submodule of Dw_j. Therefore, we have

$$D^n/K \simeq \bigoplus_{j=1}^{m} Dw_j/D(d_j w_j) \oplus Dw_{m+1} \oplus \cdots \oplus Dw_n.$$

Obviously, $Dw_j/D(d_j w_j)$ is a cyclic module, and $w_j + D(d_j w_j)$ is a generator. Denoting $y_j = w_j + D(d_j w_j)$, $1 \leq j \leq m$; $y_j = w_j$, $m + 1 \leq j \leq n$, we get

$$D^n/K \simeq Dy_1 \oplus Dy_2 \oplus \cdots \oplus Dy_n.$$

Meanwhile, a direct calculation shows that $\operatorname{ann}(y_j) = \langle d_j \rangle$, $1 \leq j \leq m$, and $\operatorname{ann}(y_j) = 0$, $m + 1 \leq j \leq n$. This completes the proof of the theorem.   □

Corollary 3.4.6. *Any finitely generated torsion-free module over a principal ideal domain must be a free module.*

Proof. Let M be a finitely generated module over a principal ideal domain. Then by Theorem 3.4.5, M can be decomposed as

$$M = Dy_1 \oplus Dy_2 \oplus \cdots \oplus Dy_s,$$

where $\text{ann}(y_1) \supseteq \text{ann}(y_2) \supseteq \cdots \supseteq \text{ann}(y_s)$. Since M is torsion-free, we have $\text{ann}(y_1) = \text{ann}(y_2) = \cdots = \text{ann}(y_s) = \{0\}$. This means that $(y_1, y_2, \ldots, y_s)$ is a base of M. Hence, M is a free module. $\square$

Problem 3.4.7. Construct an example to show that Corollary 3.4.6 is not true for modules over a general commutative ring with identity.

Now, we give some definitions related to the conclusion of Theorem 3.4.5. Let M be a finitely generated module over a principal ideal domain D. Suppose M can be decomposed as

$$M = Dy_1 \oplus Dy_2 \oplus \cdots \oplus Dy_s,$$

where $\text{ann}(y_1) \supseteq \text{ann}(y_2) \supseteq \cdots \supseteq \text{ann}(y_s)$, and $y_j \neq 0$, $j = 1, 2, \ldots, s$. If $\text{ann}(y_j) \neq 0$ for $1 \leq j \leq k$, and $\text{ann}(y_j) = 0$ for $j > k$, then we say that the rank of M is $s - k$. Moreover, for any $1 \leq j \leq k$, there exists a unique (up to a unit multiple) d_j such that $\text{ann}(y_j) = \langle d_j \rangle$, and $d_j | d_{j+1}$, $j = 1, 2, \ldots, k-1$. The elements $d_1, d_2, \ldots, d_k$ are called the **invariant factors** of M.

The careful readers may have noticed that the definition of invariant factors is not rigorous, since the decomposition of a finitely generated module into the summation of cyclic submodules is generically not unique, for example, the base of a free module is in general not unique. To make the above definition more accurate, we need to show that the annihilators in different decompositions of a finitely generated module are the same. Our next goal is to prove this assertion. For this we need to introduce the notion of p-modules and study the properties of p-modules in some detail.

Definition 3.4.8. Let N be a finitely generated module over a principal ideal domain D, and p an irreducible element in D. If for any $x \in N$, there exists a natural number e such that $p^e x = 0$, then N is called a p-**module**.

It is obvious that a p-module must be a torsion module. For a finitely generated torsion module M over D, set

$$M_p = \{x \in M | \exists e \in \mathbb{N}, p^e x = 0\}.$$

Then M_p is a submodule of M, and it is a p-module. M_p is called the p-**component** of M.

Theorem 3.4.9. *Let M be a finitely generated torsion module over D. Then there are finitely many irreducible elements $p_1, p_2, \ldots, p_m$, which are relatively prime to each other, such that*

$$M = M_{p_1} \oplus M_{p_2} \oplus \cdots \oplus M_{p_m}.$$

Moreover, for any irreducible element p which is not associated with any of p_j, $1 \leq j \leq m$, the corresponding p-component M_p is equal to zero.

Proof. We first prove a general result, namely, if Dy is a cyclic module such that the annihilator of y is $\mathrm{ann}(y) = \langle ab \rangle$, where $a, b \in R$ are relatively prime, then Dy can be decomposed as the direct sum of two submodules $Dy = Dy_1 \oplus Dy_2$, such that $\mathrm{ann}(y_1) = \langle a \rangle$, and $\mathrm{ann}(y_2) = \langle b \rangle$. To see this, set $y_1 = by$, and $y_2 = ay$. Since D is a principal ideal ring, there exist $u, v \in D$, such that $ua + vb = 1$. Then we have $y = uay + vby = vy_1 + uy_2$. Thus, $Dy = Dy_1 + Dy_2$. On the other hand, if $z \in Dy_1 \cap Dy_2$, then there exist $c_1, c_2 \in D$, such that $z = c_1 y_1 = c_1 by = c_2 y_2 = c_2 ay$. This implies that $az = bz = 0$, hence $z = (ua + vb)z = 0$. Therefore, $Dy = Dy_1 \oplus Dy_2$. Furthermore, a direct calculation shows that $\mathrm{ann}(y_1) = \langle a \rangle$, $\mathrm{ann}(y_2) = \langle b \rangle$. This completes the proof of the assertion.

Now, we turn to the proof of the theorem. By Theorem 3.4.5, M can be decomposed as

$$M = Dy_1 \oplus Dy_2 \oplus \cdots \oplus Dy_s,$$

where $y_i \neq 0$, $\mathrm{ann}(y_i) = \langle d_i \rangle$, $i = 1, 2, \ldots, s$, and $d_i | d_{i+1}$, $i = 1, 2, \ldots, s - 1$. Since M is a torsion module, for any i, $d_i \neq 0$ and d_i is not a unit. Thus, d_i can be factored as

$$d_i = \varepsilon_i p_1^{k_{i1}} p_2^{k_{i2}} \cdots p_m^{k_{im}}, \quad i = 1, 2, \ldots, s,$$

where $p_1, p_2, \ldots, p_m$ are irreducible elements which are not associated with each other, ε_i is a unit, and $k_{ij} \geq 0$. Moreover, for any $1 \leq l \leq m$, we have $k_{1l} \leq k_{2l} \leq \cdots \leq k_{sl}$.

By the above assertion, Dy_i can be decomposed as

$$Dy_i = Dy_{i1} \oplus Dy_{i2} \oplus \cdots \oplus Dy_{im},$$

where $\text{ann}(y_{ij}) = \langle p_j^{k_{ij}} \rangle$. Now, we set $M_j = Dy_{1j} \oplus Dy_{2j} \oplus \cdots \oplus Dy_{sj}$. Then M_j is a p_j-module, and $M = M_1 \oplus M_2 \oplus \cdots \oplus M_m$. It is easy to check that M_j is exactly equal to the p_j-component of M. Moreover, suppose p is an irreducible element in R and it is relatively prime to any of $p_1, p_2, \ldots, p_m$. Given $w \in M_p$, we can write $w = w_1 + w_2 + \cdots + w_m$, where $w_j \in M_j$, $j = 1, 2, \ldots, s$. Then there exists $n > 0$ such that $p^n w = p^n w_1 + \cdots + p^n w_s = 0$, hence $p^n w_1 = \cdots = p^n w_s = 0$. On the other hand, for any w_j, there exists $n_j > 0$ such that $p_j^{n_j} w_j = 0$. Since p^n is relatively prime to $p_j^{n_j}$, there exists $u_j, v_j \in D$, such that $u_j p^n + v_j p_j^{n_j} = 1$. Thus, $w_j = u_j p^n w_j + v_j p_j^{n_j} w_j = 0$. Hence, $w = 0$. Consequently, $M_p = \{0\}$. This completes the proof of the theorem. $\qquad\square$

Now, we can state the main theorem of this section.

Theorem 3.4.10. *Let M be a finitely generated module over a principal ideal domain D. Suppose*

$$M = Dy_1 \oplus Dy_2 \oplus \cdots \oplus Dy_s = Dz_1 \oplus Dz_2 \oplus \cdots \oplus Dz_t$$

are two kinds of decompositions of M, where $\text{ann}(y_1) \supseteq \text{ann}(y_2) \supseteq \cdots \supseteq \text{ann}(y_s)$, $\text{ann}(z_1) \supseteq \text{ann}(z_2) \supseteq \cdots \supseteq \text{ann}(z_t)$, and $y_i \neq 0$, $z_j \neq 0$, $1 \leq i \leq s$, $1 \leq j \leq t$. Then $s = t$, and $\text{ann}(y_i) = \text{ann}(z_i)$, $1 \leq i \leq s(= t)$.

Proof. Since the assertion is obviously true for a free module, we assume that M is not a free module. Consider the following non-empty subset of M

$$\text{Tor}(M) = \{x \in M \mid \exists a \in D, a \neq 0, ax = 0\}.$$

Then it is easy to prove that $\text{Tor}(M)$ is a submodule of M (left as an exercise). Suppose in the first decomposition, $\text{ann}(y_j) \neq \{0\}$, for $j \leq k$, but $\text{ann}(y_{k+1}) = \cdots = \text{ann}(y_s) = \{0\}$, where k is a positive integer. We now show that $\text{Tor}(M) = Dy_1 \oplus Dy_2 \oplus \cdots \oplus Dy_k$. Denote the subset of the right-hand side as M_1. Then it is obvious that $M_1 \subseteq \text{Tor}(M)$. On the other hand, if $w \in \text{Tor}(M)$, then we have a decomposition $w = w_1 + w_2 + \cdots + w_s$, where $w_j \in Dy_j$, $1 \le j \le s$. Select $b \ne 0, b \in D$ such that $bw = 0$. Then we have $0 = bw = bw_1 + bw_2 + \cdots + bw_s$. Thus, $bw_1 = bw_2 = \cdots = bw_s = 0$. This means that for $j > k$, we have $w_j = 0$. Hence, $w \in M_1$. Therefore, $\text{Tor}(M) = M_1$. Similarly, suppose in the second decomposition, $\text{ann}(z_j) \ne \{0\}$, for $j \le l$, but $\text{ann}(z_{l+1}) = \cdots = \text{ann}(z_t) = \{0\}$, where l is a positive integer. Then we also have $\text{Tor}(M) = Dz_1 \oplus \cdots \oplus Dz_l$.

The above argument shows that the free module $M/\text{Tor}(M)$ can be decomposed as

$$M/\text{Tor}(M) = Dy_{k+1} \oplus \cdots \oplus Dy_s$$
$$= Dz_{l+1} \oplus \cdots \oplus Dz_t.$$

Thus, $s - k = t - l$. This reduces the proof of the theorem to the case that M is a torsion module. Therefore, in the following we assume that M is a torsion module.

By Theorem 3.4.9, M can be decomposed into the direct sum of finitely many p-modules. Similarly, each of the submodules Dy_i, Dz_j can also be decomposed as the direct sum of some p-modules. By Theorem 3.4.9, it is easily seen that any p-component appearing in the decomposition of Dy_i or Dz_j must appear in the decomposition of M. Therefore, if M is decomposed as $M = M_{p_1} \oplus \cdots \oplus M_{p_k}$, then we have

$$M = \bigoplus_{i=1}^{s}\bigoplus_{l=1}^{k} Dy_i \cap M_{p_l} = \bigoplus_{j=1}^{t}\bigoplus_{l=1}^{k} Dz_j \cap M_{p_l}.$$

This reduces the proof of the assertion to the case of a p-module.

Now, suppose M is a p-module, and it has two kinds of decompositions $M = Dy_1 \oplus Dy_2 \oplus \cdots \oplus Dy_s = Dz_1 \oplus Dz_2 \oplus \cdots \oplus Dz_t$. Then we have $\text{ann}(y_i) = \langle p^{m_i} \rangle$, and $\text{ann}(z_j) = \langle p^{n_j} \rangle$, where $m_1 \le \cdots \le m_s$, $n_1 \le \cdots \le n_t$. Next we prove that $s = t$, and that $m_i = n_i$, $i = 1, 2, \ldots, s$.

We will apply the dimension theory of vector spaces over a general field. Notice that for any k, $p^k M = \{p^k x | x \in M\}$ is a submodule of M and $M \supseteq pM \supseteq p^2 M \supseteq \ldots$. Consider the quotient modules $N_k = p^k M / p^{k+1} M$. It is easily seen that for any $x \in M$, we have $p(p^k x + p^k M) = p^{k+1} x + p^{k+1} M = p^{k+1} M$. Thus, $p\alpha = \bar{0}$, $\forall \alpha \in N_k$. Therefore, N_k can be viewed as a $D/\langle p \rangle$-module. As $\mathbb{F} = D/\langle p \rangle$ is a field, N_k becomes a vector space over the field $\mathbb{F}$.

Now, we consider the dimensions of the vector spaces N_k. If $k \geq m_s$, then $p^k M = \{0\}$, hence, $\dim N_k = 0$. Now, suppose $k < m_s$ and l is the least index satisfying the condition $m_l > k$. Then we have $p^k M = Dp^k y_l + Dp^k y_{l+1} + \cdots + Dp^k y_s$. Therefore, $\{p^k y_l + p^{k+1} M, p^k y_{l+1} + p^{k+1} M, \ldots, p^k y_s + p^{k+1} M\}$ is a set of generators of the vector space N_k.

Next we show that the aforementioned elements are linearly independent in N_k. In fact, if $\bar{a}_j = a_j + \langle p \rangle$ are elements in the field $\mathbb{F}$, where $a_j \in D$, $l \leq j \leq s$, such that

$$\sum_{j=l}^{s} \bar{a}_j (p^k y_j + p^{k+1} M) = \left(\sum_{j=l}^{s} a_j p^k y_j \right) + p^{k+1} M = \bar{0}.$$

Then we have

$$\sum_{j=l}^{s} a_j p^k y_j \in p^{k+1} M.$$

Since l is the least index satisfying the condition $m_l > k$, we have $p^{k+1} D y_i = \{0\}$, for $i \leq l - 1$. Thus,

$$p^{k+1} M = \sum_{j=l}^{s} p^{k+1} D y_j.$$

By the properties of direct sum of submodules, we get

$$a_j p^k y_j \in p^{k+1} D y_j, \quad j \geq l.$$

This enables us to assume that

$$a_j p^k y_j = b_j p^{k+1} y_j, \quad j \geq l,$$

where $b_j \in D$. Using $p^{m_j - k - 1}$ to operate both sides of the above equation, we get

$$a_j p^{m_j - 1} y_j = 0, \quad j \geq l.$$

Since $\text{ann}(y_j) = \langle p^{m_j} \rangle$, we have $p \mid a_j$, $j \geq l$. Consequently, in the field $\mathbb{F}$ we have $\bar{a}_j = 0$, $j \geq l$. This proves our assertion.

The above argument shows that $\dim N_k$ is equal to the number of elements of the set $\{j \in \mathbb{N} | m_j > k\}$. Analogously, $\dim N_k$ is also equal to the number of elements of the set $\{j \in \mathbb{N}|n_j > k\}$. From this it follows that $|\{j \in \mathbb{N}|m_j > k\}| = |\{j \in \mathbb{N}|n_j > k\}|$, $\forall k \geq 1$. Therefore, $s = t$ and $m_j = n_j$, $\forall j$. This completes the proof of the theorem. $\square$

Now, we introduce the notion of **elementary factors**.

Theorem 3.4.11. *Let M be a finitely generated torsion module over a principal ideal module D. Then M can be decomposed as the direct sum of cyclic submodules:*

$$M = Dz_1 \oplus Dz_2 \oplus \cdots \oplus Dz_r,$$

where $\mathrm{ann}(z_i) = \langle p_i^{n_i} \rangle$, p_i *are irreducible elements, and* $n_i > 0$. *Moreover, the set* $\{p_1^{n_1}, \ldots, p_r^{n_r}\}$ *is uniquely determined by M (up to multiples of units). The elements* $p_i^{n_i}$, $1 \leq i \leq r$ *are called the elementary factors of M.*

Proof.   We first prove the existence. Decompose M into the direct sum of submodules as

$$M = Dy_1 \oplus Dy_2 \oplus \cdots \oplus Dy_s, \tag{3.8}$$

where $y_i \neq 0$, $\mathrm{ann}(y_i) = \langle d_i \rangle$, and $d_i|d_{i+1}$, $i = 1, 2, \ldots, s - 1$. Then we factor d_i into the product of irreducible elements as

$$d_i = \varepsilon_i p_{i1}^{l_{i1}} p_{i2}^{l_{i2}} \cdots p_{ir_i}^{l_{ir_i}},$$

where ε_i is a unit, and $p_{i1}, p_{i2}, \ldots, p_{ir_i}$ are irreducible elements which are not associated with each other, and $l_{ij} > 0$, $1 \leq j \leq r_i$. By the proof of the first assertion of Theorem 3.4.9, it is easy to see that Dy_i is equal to the direct sum of cyclic submodules:

$$Dy_i = Dw_{i1} \oplus Dw_{i2} \oplus \cdots \oplus Dw_{ir_i}, \tag{3.9}$$

where $\mathrm{ann}(w_{ij}) = \langle p_{ij}^{l_{ij}} \rangle$. Substituting (3.9) into (3.8), we prove the existence of the decomposition.

Now, we prove that the set $\{p_1^{n_1}, \ldots, p_r^{n_r}\}$ is uniquely determined by M. For this we only need to show that the invariant factors of M are uniquely determined by the set $\{p_1^{n_1}, \ldots, p_r^{n_r}\}$.

Divide $\{p_1^{n_1}, \ldots, p_r^{n_r}\}$ into groups, such that the powers of the associated irreducible elements are included into the same group, as follows:

$$(p_1')^{k_{11}}, (p_1')^{k_{12}}, \ldots, (p_1')^{k_{1s_1}};$$

$$(p_2')^{k_{21}}, (p_2')^{k_{22}}, \ldots, (p_2')^{k_{2s_2}};$$

$$\cdots\cdots\cdots\cdots$$

$$(p_t')^{k_{t1}}, (p_t')^{k_{t2}}, \ldots, (p_t')^{k_{ts_t}},$$

where $p_1', p_2', \ldots, p_{t'}'$ are not associated with each other, and $k_{ij} \leq k_{ij+1}$. Now, suppose $s' = \max\{s_1, \ldots, s_t\}$, and

$$d_{s'}' = (p_1')^{k_{1s_1}} (p_2')^{k_{2s_2}} \ldots (p_t')^{k_{ts_t}}.$$

Let $d_{s'-1}'$ be the product of all the powers among $(p_1')^{l_{1j}}$, $(p_2')^{l_{2j}}, \ldots, (p_{t'}')^{l_{t'j}}$ which are the second highest in each group (if for some i, the power of p_i' appears only once, then we use $p_i'^0$ instead), and define $d_{s'-2}', \ldots, d_2', d_1'$ successively. Then by the first assertion in the proof of Theorem 3.4.9, M can be decomposed as

$$M = Dy_1' \oplus Dy_2' \oplus \cdots \oplus Dy_{s'}',$$

where $\mathrm{ann}(y_i') = \langle d_i' \rangle$. This means that $d_1', d_2', \ldots, d_{s'}'$ are the invariant factors of M. Therefore, the set $\{p_1^{n_1}, \ldots, p_r^{n_r}\}$ is uniquely determined by the invariant factors of M. This completes the proof of the theorem. $\qquad\square$

Corollary 3.4.12. *Let M_1, M_2 be two finitely generated modules over a principal ideal domain. Then M_1 is isomorphic to M_2 $\Longleftrightarrow$ M_1, M_2 has the same rank and invariant factors $\Longleftrightarrow$ M_1, M_2 has the same rank and elementary factors.*

Finally, we give an example to show how to determine invariant factors by elementary factors.

Example 3.4.13. Let G be a finite abelian group. Suppose the elementary factors of G as a $\mathbb{Z}$-module are $2^4, 2^2, 2; 3^2, 3; 5$. Determine the structure of G.

Notice that there are three prime numbers appearing in the elementary factors, and 2 appears 3 times, which is the most frequent occurrence. Thus, there are three invariant factors, which are

$d_3 = 2^4 \cdot 3^2 \cdot 5 = 720$, $d_2 = 2^2 \cdot 3 = 12$, $d_1 = 2$. Therefore, G can be decomposed as

$$G = \mathbb{Z}y_1 \oplus \mathbb{Z}y_2 \oplus \mathbb{Z}y_3,$$

where $\text{ann}y_1 = \langle 2 \rangle$, $\text{ann}y_2 = \langle 12 \rangle$, and $\text{ann}y_3 = \langle 720 \rangle$. Consequently,

$$G \simeq \mathbb{Z}_2 \oplus \mathbb{Z}_{12} \oplus \mathbb{Z}_{720}.$$

Exercises

(1) Find an example of an ideal I of a ring R, such that as an R-module, I is not finitely generated.

(2) Let N be a subring of the ring $R = \mathbb{C}[x]$, and $\mathbb{C} \subsetneq N$. Show that R is finitely generated as an N-module.

(3) Suppose a submodule N of $\mathbb{Z}^4$ has a set of generators

$$h_1 = (1, 2, 1, 0), \quad h_2 = (2, 1, -1, 1), \quad h_3 = (0, 0, 1, 1).$$

Determine the rank r of N, and find a base (e_1, e_2, e_3, e_4) of $\mathbb{Z}^4$, and a base $(f_1, f_2, \ldots, f_r)$ of N such that $f_i = d_i e_i$, $i = 1, 2, \ldots, r$, and $d_i | d_{i+1}, i = 1, 2 \ldots, r - 1$.

(4) Let $R = \mathbb{Z}[x]$. Construct an example of a finitely generated R-module M, such that M cannot be decomposed into the direct sum of finitely many cyclic submodules.

(5) Let $D = \mathbb{Z}[\sqrt{-1}]$, and K be the submodule of D^3 generated by $f_1 = (1, 3, 6)$, $f_2 = (2 + 3\sqrt{-1}, -3\sqrt{-1}, 12 - 18\sqrt{-1})$, $f_3 = (2 - 3\sqrt{-1}, 6 + 9\sqrt{-1}, -18\sqrt{-1})$. Determine the invariant factors and elementary factors of the quotient module D^3/K.

(6) Let D be a principal ideal domain, and M, a finitely generated torsion module over D. For $a \in D$, denote $M(a) = \{x \in M \,|\, ax = 0\}$. Prove the following assertions:

   (i) $M(a)$ is a submodule of M. Moreover, if a is invertible, then $M(a) = \{0\}$.

   (ii) If $a \mid b$, then $M(a) \subseteq M(b)$.

   (iii) For any $a, b \in D$, $M(a) \cap M(b) = M((a, b))$, where (a, b) is a greatest common divisor of a, b.

   (iv) If a, b are relatively prime, then $M(ab) = M(a) \oplus M(b)$.

Exercises (hard)

(7) Let M be a finitely generated torsion module over a principal ideal domain D. Show that there exists $a_M \in D$, such that $M = M(a_M)$. Show also that a_M is unique up to a unit multiple.

(8) Let M be a finitely generated module over a principal ideal domain D. Show that a submodule of M is also finitely generated as a D-module.

(9) Let M be a finitely generated module over a principal ideal domain D, and N, a submodule of M. Show that $r(M) = r(N) + r(M/N)$.

(10) Let M be a torsion module over a principal ideal domain D. Show that M is a simple module if and only if $M = Dz$, and $\text{ann}(z) = \langle p \rangle$, where p is a prime element. Moreover, M is an indecomposable module if and only if $M = Dz$, and $\text{ann}(z) = \{0\}$ or $\langle p^e \rangle$, where p is a prime element, and $e \in \mathbb{N}$.

3.5 Finitely generated abelian groups

In this section, we give the first application of the theory of finitely generated modules over a principal ideal domain, namely, a classification of finitely generated abelian groups. As we mentioned before, every abelian group can be viewed as a module over the ring of integers. If G is a finitely generated abelian group, then it is a finitely generated $\mathbb{Z}$-module. Thus, we have a decomposition:

$$G = \mathbb{Z}a_1 \oplus \cdots \oplus \mathbb{Z}a_s \oplus \mathbb{Z}b_1 \oplus \cdots \oplus \mathbb{Z}b_r,$$

where $\text{ann}(a_i) = \langle d_i \rangle$, $d_i \neq 0$, $d_i | d_{i+1}$ (notice that s can be zero, in which case only finite submodules appear); and $\text{ann} b_j = \{0\}$, $j = 1, 2, \ldots, r$ (r can also be zero, in which case only infinite submodules appear). We can certainly assume that $d_i > 0$. Then it is easy to show that $\mathbb{Z}a_i \simeq \mathbb{Z}/\langle d_i \rangle = \mathbb{Z}_{d_i}$, and $\mathbb{Z}b_j \simeq \mathbb{Z}$. This proves the following:

Theorem 3.5.1. *Let G be a finitely generated abelian group. Then it can be decomposed as the direct sum of finitely many cyclic subgroups:*

$$G = \left(\bigoplus_{i=1}^{s} G_i \right) \oplus \left(\bigoplus_{j=1}^{r} G'_j \right),$$

where $G_i \simeq \mathbb{Z}_{d_i}$, $1 \leq i \leq s$, $d_j \mid d_{j+1}$, $j = 1, \ldots, s-1$, and $G'_k \simeq \mathbb{Z}$, $k = 1, 2, \ldots, r$.

The number r in the above theorem is called the **rank** of G, and the positive integers $d_1, \ldots, d_s$ are called the **invariant factors** of G. It follows that two finitely generated abelian groups are isomorphic if and only if they have the same rank and the same invariant factors. This gives a classification of finitely generated abelian groups. Next we consider the classification of finite abelian groups (which have rank 0).

Theorem 3.5.2. *Suppose a positive integer n is factored as $n = p_1^{m_1} p_2^{m_2} \cdots p_s^{m_s}$, where $p_1, p_2, \ldots, p_s$ are distinct prime numbers, and $m_i > 0$. Then up to isomorphism there are*

$$\prod_{i=1}^{s} \rho(m_i)$$

abelian groups with order n, where $\rho(m_i)$ denotes the number of partitions of m_i.

Proof. Since $|G| = n$, by Sylow's theorem, for any $1 \leq i \leq s$, there exists a subgroup G_i of G such that $|G_i| = p_i^{m_i}$. Then it follows from the commutativity of G that $G = G_1 \otimes G_2 \cdots \otimes G_s$. It is obvious that two abelian groups with the same order n are isomorphic if and only if their corresponding subgroups G_i, $i = 1, 2, \ldots, s$ are accordingly isomorphic. Thus, up to isomorphism there are

$$\prod_{i=1}^{s} \Gamma(p_i^{m_i})$$

abelian groups of order n, where $\Gamma(p_i^{m_i})$ denotes the number of abelian groups of order $p_i^{m_i}$ up to isomorphism.

Now, suppose K is an abelian group with order $p_i^{m_i}$. Then K can be decomposed as

$$K = K_1 \oplus \cdots \oplus K_t,$$

where $K_j \simeq \mathbb{Z}_{p^{e_j}}$, $e_j \in \mathbb{N}$, and $e_1 \leq \cdots \leq e_t$, $e_1 + e_2 + \cdots + e_t = m_i$. This implies that up to isomorphism there are $\rho(m_i)$ abelian groups of order $p_i^{m_i}$. $\qquad \square$

Finally, we give two examples.

Example 3.5.3. Determine the number of classes of abelian groups of order p^2 up to isomorphism, where p is a prime number.

We first show that a group of order p^2 must be abelian. In fact, if $|H| = p^2$, then the order of a non-identity element in H can only be p or p^2. If there exists an element in H with order p^2, then H must be cyclic. If there does not exist any element of H with order p^2, then all the elements in H, except for the identity, have order p. Fix $a \in H$, $a \neq e$. Then $\langle a \rangle \neq H$. Now, fix $b \notin \langle a \rangle$. Then the order of b is also equal to p, Thus, $\langle a \rangle \cap \langle b \rangle = \{e\}$, hence $|H| = |\langle a \rangle| \cdot |\langle b \rangle|$. It follows that $H = \langle a \rangle \langle b \rangle$. On the other hand, since $[H : \langle a \rangle] = [H : \langle b \rangle] = p$, and p is the least prime divisor of $|H|$, we have $\langle a \rangle \triangleleft H$, $\langle b \rangle \triangleleft H$. Therefore, $H = \langle a \rangle \otimes \langle b \rangle$. Thus, H is abelian.

The above argument has actually shown that there are only two groups of order p^2 (up to isomorphism), namely, the cyclic group $\mathbb{Z}_{p^2}$ and the abelian group $\mathbb{Z}_p \otimes \mathbb{Z}_p$. Of course, once we have proved that a group of order p^2 is abelian, we can apply Theorem 3.5.2 to get the same conclusion.

Example 3.5.4. Give the classification of abelian groups of order 100.

Notice that $100 = 2^2 \times 5^2$, and $\rho(2) = 2$. By Theorem 3.5.2, up to isomorphism, there are four abelian groups of order 100, namely, $\mathbb{Z}_4 \otimes \mathbb{Z}_{25}$, $\mathbb{Z}_4 \otimes \mathbb{Z}_5 \otimes \mathbb{Z}_5$, $\mathbb{Z}_2 \otimes \mathbb{Z}_2 \otimes \mathbb{Z}_{25}$, and $\mathbb{Z}_2 \otimes \mathbb{Z}_2 \otimes \mathbb{Z}_5 \otimes \mathbb{Z}_5$.

Exercises

(1) Classify abelian groups of order 24.
(2) Classify abelian groups of order 16.
(3) Determine the number of abelian groups of order 72 up to isomorphism.
(4) Determine the number of abelian groups of order 936 up to isomorphism.
(5) Show that there does not exist a positive integer n such that the number of abelian groups of order n up to isomorphism is equal to 13.

(6) Let α be a complex number. Denote the minimal subring of $\mathbb{C}$ containing α and $\mathbb{Z}$ as $\mathbb{Z}[\alpha]$. Show that α is an algebraic integer if and only if $\mathbb{Z}[\alpha]$ is a finitely generated abelian group.

(7) Show that the set of all the algebraic integers is a subring of the field of complex numbers.

(8) Let $G = \mathbb{Z}_{p^e}$, where p is a prime number, and $e > 0$. Denote the order of an element g in G as $o(g)$. Show that if $l \leq e$, then $|\{g \in G \,|\, o(g) \mid p^l\}| = p^l$; and if $l > e$, then $|\{g \in G \,|\, o(g) \mid p^l\}| = p^e$.

(9) Let $H_1, H_2, \ldots, H_s$ be finite abelian groups and q, a positive integer. Set $m_j = |\{h \in H_j \,|\, o(h) \mid q\}|$, $G = H_1 \otimes H_2 \otimes \cdots \otimes H_s$. Show that $|\{g \in G \,|\, o(g) \mid q\}| = m_1 m_2 \cdots m_s$.

(10) Determine how many elements there are in $\mathbb{Z}_4 \otimes \mathbb{Z}_8 \otimes \mathbb{Z}_{16}$ whose order is a divisor of 4.

(11) Let G be an abelian group of order p^n, where p is prime and $n > 0$. Show that there exists a set of generators $g_1, g_2, \ldots, g_s$ of G such that $o(g_i) = \max_{g \in G} o(g)$, $i = 1, 2, \ldots, s$.

Exercises (hard)

(12) Let G be a finite abelian group and φ a non-trivial homomorphism from G to the multiplicative group of non-zero complex numbers $\mathbb{C}^*$. Show that $\sum_{g \in G} \varphi(g) = 0$.

(13) Let G be an abelian group of odd order. Show that any homomorphism from G to the multiplicative group of non-zero rational numbers $\mathbb{Q}^*$ must be trivial.

(14) Classify abelian groups of order 8 up to isomorphism, and determine all the homomorphisms from such groups to $\mathbb{Q}^*$.

(15) Classify finite groups of order 8.

3.6 Canonical forms of linear transformations

In this section, we study canonical forms of linear transformations on a finite-dimensional vector space over a general field. Suppose V is a vector space of dimension n over a field $\mathbb{F}$, and $\mathcal{A}$ is a linear transformation on V. Our goal is to find an appropriate base of V such that the matrix of $\mathcal{A}$ with respect to this base has the simplest form.

We first deal with the general case, namely, we do not make any restriction on the field $\mathbb{F}$. As we mentioned before, the vector space V can be viewed as an $\mathbb{F}[\lambda]$-module. Since V is finite-dimensional, it is a finitely generated $\mathbb{F}[\lambda]$-module. Moreover, by the Hamilton–Cayley Theorem, we have

$$f(\lambda)v = f(\mathcal{A})(v) = 0(v) = 0, \quad \forall v \in V,$$

where $f(\lambda)$ is the characteristic polynomial of $\mathcal{A}$. Therefore, V is a torsion module. Notice that $\mathbb{F}[\lambda]$ is a Euclidean domain. Thus, the structure theory of finitely generated torsion modules over a principal ideal domain can be applied to study related problems.

First of all, fix a base $(v_1, v_2, \ldots, v_n)$ of the vector space V. Then $v_1, v_2, \ldots, v_n$ form a set of generators of V as an $\mathbb{F}[\lambda]$-module. Fix a base $(w_1, w_2, \ldots, w_n)$ of $\mathbb{F}[\lambda]^n$ as a free $\mathbb{F}[\lambda]$-module. Then there exists a surjective homomorphism φ from the $\mathbb{F}[\lambda]$-module $\mathbb{F}[\lambda]^n$ to the $\mathbb{F}[\lambda]$-module V such that $\varphi(w_i) = v_i$. Therefore, as an $\mathbb{F}[\lambda]$-module, V is isomorphic to the quotient module $\mathbb{F}[\lambda]^n/K$, where $K = \operatorname{Ker}\varphi$ is a submodule of $\mathbb{F}[\lambda]^n$. By Theorem 3.4.1, K is also a free module. Now, we give a base of K.

Theorem 3.6.1. *Suppose V is a vector space of dimension n over a field $\mathbb{F}$, and $\mathcal{A}$ is a linear transformation on V whose matrix with respect to a base $(v_1, v_2, \ldots, v_n)$ is $A = (a_{ij})_{n \times n}$. Let φ be the homomorphism of modules as above, and $K = \operatorname{Ker}\varphi$. Then the elements*

$$u_i = \lambda w_i - \sum_{j=1}^{n} a_{ji} w_j, \quad i = 1, 2, \ldots, n,$$

form a base of K.

Proof. This will be proved in three steps:

(1) $u_i \in K$, $i = 1, 2, \ldots, n$. In fact,

$$\varphi(u_i) = \varphi(\lambda w_i) - \varphi\left(\sum_{j=1}^{n} a_{ji} w_j\right) = \lambda \varphi(w_i) - \sum_{j=1}^{n} a_{ji} \varphi(w_j)$$

$$= \lambda v_i - \sum_{j=1}^{n} a_{ji} v_j = \mathcal{A}(v_i) - \sum_{j=1}^{n} a_{ji} v_j$$

$$= 0.$$

The reason for the last equality to hold is that the matrix of $\mathcal{A}$ with respect to the base $(v_1, \ldots, v_n)$ is exactly A.

(2) $u_1, \ldots, u_n$ form a set of generators of K. Given $w \in K$, since $(w_1, \ldots, w_n)$ is a base of the free module $\mathbb{F}[\lambda]^n$, there exists $g_1(\lambda), \ldots, g_n(\lambda) \in \mathbb{F}[\lambda]$ such that

$$w = \sum_{j=1}^{n} g_j(\lambda) w_j.$$

Write $g_j(\lambda)$ as $g_j(\lambda) = h_j(\lambda)\lambda + c_j$, where $h_j(\lambda) \in \mathbb{F}[\lambda]$, $c_j \in \mathbb{F}$. Then we have

$$g_j(\lambda) w_j = h_j(\lambda)\lambda w_j + c_j w_j$$

$$= h_j(\lambda)\left(u_j + \sum_{k=1}^{n} a_{kj} w_k \right) + c_j w_j$$

$$= h_j(\lambda) u_j + \sum_{k=1}^{n} a_{kj} h_j(\lambda) w_k + c_j w_j.$$

Notice that if $\deg g_j(\lambda) \geq 1$, then $\deg h_j(\lambda) = \deg g_j(\lambda) - 1$. This process applied repeatedly shows that there are polynomials $f_k(\lambda) \in \mathbb{F}[\lambda]$ and $b_k \in \mathbb{F}$, $k = 1, \ldots, n$, such that

$$\sum_{j=1}^{n} g_j(\lambda) w_j = \sum_{k=1}^{n} f_k(\lambda) u_k + \sum_{k=1}^{n} b_k w_k.$$

Then it follows from $\varphi(u_k) = 0$ that

$$0 = \varphi(w) = \sum_{k=1}^{n} f_k(\lambda)\varphi(u_k) + \sum_{k=1}^{n} b_k \varphi(w_k) = \sum_{k=1}^{n} b_k v_k = 0.$$

Since $(v_1, \ldots, v_n)$ is a base of V as a vector space over $\mathbb{F}$, we conclude that $b_1 = b_2 = \cdots = b_n = 0$. Thus,

$$w = \sum_{j=1}^{n} f_j(\lambda) u_j.$$

Therefore, $\{u_1, \ldots, u_n\}$ is a set of generators of K.

(3) $u_1, \ldots, u_n$ are linearly independent in the free module $\mathbb{F}[\lambda]^n$. Suppose

$$\sum_{j=1}^{n} l_j(\lambda) u_j = 0,$$

where $l_1(\lambda), \ldots, l_n(\lambda) \in \mathbb{F}[\lambda]$. Then we have

$$\sum_{j=1}^{n} l_j(\lambda) \left(\lambda w_j - \sum_{k=1}^{n} a_{kj} w_k \right) = 0.$$

This equality can be written in matrix form as

$$(w_1, , \ldots, w_n) \begin{pmatrix} \lambda l_1(\lambda) \\ \lambda l_2(\lambda) \\ \vdots \\ \lambda l_n(\lambda) \end{pmatrix} = (w_1, \ldots, w_n) A \begin{pmatrix} l_1(\lambda) \\ l_2(\lambda) \\ \vdots \\ l_n(\lambda) \end{pmatrix}.$$

Since $(w_1, \ldots, w_n)$ is a base of $\mathbb{F}[\lambda]^n$, they are linearly independent. Therefore, we have

$$\lambda l_j(\lambda) = \sum_{k=1}^{n} a_{jk} l_k(\lambda), \quad j = 1, \ldots, n.$$

This implies that any of $l_1(\lambda), l_2(\lambda), \ldots, l_n(\lambda)$ must be equal to 0. In fact, if at least one of them is non-zero, then we can select $l_q(\lambda)$ from $l_1(\lambda), \ldots, l_n(\lambda)$ such that the degree of $l_q(\lambda)$ attains the maximum. Then $\lambda l_q(\lambda) \neq 0$, and for any $1 \leq k \leq n$, either $l_k(\lambda)$ is equal to 0 or its degree is less than $\deg \lambda l_q(\lambda)$. But $\lambda l_q(\lambda)$ can be expressed as an $\mathbb{F}$-combination of $l_1(\lambda), \ldots, l_n(\lambda)$, which is absurd. Therefore, $u_1, \ldots, u_n$ are linearly independent.

The proof of the theorem is now completed. $\qquad\square$

Next we apply Theorem 3.6.1 to study the structure of V as an $\mathbb{F}[\lambda]$-module, and find a base of V such that the matrix of $\mathcal{A}$ with respect to this base has the simplest form. Notice that the matrix of the base $(u_1, u_2, \ldots, u_n)$ of K with respect to the base $(w_1, w_2, \ldots, w_n)$ of $\mathbb{F}[\lambda]^n$ is equal to $\lambda I_n - A$. Thus, to determine the structure of the module V, we only need to determine the invariant factors or the elementary factors of the matrix $\lambda I_n - A$. This is the theoretical basis of the Jordan canonical theory in linear algebra.

Now, we study the matrix $\lambda I_n - A$. The nth determinant factor $\Delta_n(\lambda I_n - A)$ of this matrix is just the characteristic polynomial of A (or $\mathcal{A}$), which is a monic polynomial with degree n. In particular, the rank of the matrix $\lambda I_n - A$ is equal to n. Thus, the normal form of the matrix $\lambda I_n - A$ can be written as $\text{diag}(1, \ldots, 1, d_1(\lambda), \ldots, d_s(\lambda))$. Therefore, as an $\mathbb{F}[\lambda]$-module, V can be written as the direct sum of the cyclic submodules:

$$V = \mathbb{F}[\lambda]x_1 \oplus \mathbb{F}[\lambda]x_2 \oplus \cdots \oplus \mathbb{F}[\lambda]x_s,$$

where $\text{ann}(x_j) = \langle d_j(\lambda) \rangle$.

Now, let us study the cyclic submodule $V_j = \mathbb{F}[\lambda]x_j$ in the above decomposition. As a submodule, V_j is an invariant subspace of the linear transformation $\mathcal{A}$. Therefore, we can denote $\mathcal{A}_j = \mathcal{A}|_{V_j}$, and study the canonical forms of the linear transformation $\mathcal{A}_j$ on V_j. For the sake of simplicity, we can assume that $s = 1$, and $V = \mathbb{F}[\lambda]x$. In this case, $\text{ann}(x) = \langle f(\lambda) \rangle$, where $f(\lambda)$ is the characteristic polynomial of $\mathcal{A}$. Notice that $f(\lambda)$ is the non-zero polynomial with the lowest degree satisfying the condition $f(\lambda)x = 0$. Thus, $x, \lambda x, \ldots, \lambda^{n-1}x$ are linearly independent in V, hence they form a base of V. Now, suppose $f(\lambda) = \lambda^n + a_1\lambda^{n-1} + \cdots + a_{n-1}\lambda + a_n$. Then we have $\mathcal{A}x = \lambda x$, $\mathcal{A}(\lambda x) = \lambda^2 x, \ldots, \mathcal{A}(\lambda^{n-2}x) = \lambda^{n-1}x$, and

$$\mathcal{A}(\lambda^{n-1}x) = \lambda^n x = -a_1\lambda^{n-1}x - \cdots - a_{n-1}\lambda x - a_n x.$$

This implies that the matrix of $\mathcal{A}$ with respect to the base $(x, \lambda x, \ldots, \lambda^{n-1}x)$ is

$$\begin{pmatrix} 0 & 0 & 0 & \cdots & 0 & -a_n \\ 1 & 0 & 0 & \cdots & 0 & -a_{n-1} \\ 0 & 1 & 0 & \cdots & 0 & -a_{n-2} \\ \vdots & \vdots & \vdots & \vdots & \vdots & \vdots \\ 0 & 0 & 0 & \cdots & 1 & -a_1 \end{pmatrix}. \tag{3.10}$$

The above argument can be applied to the general case to get the following:

Theorem 3.6.2. *Suppose V is a finite-dimensional vector space over a field $\mathbb{F}$, and $\mathcal{A}$ is a linear transformation on V. Then there exists a base $(v_1, v_2, \ldots, v_n)$ of V such that the matrix of $\mathcal{A}$ with respect to this base has the form of a diagonal block matrix as follows:*

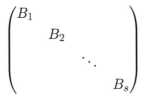

where $B_j, j = 1, 2, \ldots, s$ has the form (3.10).

In the literature, the matrix in Theorem 3.6.2 is usually called the **rational canonical form** of $\mathcal{A}$. For a general field $\mathbb{F}$, the rational canonical form is the simplest. However, if $\mathbb{F}$ is **algebraically closed**, i.e., every non-constant polynomial over $\mathbb{F}$ has a root in $\mathbb{F}$, then there is a simpler canonical form, namely, the **Jordan canonical form**.

In Theorem 3.6.2, we consider the decomposition of the module V into the sum of cyclic submodules with respect to the elementary factors. Then each block B_i corresponds to a polynomial of the form $p_i^{r_i}(\lambda)$, where $p_i(\lambda)$ is an irreducible polynomial, $r_i \geq 1$. If we further assume that $p_i(\lambda) = (\lambda - \lambda_i)$ (which is automatically true in case that $\mathbb{F}$ is algebraically closed), then we can select a base of the submodule $V_i = \mathbb{F}[x]x_i$ as

$$\varepsilon_1 = x_i, \varepsilon_2 = (\lambda - \lambda_i)x_i, \ldots, \varepsilon_{r_i} = (\lambda - \lambda_i)^{r_i-1}x_i.$$

Then we have

$$\mathcal{A}_i(\varepsilon_1) = \lambda x_i = \lambda_i x_i + (\lambda - \lambda_i)x_i = \lambda_i \varepsilon_1 + \varepsilon_2,$$
$$\mathcal{A}_i(\varepsilon_k) = \lambda(\lambda - \lambda_i)^{k-1}x_i = \lambda_i(\lambda - \lambda_i)^{k-1}x_i + (\lambda - \lambda_i)^k x_i$$
$$= \lambda_i \varepsilon_k + \varepsilon_{k+1}, \quad k = 1, \ldots, r_i - 2,$$
$$\mathcal{A}_i(\varepsilon_{r_i}) = \lambda(\lambda - \lambda_i)^{r_i-1}x_i = \lambda_i(\lambda - \lambda_i)^{r_i-1}x_i + (\lambda - \lambda_i)^{r_i}x_i$$
$$= \lambda_i(\lambda - \lambda_i)^{r_i-1}x_i = \lambda_i \varepsilon_{r_i}.$$

Thus, the matrix of $\mathcal{A}_i$ with respect to the base $(\varepsilon_1, \ldots, \varepsilon_{r_i})$ is

$$\begin{pmatrix} \lambda_i & 0 & 0 & \cdots & 0 & 0 \\ 1 & \lambda_i & 0 & \cdots & 0 & 0 \\ 0 & 1 & \lambda_i & \cdots & 0 & 0 \\ \vdots & \vdots & \vdots & \vdots & \vdots & \vdots \\ 0 & 0 & 0 & \cdots & 1 & \lambda_i \end{pmatrix}.$$

The matrix of the above form is called a **Jordan block**.

In summary, we have proved the following:

Theorem 3.6.3. *Suppose $\mathbb{F}$ is an algebraically closed field, V is an n-dimensional vector space over $\mathbb{F}$, and $\mathcal{A}$ is a linear transformation on V. Then there exists a base $(\alpha_1, \alpha_2, \ldots, \alpha_n)$ of V such that the matrix of $\mathcal{A}$ with respect to this base has the diagonal block form as follows:*

$$\begin{pmatrix} J_1 & & & \\ & J_2 & & \\ & & \ddots & \\ & & & J_s \end{pmatrix},$$

where $J_1, J_2, \ldots, J_s$ are Jordan blocks.

Since the field of complex numbers is algebraically closed, the theory of Jordan canonical forms of linear transformations on complex vector spaces in linear algebra is just a special case of the above theorem.

We leave as an exercise for readers to write down the above theorem in the matrix form, which gives the canonical forms of square matrices over an algebraically closed field. In the following, we give two examples to illustrate the main results of this section.

Example 3.6.4. Let V be a 4-dimensional vector space over $\mathbb{Q}$ with a base $(\alpha_1, \alpha_2, \alpha_3, \alpha_4)$. Suppose $\mathcal{A}$ is a linear transformation on V

whose matrix with respect to the base $(\alpha_1, \alpha_2, \alpha_3, \alpha_4)$ is

$$A = \begin{pmatrix} -1 & 2 & -1 & 0 \\ -2 & -1 & 0 & -1 \\ 0 & 0 & -1 & 2 \\ 0 & 0 & -2 & -1 \end{pmatrix}.$$

Then

$$M(\lambda) = \lambda I_4 - A = \begin{pmatrix} \lambda+1 & -2 & 1 & 0 \\ 2 & \lambda+1 & 0 & 1 \\ 0 & 0 & \lambda+1 & -2 \\ 0 & 0 & 2 & \lambda+1 \end{pmatrix}.$$

To determine the rational canonical form of $\mathcal{A}$, we need to calculate the invariant factors of $M(\lambda)$. It is obvious that the first and second determinant factors of $M(\lambda)$ are $\Delta_1 = \Delta_2 = 1$. Notice that the two 3-minors

$$M(\lambda) \begin{pmatrix} 123 \\ 234 \end{pmatrix} = 4(\lambda+1)^2,$$

$$M(\lambda) \begin{pmatrix} 123 \\ 134 \end{pmatrix} = 4 - (\lambda+1)^2,$$

are relatively prime. Thus, $\Delta_3 = 1$. Moreover, $\Delta_4 = \det M(\lambda) = ((\lambda+1)^2 + 4)^2$. Therefore, the invariant factors of $M(\lambda)$ are $d_1 = d_2 = d_3 = 1$, $d_4 = ((\lambda+1)^2 + 4)^2 = \lambda^4 + 4\lambda^3 + 14\lambda^2 + 20\lambda + 25$. By Theorem 3.6.1, there exists a base $(\beta_1, \beta_2, \beta_3, \beta_4)$ of V such that the matrix of $\mathcal{A}$ with respect to $(\beta_1, \beta_2, \beta_3, \beta_4)$ is

$$\begin{pmatrix} 0 & 0 & 0 & -25 \\ 1 & 0 & 0 & -20 \\ 0 & 1 & 0 & -14 \\ 0 & 0 & 1 & -4 \end{pmatrix}.$$

This is the rational canonical form of the linear transformation $\mathcal{A}$.

Example 3.6.5. Let us consider Example 3.6.4 for the case where the base field is $\mathbb{R}$. If we consider the rational canonical form, we will get the same result as the previous example. However, there is another kind of canonical form for linear transformations on a real vector space, which is very useful and convenient. Consider the matrix

$$B = \begin{pmatrix} B_2 & 0 \\ I_2 & B_2 \end{pmatrix},$$

where $B_2 = \begin{pmatrix} 0 & -5 \\ 1 & -2 \end{pmatrix}$. A direct calculation shows that the invariant factors of $\lambda I_4 - B$ are $d_1 = d_2 = d_3 = 1$, $d_4 = ((\lambda + 1)^2 + 2)^2$. Therefore, there exists a base $(\gamma_1, \gamma_2, \gamma_3, \gamma_4)$ of V such that the matrix of $\mathcal{A}$ with respect to $(\gamma_1, \gamma_2, \gamma_3, \gamma_4)$ is B.

This example can be generalized to any linear transformation on a finite-dimensional real vector space. Suppose V is an n-dimensional real vector space, and $\mathcal{A}$ is a linear transformation on V. If the determinant factors of $\mathcal{A}$ are

$$(\lambda - \lambda_1)^{m_1}, \ldots, (\lambda - \lambda_s)^{m_s}, (\lambda^2 - a_1\lambda + b_1)^{l_1}, \ldots, (\lambda^2 - a_t\lambda + b_t)^{l_t},$$

where $\lambda_i, a_k, b_k \in \mathbb{R}$, $m_i, l_j > 0$, $1 \leq i \leq s$, $1 \leq j, k \leq t$, and $a_k^2 - 4b_k < 0$, $1 \leq k \leq t$, then there exists a base of V such that the matrix of $\mathcal{A}$ with respect to this base has the form

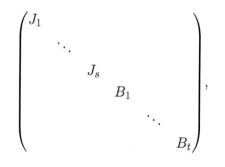

where

$$J_i = \begin{pmatrix} \lambda_i & & & \\ 1 & \lambda_i & & \\ & \ddots & \ddots & \\ & & 1 & \lambda_i \end{pmatrix} \in \mathbb{R}^{m_i \times m_i}$$

are Jordan blocks, and

$$B_i = \begin{pmatrix} A_i & & & \\ I_2 & A_i & & \\ & \ddots & \ddots & \\ & & I_2 & A_i \end{pmatrix} \in \mathbb{R}^{2l_i \times 2l_i},$$

here $A_i = \begin{pmatrix} 0 & -b_i \\ 1 & a_i \end{pmatrix}$.

The proof of the above result will be left as an exercise for readers.

Problem 3.6.6. Try to find a base of V such that the matrix of $\mathcal{A}$ with respect to this base has the aforementioned canonical form.

Exercises

(1) Let $\mathcal{A}$ be a linear transformation on a 3-dimensional complex vector space V. Suppose the matrix of $\mathcal{A}$ with respect to a base is

$$\begin{pmatrix} 1 & 3 & 5 \\ 0 & 1 & 3 \\ 0 & 0 & 3 \end{pmatrix}.$$

Determine whether V is a cyclic module or a $\mathbb{C}[\lambda]$-module.

(2) Prove the assertions in Example 3.6.5.

(3) Suppose $\mathcal{A}$ is a linear transformation on an n-dimensional vector space V over a field $\mathbb{F}$, and there exists a positive integer k such that $\mathcal{A}^k = 0$. Show that there exists a base of V such that the matrix of $\mathcal{A}$ with respect to this base has the form of a diagonal block matrix $\mathrm{diag}(N_1, N_2, \ldots, N_s)$, where N_i, $i = 1, 2, \ldots, s$ are Jordan blocks with 0s on the diagonal.

(4) Let $\mathbb{F}$ be a field. Show that two $n \times n$ matrices are similar if and only if the matrices $\lambda I_n - A$ and $\lambda I_n - B$ in $F[\lambda]^{n \times n}$ have the same invariant factors.

(5) Show that for any $n \times n$ matrix A over a field $\mathbb{F}$, A is similar to A'.

(6) Suppose $\mathcal{A}$ is a linear transformation on a 7-dimensional complex vector space V, and satisfies the conditions $(\mathrm{id} - \mathcal{A})^4 = 0$

but $(\mathrm{id} - \mathcal{A})^3 \neq 0$. Determine all the possible Jordan canonical forms of $\mathcal{A}$.

(7) Let V be a 5-dimensional complex vector space, and $\mathcal{A}$, a linear transformation on V satisfying the condition $(\mathcal{A} - 2\mathrm{id})^3 = 0$. Determine all the possible Jordan canonical forms of $\mathcal{A}$.

(8) Determine up to similarity how many complex $(2n+1) \times (2n+1)$ complex matrices there are which satisfy the conditions $(A - I_n)^{n+1} = 0$ and $(A - I_n)^n \neq 0$.

(9) Show that any Jordan block matrix over a field $\mathbb{F}$ can be written as the product of two symmetric matrices.

(10) Let $\mathbb{F}$ be an algebraically closed field. Show that every square matrix over $\mathbb{F}$ can be written as the product of two symmetric matrices.

(11) Let $\mathcal{A}$ be a linear transformation on a finite-dimensional real vector space V. Show that there exists a base of V such that the matrix of $\mathcal{A}$ with respect to this base has the form of a block matrix as

$$\begin{pmatrix} B_1 & B_2 \\ 0 & B_3 \end{pmatrix},$$

where B_1 is a square matrix of order 1 or 2.

Exercises (hard)

(12) (Weyr). Show that two complex $n \times n$ matrices A, B are similar to each other if and only if for any $a \in \mathbb{C}$ and $k \in \mathbb{N}$, $r((aI_n - A)^k) = r((aI_n - B)^k)$.

(13) Let V be an n-dimensional vector space over a field $\mathbb{F}$. Given a set of vectors $\gamma_1, \gamma_2, \ldots, \gamma_s$, we set

$$W_{\{\gamma_1, \gamma_2, \ldots, \gamma_s\}} = \left\{ (a_1, a_2, \ldots, a_s) \in F^s \mid \sum_{j=1}^{s} a_i \gamma_i = 0 \right\}.$$

Now, suppose $\{\alpha_1, \alpha_2, \ldots, \alpha_t\}$; $\{\beta_1, \beta_2, \ldots, \beta_t\}$ are two sets of vectors in V. Show that there exists a linear transformation $\mathcal{A}$ on V such that $\mathcal{A}(\alpha_j) = \beta_j$, $j = 1, 2, \ldots, t$ if and only if $W_{\{\alpha_1, \ldots, \alpha_t\}} \subseteq W_{\{\beta_1, \ldots, \beta_t\}}$. Furthermore, there exists a linear automorphism $\mathcal{A}$ of V such that $\mathcal{A}(\alpha_j) = \beta_j$, $j = 1, 2, \ldots, t$ if and only if $W_{\{\alpha_1,\ldots,\alpha_t\}} = W_{\{\beta_1,\ldots,\beta_t\}}$.

3.7 Summary of the chapter

In this chapter, we have introduced the fundamental theory of modules. The core content of this chapter is the structure theory of finitely generated modules over a principal ideal domain. The last two sections of this chapter reveal the huge power of module theory. In fact, many complicated and abstract mathematical objects can be described using the terminologies of modules, such as representations of groups, representations of algebras, etc. The module theory also has important applications in differential geometry and topology. Thus, the content of this chapter is now the prerequisite for anyone who is devoted to the study of pure mathematics.

Further topics in module theory include more structure theory and the study on special modules, such as injective modules, projective modules, Noether modules, and Artin modules, etc. Interested readers are suggested to read the references presented in this book.

Chapter 4

Fields

In this chapter, we will introduce the basic ideas of Galois theory and its applications in straightedge and compass construction and in solving equations by radicals. The concept of fields originated in the work of Abel and Galois on solving algebraic equations by radicals in the early 19th century. In 1871, Dedekind first introduced the concept of number fields. Ten years later, Kronecker defined rational function fields. In 1893, Weber gave the abstract definition of fields. In 1910, Steinitz studied the properties of fields and introduced the notions of prime and perfect fields. From 1928 to 1942, Artin systematically studied the relationship between groups and fields, and rewrote Galois theory with modern methods.

4.1 Field extensions

Recalled that a field $\mathbb{F}$ is a commutative division ring, i.e., a non-zero domain such that any non-zero element in $\mathbb{F}$ has an inverse. In linear algebra, we have known many number fields, including the field of rational numbers $\mathbb{Q}$, the field of real numbers $\mathbb{R}$, the field of complex numbers $\mathbb{C}$, $\mathbb{Q}(\sqrt{2})$, and $\mathbb{Q}(\sqrt{-1})$, etc.

By Theorem 2.6.14, if R is a commutative ring with identity, and A is a maximal ideal of R, then the quotient ring R/A is a field.

This gives us an efficient way to get new fields. For instance, in Example 2.2.21, we have seen a field with only p elements:

$$\mathbb{Z}_p = \{\overline{0}, \overline{1}, \ldots, \overline{p-1}\},$$

which is also the quotient ring of $\mathbb{Z}$ by the maximal ideal $p\mathbb{Z}$, usually denoted by $\mathbb{F}_p$. In particular, $\mathbb{F}_2$ is the field with the least number of elements. A field with a finite number of elements is called a **finite field** or a **Galois field** (see Section 4.7 for more details). A similar method may be applied to other rings.

Example 4.1.1. If $\mathbb{F}$ is a number field, and $f(x) \in \mathbb{F}[x]$ is an irreducible polynomial, then $\langle f(x) \rangle$ is a maximal ideal. Thus, $\mathbb{F}[x]/\langle f(x) \rangle$ is a field.

Once we have enough examples of fields in hand, we face a fundamental problem: are there any relationship among those different-looking fields? The key to answering the question is the following.

Definition 4.1.2. Let $\mathbb{F}, \mathbb{E}$ be two fields. A map $\varphi : \mathbb{F} \to \mathbb{E}$ is called a **field homomorphism** if φ is a ring homomorphism, i.e.,

$$\varphi(\alpha + \beta) = \varphi(\alpha) + \varphi(\beta), \quad \varphi(\alpha\beta) = \varphi(\alpha)\varphi(\beta), \quad \alpha, \beta \in \mathbb{F},$$

and $\varphi(1_\mathbb{F}) = 1_\mathbb{E}$, where $1_\mathbb{F}, 1_\mathbb{E}$ are identities of $\mathbb{F}, \mathbb{E}$ respectively. Denote by $\mathrm{Hom}\,(\mathbb{F}, \mathbb{E})$ the set of all field homomorphisms from $\mathbb{F}$ to $\mathbb{E}$. If a field homomorphism φ is also a bijection, φ is called a **field isomorphism**.

A field homomorphism (or isomorphism) from $\mathbb{E}$ to itself is call an **endomorphism** (or **automorphism**) of $\mathbb{E}$. Denote by $\mathrm{Aut}\,(\mathbb{E})$ the set of all automorphisms of $\mathbb{E}$.

Note that the composition of two field homomorphisms is also a field homomorphism, i.e., if $\varphi : \mathbb{F} \to \mathbb{E}$, $\psi : \mathbb{E} \to \mathbb{K}$ are field homomorphisms, so is $\psi \circ \varphi$. In particular, it is easy to check that $\mathrm{Hom}\,(\mathbb{E}, \mathbb{E})$ is a monoid, and $\mathrm{Aut}\,(\mathbb{E})$ is a group, called the **automorphism group** of $\mathbb{E}$.

Let's look at some examples first.

Example 4.1.3. Let $\mathbb{F} = \mathbb{Q}(\sqrt{2})$, and let $\varphi : \mathbb{F} \to \mathbb{C}$ be a field homomorphism. Then $\varphi^2(\sqrt{2}) = \varphi((\sqrt{2})^2) = \varphi(2) = 2$. Thus, $\varphi(\sqrt{2}) = \pm\sqrt{2}$. It is easy to check that $\varphi(a + b\sqrt{2}) = a \pm b\sqrt{2}$ are homomorphisms with the same image $\mathbb{F}$. Therefore, $\text{Hom}\,(\mathbb{Q}(\sqrt{2}), \mathbb{C})$ has two elements and $\text{Aut}\,(\mathbb{Q}(\sqrt{2}))$ is a group of order 2.

Example 4.1.4. Define $\varphi : \mathbb{Q}[x] \to \mathbb{Q}(\sqrt[3]{2})$ by $\varphi(f(x)) = f(\sqrt[3]{2})$. It is easy to check that φ is a ring epimorphism with kernel $\text{Ker}\,\varphi = \langle x^3 - 2 \rangle$. Since $x^3 - 2$ is irreducible, $\langle x^3 - 2 \rangle$ is a maximal ideal of $\mathbb{Q}[x]$. It follows that $\mathbb{F} = \mathbb{Q}[x]/\langle x^3 - 2 \rangle$ is a field, and φ induces a field isomorphism $\overline{\varphi} : \mathbb{F} \to \mathbb{Q}(\sqrt[3]{2})$, which is the only element in $\text{Hom}\,(\mathbb{F}, \mathbb{Q}(\sqrt[3]{2}))$.

Example 4.1.5. Let $\mathbb{F} = \mathbb{Q}(\sqrt[3]{2})$, and let $\varphi : \mathbb{F} \to \mathbb{C}$ be a field homomorphism. Then $\varphi(\sqrt[3]{2})^3 = \varphi((\sqrt[3]{2})^3) = 2$. Thus, $\varphi(\sqrt[3]{2})$ is one of the cubic roots of 2: $\sqrt[3]{2}, \sqrt[3]{2}\omega, \sqrt[3]{2}\omega^2$, where $\omega = \frac{-1+\sqrt{-3}}{2}$ is a cubic root of unity. Since $\sqrt[3]{2}$ is a root of the irreducible rational polynomial $x^3 - 2$, by Example 4.1.4, one gets that $\mathbb{Q}[x]/\langle x^3 - 2 \rangle$ is isomorphic to any of $\mathbb{Q}(\sqrt[3]{2})$, $\mathbb{Q}(\sqrt[3]{2}\omega)$ and $\mathbb{Q}(\sqrt[3]{2}\omega^2)$. Therefore, three different values of $\varphi(\sqrt[3]{2})$ define three different field isomorphisms. In particular, if $\varphi(\sqrt[3]{2}) = \sqrt[3]{2}$, the corresponding field isomorphism is the identity automorphism of $\mathbb{Q}(\sqrt[3]{2})$, which is the only element in $\text{Aut}\,(\mathbb{Q}(\sqrt[3]{2}))$. In conclusion, there are three elements in $\text{Hom}\,(\mathbb{Q}(\sqrt[3]{2}), \mathbb{C})$ and $\text{Aut}\,(\mathbb{Q}(\sqrt[3]{2}))$ is a trivial group.

By the Fundamental Theorem of Homomorphisms of rings (Theorem 2.4.5), the kernel $\text{Ker}\,\varphi$ of the homomorphism $\varphi : \mathbb{F} \to \mathbb{E}$ is an ideal of $\mathbb{F}$, which must be $\mathbb{F}$ or $\{0\}$. Since $\varphi(1_{\mathbb{F}}) = 1_{\mathbb{E}}$, we have $\text{Ker}\,\varphi = \{0\}$, which implies.

Lemma 4.1.6. *A field homomorphism* $\varphi : \mathbb{F} \to \mathbb{E}$ *must be injective.*

Therefore, a surjective homomorphism of fields is an isomorphism. Furthermore, it is easy to see that $\varphi(\mathbb{F})$, as a subset of $\mathbb{E}$, is a field isomorphic to $\mathbb{F}$. We introduce the following definition.

Definition 4.1.7. Let $\mathbb{E}$ be a field, $\mathbb{F} \subseteq \mathbb{E}$. If $\mathbb{F}$ is a field under the operations of $\mathbb{E}$, then we call $\mathbb{F}$ a **subfield** of $\mathbb{E}$, and $\mathbb{E}$ an **extension** of $\mathbb{F}$, denoted by $\mathbb{E}/\mathbb{F}$.

Any number field is a subfield of $\mathbb{C}$ and is an extension of $\mathbb{Q}$. That is to say, $\mathbb{Q}$ is the smallest number field. In general, for any field $\mathbb{E}$, the intersection of some of its subfields is still a subfield. Therefore, the intersection of all subfields of $\mathbb{E}$ is the unique smallest subfield of

$\mathbb{E}$, and it has no proper subfield, called a **prime field**. Hence, any field contains a prime field as its subfield, or, in other words, any field is an extension of a prime field.

Let $\mathbb{E}$ be a field with identity 1, which must be the identity of the prime field of $\mathbb{E}$. Consider the ring homomorphism

$$\varphi : \mathbb{Z} \to \mathbb{E}, \quad \varphi(n) = n \cdot 1.$$

Since $\mathbb{E}$ is a domain, $\operatorname{Ker}\varphi$ must be a prime ideal of $\mathbb{Z}$, and it is of the form $\langle p \rangle$, where $p = 0$ or p is a prime. If p is a prime, then $\langle p \rangle$ is also a maximal ideal of $\mathbb{Z}$. Then $\operatorname{Im}\varphi \simeq \mathbb{Z}/\langle p \rangle$ is a field. If $p = 0$, then φ is injective and can be extended to a map

$$\overline{\varphi} : \mathbb{Q} \to \mathbb{E}, \quad \overline{\varphi}(m/n) = (m \cdot 1)(n \cdot 1)^{-1}.$$

It is easy to check that $\overline{\varphi}$ is well-defined and is a field homomorphism. Therefore, each field contains a subfield isomorphic to $\mathbb{Q}$ or $\mathbb{F}_p$. In particular, we have the following.

Theorem 4.1.8. *Every prime field is isomorphic to $\mathbb{Q}$ or $\mathbb{F}_p$, for some prime p.*

In Chapter 2, we introduced the notion of characteristics for rings without zero divisors (hence for fields). Thus, the above theorem implies

Lemma 4.1.9. *If the prime field of $\mathbb{F}$ is isomorphic to $\mathbb{Q}$ (resp. $\mathbb{F}_p$), then the **characteristic** of $\mathbb{F}$ is zero (resp. p).*

Any field homomorphism $\sigma : \mathbb{F} \to \mathbb{E}$ induces an isomorphic between the prime fields of $\mathbb{F}$ and $\mathbb{E}$. Thus, $\mathbb{F}, \mathbb{E}$ share the same characteristic. Without loss of generality, we may assume that $\mathbb{F}, \mathbb{E}$ are extensions of the same prime field Π. Since $\varphi(1) = 1$, we have the following.

Corollary 4.1.10. *Let $\mathbb{E}, \mathbb{F}$ be the extensions of a prime field Π, and let $\varphi : \mathbb{F} \to \mathbb{E}$ be a field homomorphism. Then $\varphi(a) = a$, for any $a \in \Pi$, i.e., $\varphi|_\Pi = \operatorname{id}_\Pi$.*

It is easy to see that $\mathbb{E}, \mathbb{F}$ can be regarded as Π-modules or Π-vector spaces. Then a field homomorphism $\varphi : \mathbb{F} \to \mathbb{E}$ is a special linear map between vector spaces. In general, we have the following.

Theorem 4.1.11. *Let $\mathbb{E}/\mathbb{F}$ be an extension. Then $\mathbb{E}$ is an $\mathbb{F}$-vector space.*

The proof is straightforward and left to the reader. This viewpoint is crucial since results in vector spaces may be applied directly to study field theory.

Definition 4.1.12. Let $\mathbb{E}/\mathbb{F}$ be a field extension. The **degree** of $\mathbb{E}$ over $\mathbb{F}$, denoted by $[\mathbb{E} : \mathbb{F}]$, is the dimension of the vector space $\mathbb{E}$ over $\mathbb{F}$. If $[\mathbb{E} : \mathbb{F}]$ is finite, $\mathbb{E}/\mathbb{F}$ will be called a **finite extension**, otherwise, $\mathbb{E}/\mathbb{F}$ is called an **infinite extension**.

For example, $\mathbb{C}/\mathbb{R}$, $\mathbb{Q}(\sqrt{-1})/\mathbb{Q}$ are quadratic extensions (i.e., extensions of degree 2), and $\mathbb{R}/\mathbb{Q}$ is an infinite extension.

Theorem 4.1.13. *If $\mathbb{K}/\mathbb{F}$ and $\mathbb{E}/\mathbb{K}$ are extensions, then the extension $\mathbb{E}/\mathbb{F}$ is finite if and only if both $\mathbb{E}/\mathbb{K}$ and $\mathbb{K}/\mathbb{F}$ are finite. In this case, one has*

$$[\mathbb{E} : \mathbb{F}] = [\mathbb{E} : \mathbb{K}][\mathbb{K} : \mathbb{F}].$$

Proof. If $[\mathbb{E} : \mathbb{F}]$ is finite, then $[\mathbb{K} : \mathbb{F}]$ is finite since $\mathbb{K}$ is an $\mathbb{F}$-subspace of $\mathbb{E}$. Furthermore, an $\mathbb{F}$-basis of $\mathbb{E}$ is a set of $\mathbb{K}$-generators of $\mathbb{E}$, which implies that $[\mathbb{E} : \mathbb{K}] \leq [\mathbb{E} : \mathbb{F}] < \infty$.

Suppose $[\mathbb{K} : \mathbb{F}]$ and $[\mathbb{E} : \mathbb{K}]$ are finite and $a_1, a_2, \ldots, a_m$ is an $\mathbb{F}$-base for $\mathbb{K}$, $b_1, b_2, \ldots, b_n$ an $\mathbb{K}$-base for $\mathbb{E}$. For any $\alpha \in \mathbb{E}$, there exist $x_1, x_2, \ldots, x_n \in \mathbb{K}$ such that $\alpha = \sum_{j=1}^{n} x_j b_j$. Each x_j is a linear combination of $a_1, a_2, \ldots, a_m$, say, $x_j = \sum_{i=1}^{m} y_{ij} a_i$, $y_{ij} \in \mathbb{F}$. Thus, $\alpha = \sum_{i=1}^{m} \sum_{j=1}^{n} y_{ij} a_i b_j$, which implies that $a_i b_j$, $1 \leq i \leq m$, $1 \leq j \leq n$, is a set of generators of $\mathbb{E}$ over $\mathbb{F}$. Hence $[\mathbb{E} : \mathbb{F}]$ is finite. Furthermore, if $\alpha = 0$, then $x_j = 0$, which implies that $y_{ij} = 0$. Thus, $a_i b_j$, $1 \leq i \leq m$, $1 \leq j \leq n$, is an $\mathbb{F}$-basis of $\mathbb{E}$. The last assertion follows. $\square$

The field $\mathbb{K}$ is said to be an **intermediate field** of the extension $\mathbb{E}/\mathbb{F}$, or $\mathbb{K}/\mathbb{F}$ is said to be a **subextension** of $\mathbb{E}/\mathbb{F}$.

Remark 4.1.14.

(1) Similar results appear in the theory of vector spaces; for example, an n-dimensional complex vector space can be regarded as a $2n$-dimensional real vector space.
(2) If any of the extensions in the theorem is infinite, then the equality still holds.

Exercises

(1) Let $\mathbb{F}$ be a (non-zero) commutative ring with identity. Prove that $\mathbb{F}$ if a field if and only if $\mathbb{F}$ has only trivial ideals.

(2) Determine whether the following quotient rings are fields and find their characteristics.

(i) $\mathbb{Z}[\sqrt{-1}]/\langle 7 \rangle$; (ii) $\mathbb{Z}[\sqrt{-1}]/\langle 5 \rangle$; (iii) $\mathbb{Z}[\sqrt{-1}]/\langle 2 + \sqrt{-1} \rangle$.

(3) Assume that $\mathbb{F}$ is a field of characteristic $p > 0$. Show that:

(i) the map $\mathrm{Fr}(x) = x^p$ is an endomorphism of $\mathbb{F}$;
(ii) $(a_1 + a_2 + \cdots + a_r)^p = a_1^p + a_2^p + \cdots + a_r^p$;
(iii) for any $n \in \mathbb{N}$, $(a \pm b)^{p^n} = a^{p^n} \pm b^{p^n}$;
(iv) $(a - b)^{p-1} = \sum_{i=0}^{p-1} a^i b^{p-1-i}$.

(4) Let $\mathbb{E}$ be a finite field and $\mathbb{F}_p$ the prime field of $\mathbb{E}$. Show that $\mathbb{E}/\mathbb{F}_p$ is finite. Set $n = [\mathbb{E} : \mathbb{F}_p]$. Determine the number of elements in $\mathbb{E}$.

(5) Determine the degree of each field extension:
(i) $(\mathbb{Q}(\sqrt{2} + \sqrt{3}) : \mathbb{Q})$; (ii) $(\mathbb{Q}(\sqrt{2} + \sqrt{3}) : \mathbb{Q}(\sqrt{3}))$.

(6) Prove $\mathbb{Q}(\sqrt{2} + \sqrt{3}) = \mathbb{Q}(\sqrt{2}, \sqrt{3})$.

(7) Are the two fields $\mathbb{Q}(\sqrt{5})$ and $\mathbb{Q}(\sqrt{-5})$ isomorphic?

(8) Let $\alpha \in \mathbb{C}$ be a root of $x^3 - 3x^2 + 15x + 6$.

(i) Show that $[\mathbb{Q}(\alpha) : \mathbb{Q}] = 3$.
(ii) Write $\alpha^4, (\alpha-2)^{-1}, (\alpha^2 - \alpha + 1)^{-1}$ as the linear combinations of $1, \alpha, \alpha^2$ with coefficients in $\mathbb{Q}$.

(9) Let Π be a prime field and let $\mathbb{F}, \mathbb{E}$ be field extensions of Π. Show that $f(k\alpha) = kf(\alpha)$, for any $f \in \mathrm{Hom}\,(\mathbb{F}, \mathbb{E})$, $k \in \Pi$, $\alpha \in \mathbb{F}$.

(10) Let $\mathbb{K}/\mathbb{F}$ be a finite extension and let D be a subring of $\mathbb{E}$ containing $\mathbb{F}$. Show that D is a field.

(11) Show that $\mathbb{R}/\mathbb{Q}$ is infinite.

(12) Let $\mathbb{E}$ be the field of fractions of a domain R. For each automorphism σ of R, define $\tilde{\sigma}(ab^{-1}) = \sigma(a)(\sigma(b))^{-1}$. Show that $\tilde{\sigma}$ is an automorphism of $\mathbb{E}$.

(13) Show that $x^3 + x + 1 \in \mathbb{F}_2[x]$ is irreducible, and the field $\mathbb{E} = \mathbb{F}_2[x]/\langle x^3 + x + 1 \rangle$ has 8 elements. Determine the multiplication table for $\mathbb{E}^*$.

Exercises (hard)

(14) Show that $\mathrm{Aut}\,(\mathbb{R})$ is a trivial group.

(15) Let $\alpha \in \mathbb{C}$ be a root of an irreducible polynomial $f(x) \in \mathbb{Q}[x]$ of degree n.

    (i) Show that

$$\mathbb{Q}(\alpha) = \{a_0 + a_1\alpha + \cdots + a_{n-1}\alpha^{n-1} | a_0, a_1, \ldots, a_{n-1} \in \mathbb{Q}\}$$

       is a field, and $[\mathbb{Q}(\alpha) : \mathbb{Q}] = n$.

    (ii) Determine $\mathrm{Hom}\,(\mathbb{Q}(\alpha), \mathbb{C})$.

4.2   Algebraic extensions

The theory of field extension arises in the process of finding the (radical) solution of an algebraic equation. Let's first consider a fundamental problem: given a field extension $\mathbb{E}/\mathbb{F}$, which elements in $\mathbb{E}$ are the roots of some non-zero polynomials $f(x) \in \mathbb{F}[x]$? The key ingredient to solving this problem is still applying homomorphisms.

Let $\mathbb{E}/\mathbb{F}$ be a field extension. For any $\alpha \in \mathbb{E}$, consider the map

$$\varphi : \mathbb{F}[x] \to \mathbb{E}, \quad g(x) \mapsto g(\alpha). \tag{4.1}$$

It is straightforward to check that φ is not only a linear map between $\mathbb{F}$-vector spaces but also a ring homomorphism. It follows that $\mathrm{Ker}\,\varphi$ is an ideal of $\mathbb{F}[x]$, where $\mathrm{Ker}\,\varphi$ is the subset of $\mathbb{F}[x]$ consisting of all **annihilating polynomials** of α, i.e., polynomials having α as a root. By the Fundamental Theorem of Homomorphisms of rings, we have $\mathbb{F}[x]/\mathrm{Ker}\,\varphi \simeq \mathrm{Im}\,\varphi$. Since $\mathrm{Im}\,\varphi$ is a domain (as a subring of the field $\mathbb{E}$), $\mathrm{Ker}\,\varphi$ is a prime ideal of $\mathbb{F}[x]$ and must be of the form $\mathrm{Ker}\,\varphi = \langle f_\alpha(x) \rangle$, where $f_\alpha(x)$ is zero or a monic irreducible polynomial in $\mathbb{F}[x]$. If $f_\alpha(x) \neq 0$, then α is called an **algebraic number** over $\mathbb{F}$, and $f_\alpha(x)$ is called the **minimal polynomial** of α over $\mathbb{F}$, also denoted by $\mathrm{Irr}(\alpha, \mathbb{F})$. The minimal polynomial of α is the unique monic annihilating polynomial of α with the lowest degree, and its degree is called the **degree** of α over $\mathbb{F}$, denoted by $\deg(\alpha, \mathbb{F})$. If $f_\alpha(x) = 0$, then we say that α is **transcendental** over $\mathbb{F}$. In other words, an algebraic element over $\mathbb{F}$ is a root of some non-zero polynomial in $\mathbb{F}[x]$, and a transcendental element over $\mathbb{F}$ is not a root of any non-zero polynomial in $\mathbb{F}[x]$. By (4.1) we have a field homomorphism

$$\begin{cases} \varphi : \mathbb{F}(x) \to \mathbb{E}, & \text{if } \alpha \text{ is transcendental over } \mathbb{F}, \\ \varphi : \mathbb{F}[x]/\langle f_\alpha(x) \rangle \to \mathbb{E}, & \text{if } \alpha \text{ is algebraic over } \mathbb{F}. \end{cases}$$

It is easy to see that 2, $\sqrt{2}$, $\sqrt[3]{2}$, $\sqrt{-1}$ are all algebraic numbers over $\mathbb{Q}$. In 1874, Cantor proved that algebraic numbers over $\mathbb{Q}$ are countable (i.e., there is a one-to-one correspondence between the set of algebraic numbers over $\mathbb{Q}$ and $\mathbb{N}$), and there are more transcendental numbers than algebraic ones. But it is quite difficult to show which number is transcendental. In 1844, the French mathematician Liouville first proved that a class of numbers (now called Liouville numbers) are transcendental. Subsequently, Hermite proved the transcendence of e in 1873; Lindemann proved the transcendence of π in 1882. In 1900, Hilbert incorporated the problem of proving the transcendence of some numbers into one of his famous 23 problems: if α, β are algebraic with $\alpha \neq 0, 1$ and $\beta \notin \mathbb{Q}$, then α^β is transcendental. This problem was answered affirmatively by Aleksandr Gelfond and Theodor Schneider independently in 1934.

In any field extension $\mathbb{E}/\mathbb{F}$, algebraic elements over $\mathbb{F}$ always exist; for example, every element in $\mathbb{F}$ is algebraic. However, transcendental elements do not necessarily exist. In general, we have the following conclusion.

Lemma 4.2.1. *Let $\mathbb{E}/\mathbb{F}$ be a finite extension. Then any element in $\mathbb{E}$ is algebraic over $\mathbb{F}$.*

Proof. If $[\mathbb{E} : \mathbb{F}] = n$, then for any $\alpha \in \mathbb{E}$, $1, \alpha, \ldots, \alpha^n$ are linearly dependent, i.e., there exist $a_0, a_1, \ldots, a_n \in \mathbb{F}$, not all zero, such that $a_0 + a_1\alpha + \cdots + a_n\alpha^n = 0$. Therefore, α is a root of the non-zero polynomial $a_0 + a_1 x + \cdots + a_n x^n \in \mathbb{F}[x]$, hence it is algebraic over $\mathbb{F}$. $\square$

For convenience, we introduce the following.

Definition 4.2.2. Let $\mathbb{E}/\mathbb{F}$ be a field extension. If every element in $\mathbb{E}$ is algebraic over $\mathbb{F}$, then $\mathbb{E}/\mathbb{F}$ is called an **algebraic extension**, or $\mathbb{E}$ is algebraic over $\mathbb{F}$. Otherwise, $\mathbb{E}/\mathbb{F}$ is called a **transcendental extension**.

Finite extensions are algebraic. In particular, for any irreducible polynomial $f(x) \in \mathbb{F}[x]$, $\mathbb{F}[x]/\langle f(x) \rangle$ is an algebraic extension of $\mathbb{F}$;

while $\mathbb{R}/\mathbb{Q}$ and $\mathbb{C}/\mathbb{Q}$ are both transcendental extensions. In general, to check whether an extension is algebraic, it is not necessary to check that every element of the extension is algebraic. We need some effective criteria. The key point is that, for any extension $\mathbb{E}/\mathbb{F}$, the set of all algebraic elements is a subfield of $\mathbb{E}$. In other words, the sum, difference, product, and quotient (with a non-zero denominator) of two algebraic elements over $\mathbb{F}$ are still algebraic. First, we introduce a definition.

Definition 4.2.3. Let $\mathbb{E}/\mathbb{F}$ be a field extension and $S \subseteq \mathbb{E}$. The subfield of $\mathbb{E}/\mathbb{F}$ generated by S is the intersection of all subfields of $\mathbb{E}$ containing $\mathbb{F} \cup S$, or, in other words, the smallest subfield of $\mathbb{E}$ containing $\mathbb{F} \cup S$.

Let's look at the structure of $\mathbb{F}(S)$. Denote by $\mathbb{F}[S]$ the set of all finite sums of the following forms

$$\sum_{i_1,i_2,\ldots,i_n \geq 0} a_{i_1 i_2 \ldots i_n} \alpha_1^{i_1} \alpha_2^{i_2} \cdots \alpha_n^{i_n},$$

where $a_{i_1 i_2 \ldots i_n} \in \mathbb{F}$, $\alpha_j \in S$ for $j = 1, 2, \ldots, n$. It is easy to check that $\mathbb{F}[S]$ is a subring of $\mathbb{E}$. Thus, $\mathbb{F}[S]$ is a domain. We claim that $\mathbb{F}(S)$ is the fractional field of $\mathbb{F}[S]$. In fact, the fractional field of $\mathbb{F}[S]$ contains $\mathbb{F} \cup S$, and hence contains $\mathbb{F}[S]$. Moreover, any field containing $\mathbb{F} \cup S$ must contain $\mathbb{F}[S]$. Since the fractional field of a domain is the smallest field containing the domain, it must be a subfield of $\mathbb{F}(S)$. Thus, $\mathbb{F}(S)$ is exactly the fractional field of $\mathbb{F}[S]$. If $S = \{\alpha_1, \alpha_2, \ldots, \alpha_n\}$, we put $\mathbb{F}(S) = \mathbb{F}(\alpha_1, \alpha_2, \ldots, \alpha_n)$. In particular, if $S = \{\alpha\}$, then $\mathbb{F}(\alpha) = \left\{ \frac{f(\alpha)}{g(\alpha)} \middle| f(x), g(x) \in \mathbb{F}[x], g(\alpha) \neq 0 \right\}$, which is called a **simple extension**.

Proposition 4.2.4. *Let $\mathbb{E}/\mathbb{F}$ be a field extension, $S \subseteq \mathbb{E}$.*

(1) $\mathbb{F}(S) = \bigcup_{S' \subseteq S} \mathbb{F}(S')$, *where the sum is taken over all finite subsets S' of S.*
(2) *If $S = S_1 \cup S_2$, then $\mathbb{F}(S) = \mathbb{F}(S_1)(S_2) = \mathbb{F}(S_2)(S_1)$.*

Proof. (1) Since $\mathbb{F}(S') \subseteq \mathbb{F}(S)$ for any finite subset $S' \subseteq S$, $\bigcup_{S' \subseteq S} \mathbb{F}(S') \subseteq \mathbb{F}(S)$. Conversely, for any $a \in \mathbb{F}(S)$, there exist $f, g \in \mathbb{F}[S]$ with $g \neq 0$, such that $a = fg^{-1}$. Since the expressions of

f and g are in the form of finite sums, there is a finite subset S' of S such that $f, g \in \mathbb{F}[S']$; hence $a \in \mathbb{F}(S')$.

(2) Just need to prove that $\mathbb{F}(S) = \mathbb{F}(S_1)(S_2)$. Since the field $\mathbb{F}(S_1)(S_2)$ contains $\mathbb{F}$, S_1, S_2, and $\mathbb{F}(S)$ is the smallest subfield of $\mathbb{E}$ containing $\mathbb{F}$, S, we have $\mathbb{F}(S) \subseteq \mathbb{F}(S_1)(S_2)$.

On the other hand, since $\mathbb{F}(S_1)(S_2)$ is the smallest subfield of $\mathbb{E}$ containing $\mathbb{F}(S_1)$ and S_2, and the field $\mathbb{F}(S)$ contains both $\mathbb{F}(S_1)$ and S_2, we have $\mathbb{F}(S_1)(S_2) \subseteq \mathbb{F}(S_1 \cup S_2)$. Then (2) follows. □

By induction, one may quickly get the following conclusion.

Corollary 4.2.5. $\mathbb{F}(\alpha_1, \alpha_2, \ldots, \alpha_n) = \mathbb{F}(\alpha_1)(\alpha_2) \cdots (\alpha_n)$.

For any field extension $\mathbb{E}/\mathbb{F}$, $\mathbb{E}$ is always generated over $\mathbb{F}$ by some subset $S \subseteq \mathbb{E}$. If $\mathbb{E} = \mathbb{F}(S)$ is an algebraic extension, then every element in S is algebraic over $\mathbb{F}$. We will find that the converse also holds. We first consider the case when S is finite.

Theorem 4.2.6. *If* $\mathbb{E} = \mathbb{F}(\alpha_1, \alpha_2, \ldots, \alpha_n)$, *then the following assertions are equivalent*:

(1) $\mathbb{E}/\mathbb{F}$ *is an algebraic extension*;
(2) $\alpha_1, \alpha_2, \ldots, \alpha_n$ *are algebraic over* $\mathbb{F}$;
(3) $\mathbb{E}/\mathbb{F}$ *is a finite extension*.

Proof. It is easy to see that $(3) \Rightarrow (1) \Rightarrow (2)$. Noting that $\mathbb{F}(\alpha_1, \alpha_2, \ldots, \alpha_n) = \mathbb{F}(\alpha_1)(\alpha_2) \cdots (\alpha_n)$, by Theorem 4.1.13, we have

$$[\mathbb{E} : \mathbb{F}] = [\mathbb{F}(\alpha_1) : \mathbb{F}][\mathbb{F}(\alpha_1, \alpha_2) : \mathbb{F}(\alpha_1)] \cdots [\mathbb{E} : \mathbb{F}(\alpha_1, \alpha_2, \ldots, \alpha_{n-1})],$$

which implies that $[\mathbb{E} : \mathbb{F}]$ is finite. □

In particular, for algebraic elements α, β in the extension $\mathbb{E}/\mathbb{F}$, $\mathbb{F}(\alpha, \beta)$ is algebraic over $\mathbb{F}$. Thus, we have the following conclusion.

Corollary 4.2.7. *The sum, difference, product, and quotient (with a non-zero denominator) of two algebraic elements over* $\mathbb{F}$ *are still algebraic over* $\mathbb{F}$.

Problem 4.2.8. Let $\alpha, \beta \in \mathbb{E}$ be algebraic over $\mathbb{F}$ with minimal polynomials $\mathrm{Irr}(\alpha, \mathbb{F}), \mathrm{Irr}(\beta, \mathbb{F})$, respectively. How to determine the minimal polynomials or annihilating polynomials of $\alpha + \beta$, $\alpha\beta$?

By corollary 4.2.7 we have the following.

Corollary 4.2.9. *Let $\mathbb{E}/\mathbb{F}$ be a field extension, $S \subseteq \mathbb{E}$ with $\mathbb{E} = \mathbb{F}(S)$. If every element in S is algebraic over $\mathbb{F}$, then $\mathbb{E}/\mathbb{F}$ is an algebraic extension.*

Now, we are ready to prove the following important result in field theory, i.e., transitivity of algebraic extensions.

Theorem 4.2.10. *If $\mathbb{E}/\mathbb{K}, \mathbb{K}/\mathbb{F}$ are algebraic extensions, so is $\mathbb{E}/\mathbb{F}$.*

Proof. We need to show that any element $\alpha \in \mathbb{E}$ is algebraic over $\mathbb{F}$. Since α is algebraic over $\mathbb{K}$, there exist $a_1, a_2, \ldots, a_n \in \mathbb{K}$, not all zero, such that

$$a_1 + a_2\alpha + \cdots + a_n\alpha^{n-1} = 0.$$

Hence, α is algebraic over $\mathbb{F}(a_1, a_2, \ldots, a_n)$. Furthermore, $\mathbb{K}/\mathbb{F}$ is an algebraic extension, thus $\mathbb{F}(a_1, a_2, \ldots, a_n)/\mathbb{F}$ is a finite extension, which implies that $[\mathbb{F}(a_1, a_2, \ldots, a_n, \alpha) : \mathbb{F}] < \infty$. Therefore, α is algebraic over $\mathbb{F}$. □

For any extension $\mathbb{E}/\mathbb{F}$, denote by $\mathbb{E}_0$ the set of elements in $\mathbb{E}$ which are algebraic over $\mathbb{F}$. By Corollary 4.2.7, $\mathbb{E}_0$ is a field, called the **algebraic closure** of $\mathbb{F}$ in $\mathbb{E}$. By the transitivity of algebraic extensions, we have the following.

Theorem 4.2.11. *Let $\mathbb{E}$ be an extension of $\mathbb{F}$, and $\mathbb{E}_0$ the algebraic closure of $\mathbb{F}$ in $\mathbb{E}$. Then $\mathbb{E}_0$ is the largest algebraic extension of $\mathbb{F}$ in $\mathbb{E}$, in the sense that, for any $\delta \in \mathbb{E} \setminus \mathbb{E}_0$, δ is transcendental over $\mathbb{E}_0$.*

Example 4.2.12. Denote by $\overline{\mathbb{Q}}$ the algebraic closure of $\mathbb{Q}$ in $\mathbb{C}$. $\overline{\mathbb{Q}}$ is the union of all algebraic extensions of $\mathbb{Q}$ in $\mathbb{C}$, and it consists of all complex roots of every non-zero rational polynomial. Furthermore, every polynomial with coefficients in $\overline{\mathbb{Q}}$ splits in $\overline{\mathbb{Q}}$. The field $\overline{\mathbb{Q}}$ is a very crucial subject in number theory.

We call $\mathbb{E}/\mathbb{F}$ a **purely transcendental extension**, if any $\alpha \in \mathbb{E} \setminus \mathbb{F}$ is transcendental over $\mathbb{F}$. As the above theorem suggests, we may study the field extension $\mathbb{E}/\mathbb{F}$ in two steps: first consider the algebraic extension $\mathbb{E}_0/\mathbb{F}$, then consider the purely transcendental extension $\mathbb{E}/\mathbb{E}_0$. The most fundamental extensions are the simple ones: extensions generated by one element α over some field $\mathbb{F}$. If α is algebraic over $\mathbb{F}$, $\mathbb{F}(\alpha)$ is called a **simple algebraic extension** of $\mathbb{F}$; otherwise, $\mathbb{F}(\alpha)$ is called **simple transcendental extension** of $\mathbb{F}$. Applying the homomorphism in (4.1), we have the following.

Theorem 4.2.13.

(1) *Any simple algebraic extension $\mathbb{F}(\alpha)$ of $\mathbb{F}$ is isomorphic to $\mathbb{F}[x]/\langle f_\alpha(x) \rangle$.*
(2) *Any simple transcendental extension of $\mathbb{F}$ is isomorphic to $\mathbb{F}(x)$.*

One can see from this theorem that a simple transcendental extension of a field $\mathbb{F}$ is unique up to isomorphism, while simple algebraic extensions are much more complicated. In addition, a simple algebraic extension of $\mathbb{F}$ corresponds to a monic irreducible polynomial in $\mathbb{F}[x]$. It is worth noting that such a monic irreducible polynomial is not unique. For example, it is easy to check that $\mathbb{Q}(\sqrt{2}) = \mathbb{Q}(\sqrt{2}+1)$, while $\mathrm{Irr}(\sqrt{2}, \mathbb{Q}) = x^2 - 2$, $\mathrm{Irr}(\sqrt{2}+1, \mathbb{Q}) = x^2 - 2x - 1$.

Exercises

(1) For each $\alpha \in \mathbb{C}$, determine $\mathrm{Irr}(\alpha, \mathbb{Q})$:

    (i) $1 + \sqrt{-1}$;   (ii) $\sqrt{2} + \sqrt{-1}$;

    (iii) $\sqrt{1 + \sqrt[3]{2}}$;   (iv) $\sqrt{\sqrt[3]{2} - \sqrt{-1}}$.

(2) Determine the degree of the following extensions:

    (i) $[\mathbb{Q}(\sqrt{2}, \sqrt{-1}) : \mathbb{Q}]$;   (ii) $[\mathbb{Q}(\sqrt{2}, \sqrt[3]{2}) : \mathbb{Q}]$;

    (iii) $[\mathbb{Q}(\sqrt[3]{2}, \sqrt[3]{6}, \sqrt[3]{24}) : \mathbb{Q}]$;   (iv) $[\mathbb{Q}(\sqrt{2}, \sqrt{6}) : \mathbb{Q}(\sqrt{3})]$.

(3) Determine the minimal polynomials of $\sqrt{2} + \sqrt{3}$ over $\mathbb{Q}$, $\mathbb{Q}(\sqrt{2})$, $\mathbb{Q}(\sqrt{6})$, respectively.

(4) Let $\mathbb{E}/\mathbb{F}$ be a field extension of degree p, where p is a prime. Show that $\mathbb{E} = \mathbb{F}(\alpha)$, for any $\alpha \in \mathbb{E} \setminus \mathbb{F}$.

(5) Let $\mathbb{F}(\alpha)$ be a simple algebraic extension, $\deg(\alpha, \mathbb{F}) = n$. Show that $\mathbb{F}(\alpha)$ is an n-dimensional $\mathbb{F}$-vector space, and $1, \alpha, \ldots, \alpha^{n-1}$ form a basis of $\mathbb{F}(\alpha)$.

(6) Let n be a positive integer. Show that there exists an algebraic extension $\mathbb{K}/\mathbb{Q}$ of degree n.

(7) Let $\mathbb{E}$ be an extension of $\mathbb{F}$, and let $\alpha, \beta \in \mathbb{E}$ be algebraic over $\mathbb{F}$ such that $\deg(\alpha, \mathbb{F})$ and $\deg(\beta, \mathbb{F})$ are coprime.

Show that $\mathrm{Irr}(\alpha, \mathbb{F})$ is irreducible in $\mathbb{F}(\beta)[x]$, and $[\mathbb{F}(\alpha, \beta) : \mathbb{F}] = \deg(\alpha, \mathbb{F}) \deg(\beta, \mathbb{F})$.

(8) Let $\mathbb{E}$ be an extension of $\mathbb{F}$, and let $\alpha \in \mathbb{E}$ be algebraic over $\mathbb{F}$ such that $\deg(\alpha, \mathbb{F})$ is odd. Show that $\mathbb{F}(\alpha^2) = \mathbb{F}(\alpha)$.

(9) For $a, b \in \mathbb{C}$ with $a^2, b^2 \in \mathbb{Q}$ and $a^2 \neq b^2$, show that

$$Q(a, b) = Q(a + b).$$

(10) Let $\mathbb{E} = \mathbb{F}_p(t)$ be a simple transcendental extension of $\mathbb{F}_p$.

(i) Show that $x^p - t \in \mathbb{E}[x]$ is irreducible.
(ii) Let θ be a root of $x^p - t$. Write $x^p - t$ as the product of irreducible polynomials over $\mathbb{E}(\theta)$.

(11) Let $\mathbb{E}$ be an extension of $\mathbb{F}$, $u \in \mathbb{E}$, $\mathrm{Irr}(u, \mathbb{F}) = x^n - a$. For any divisor m of n, find the minimal polynomial of u^m over $\mathbb{F}$.

(12) Let $\mathbb{E}/\mathbb{F}$ be an algebraic extension, and let D be a subring of $\mathbb{E}$ containing $\mathbb{F}$. Show that D is a field. In particular, if $\alpha_1, \alpha_2, \ldots, \alpha_n$ is algebraic over $\mathbb{F}$, show that $\mathbb{F}[\alpha_1, \alpha_2, \ldots, \alpha_n] = \mathbb{F}(\alpha_1, \alpha_2, \ldots, \alpha_n)$.

(13) Let $\mathbb{F}(x)$ be a simple transcendental extension, $u \in \mathbb{F}(x) \setminus \mathbb{F}$. Show that x is algebraic over $\mathbb{F}(u)$.

Exercises (hard)

(14) Let $\mathbb{E}$ be an extension of $\mathbb{F}$.

(i) Show that $\alpha \in \mathbb{E}$ is algebraic over $\mathbb{F}$ if and only if α is an eigenvalue of some matrix over $\mathbb{F}$;
(ii) If $\alpha, \beta \in \mathbb{E}$ are algebraic over $\mathbb{F}$, then $\alpha \pm \beta, \alpha\beta, \alpha/\beta$ $(\beta \neq 0)$ are all eigenvalue of some matrices over $\mathbb{F}$, hence, they are algebraic over $\mathbb{F}$.

(15) Show that the simple transcendental extension $\mathbb{F}(x)$ of $\mathbb{F}$ is purely transcendental.

4.3 Straightedge and compass construction

Before going further, let's use what we learned in field theory to solve the famous Greek construction problems: trisecting any angle, squaring a circle, and duplicating a cube with only straightedge and compass. Those problems stood unsolved for more than two thousand years before Galois theory emerged.

In ancient Greece, Pythagoras and his followers believed that "all are (rational) numbers". However, Hippasus, one of Pythagoras' followers, discovered that the diagonal length of a unit square is not a ratio of two integers. This discovery caused great panic, which was the so-called first mathematical crisis. But the ancient Greeks found it easy to construct a unit square and the diagonal length with straightedge and compass.

As the term implies, the straightedge and compass construction refers to drawing pictures with only straightedge and compass. The straightedge is assumed to be infinite in length, and the compass can draw a circle with any given radius. The straightedge and compass construction is carried out on a plane, which can be identified with the complex number field $\mathbb{C}$ by the Cartesian coordinate system. In this way, any point on the plane corresponds to a unique coordinate (a, b) $(a, b \in \mathbb{R})$, which in turn corresponds to a complex number $a + b\sqrt{-1}$. A point in the plane is **constructible** if one can construct it with straightedge and compass in a finite number of steps, starting from a given unit length. A complex number is **constructible** if it corresponds to a constructible point. The natural question is, are all complex numbers constructible?

Theorem 4.3.1. *The set of all constructible numbers is a number field, i.e., if a, b are constructible, then so are $a + b$, $-a$, ab and a/b $(b \neq 0)$. Furthermore, if a is constructible, then so is $\sqrt{a}$.*

Notice that a complex number $x + y\sqrt{-1}$ is constructible if and only if $x, y \in \mathbb{R}$ are both constructible. We need to prove the theorem for the case of real constructible numbers, which reduces the problem to exercises in plane geometry. We leave it to the reader.

It is easy to see that all numbers in $\mathbb{Q}$, $\mathbb{Q}(\sqrt{-1})$ or $\mathbb{Q}(\sqrt{2})$ are constructible. What about other numbers, say, $\sqrt[3]{2}$, π? To answer this question, we must understand what a straightedge and compass can do. A straightedge can be used to draw a straight line determined by any two given points, while a compass can be used to draw a circle with any given point as the center and the distance between any two given points as the radius. The new points we get are the intersections of these lines and circles.

Suppose that all numbers in a number field $\mathbb{K}$ are constructible. We may assume $\sqrt{-1} \in \mathbb{K}$. Then $\mathbb{K} = \{a + b\sqrt{-1} \mid a, b \in \mathbb{F}\}$, where $\mathbb{F} = \mathbb{K} \cap \mathbb{R}$. To get new constructible points from $\mathbb{K}$, we may draw a line determined by two points in $\mathbb{K}$, and its equation is

$$ax + by + c = 0, \quad a, b, c \in \mathbb{F}.$$

Moreover, we may also make a circle whose equation is

$$(x - d)^2 + (y - e)^2 = r^2, \quad d, e, r \in \mathbb{F}.$$

The possible new constructible point is the intersection of some of these lines and circles. There are three cases.

(1) The intersection of two straight lines. Using Cramer's rule, we can see that the coordinates of the point of intersection are still in $\mathbb{F}$; that is, the point of intersection belongs to $\mathbb{K}$.

(2) The intersections of a straight line and a circle. The coordinates (x, y) of the point of intersection are the solution to the following equations

$$\begin{cases} ax + by + c = 0, \\ (x - d)^2 + (y - e)^2 = r^2. \end{cases}$$

Then x is the solution of a quadratic equation, which implies $[\mathbb{F}(x) : \mathbb{F}] \leq 2$. Noting that $y \in \mathbb{F}(x)$, we have $x + y\sqrt{-1} \in \mathbb{K}(x)$.

(3) The intersections of two circles. The coordinates of the point of the intersection are the solution to the equations

$$\begin{cases} (x - d_1)^2 + (y - e_1)^2 = r_1^2, \\ (x - d_2)^2 + (y - e_2)^2 = r_2^2. \end{cases}$$

The difference of the above two equations gives a linear equation, which reduces this case to case (2).

The above discussion shows that any new constructible number obtained from $\mathbb{K}$ by the intersection of a line and a circle lies in some quadratic extension of $\mathbb{K}$. Thus, we have the following theorem.

Theorem 4.3.2. *A complex number α is constructible if and only if there is a sequence of fields*

$$\mathbb{Q} = \mathbb{K}_0 \subset \mathbb{K}_1 \subset \cdots \subset \mathbb{K}_r,$$

such that $\alpha \in \mathbb{K}_r$ and $\mathbb{K}_i / \mathbb{K}_{i-1}$ ($i = 1, 2, \ldots, r$) are quadratic extensions.

Proof. We only need to prove the sufficiency. Assume that all numbers in $\mathbb{K}_i$ are constructible. For any $\alpha_i \in \mathbb{K}_{i+1} \setminus \mathbb{K}_i$, set $\mathrm{Irr}(\alpha_i, \mathbb{K}_i) = x^2 + b_i x + c_i$. Then $\alpha_i = \frac{-b_i \pm \sqrt{b_i^2 - 4c_i}}{2}$. By Theorem 4.3.1, α_i is constructible. Thus, any number including α in $\mathbb{K}_r$ is constructible by induction. $\qquad\square$

A simple but frequently used result is the following.

Corollary 4.3.3. *A constructible number z is an algebraic number over $\mathbb{Q}$, and $\mathrm{Irr}(z, \mathbb{Q})$ is a polynomial of degree 2^l for some $l \in \mathbb{N}$.*

Proof. Any constructible number is contained in some finite extension of $\mathbb{Q}$ and must be algebraic over $\mathbb{Q}$. If $\deg \mathrm{Irr}(z, \mathbb{Q}) = n$, then $[\mathbb{Q}(z) : \mathbb{Q}] = n$. By the above theorem, the field K_r containing z must contain $\mathbb{Q}(z)$. Since $[\mathbb{K}_r : \mathbb{Q}] = 2^r$, we have $n \mid 2^r$. $\qquad\square$

We can use the above discussion to give negative answers to the three ancient Greek problems.

Trisection of any angle. We will show that $60°$ can not be trisected. By the formula $\cos 3\theta = 4\cos^3 \theta - 3\cos \theta$, one can easily see that $\cos 20°$ is a root of the irreducible rational polynomial $4x^3 - 3x - \frac{1}{2}$. Then $\cos 20°$ is not constructible since it is contained in an extension of $\mathbb{Q}$ of degree 3. Of course, we may trisect some special angles, say, $90°$, $45°$, etc.

Duplication of the cube. This problem is essential to make a regular cube with the side length being $\sqrt[3]{2}$. However, $\sqrt[3]{2}$ is not constructible since its minimum polynomial is $x^3 - 2$.

Squaring the circle. It is to construct a square whose area is the same as that of the unit circle. The square's side length is $\sqrt{\pi}$, which is not constructible since it is transcendental over $\mathbb{Q}$.

The general problem of straightedge and compass construction is constructing a specified figure from a known one. Some particular points usually determine the known figure; for example, a circle is determined by its center and a point on it.

Given a planar point set $\mathcal{S} = \{p_1, p_2, \ldots, p_{n+2}\}$, $n \in \mathbb{N}$, we establish a Cartesian coordinate system with the point p_1 as the origin so that the coordinate of p_2 is $(1, 0)$. Thus, we get a set of numbers $\mathcal{S} = \{0, 1, z_1, \ldots, z_n\}$, where z_i are the complex numbers corresponding to p_{i+2}. The real and imaginary parts of these complex numbers, or $\bar{z}_1, \ldots, \bar{z}_n$, are given. Therefore, we start with the number field

$$\mathbb{F} = \mathbb{Q}(z_1, \ldots, z_n, \bar{z}_1, \ldots, \bar{z}_n).$$

Similar to the above discussion, we can quickly draw the following conclusion.

Theorem 4.3.4. *Let* $z_i \in \mathbb{C}$, $i = 1, \ldots, n$, $\mathbb{F} = \mathbb{Q}(z_1, \ldots, z_n, \bar{z}_1, \ldots, \bar{z}_n)$. *Then a complex number* z *is constructible from the set* $\mathcal{S} = \{0, 1, z_1, \ldots, z_n\}$ *if and only if there is a sequence of quadratic extensions*

$$\mathbb{F} = \mathbb{F}_1 \subset \mathbb{F}_2 \subset \cdots \subset \mathbb{F}_t = \mathbb{K}$$

such that $z \in \mathbb{K}$.

Exercises

(1) Show that the regular pentagon can be constructed with straight-edge and compass. How many ways can you find to realize the construction?

(2) Is regular 9-gon constructible with straightedge and compass?

(3) Prove Theorem 4.3.1.

(4) Let $n = pq$, where p and q are coprime positive integers. Show that if the regular p-gon and q-gon can be constructed with straightedge and compass, then so is the regular n-gon.

(5) Show that $\arccos \frac{6}{11}$ can not be trisected with straightedge and compass. What about $\arccos \frac{11}{16}$?

Exercises (hard)

(6) Show that a regular heptadecagon (17-gon) can be constructed with straightedge and compass.

4.4 Splitting fields

Now, we go back to the exploration of solving algebraic equations. We will gradually understand what the great mathematicians did in history: conceiving new mathematical concepts, finding new methods, and proposing and solving new problems. Since equations with rational coefficients may not have rational roots, irrational numbers and real numbers were introduced. Furthermore, equations with real coefficients may not have real roots, thus imaginary numbers and complex numbers were introduced. The fundamental theorem of algebra tells us that a non-constant polynomial with complex coefficients must have at least one complex root. More generally, we have the following result.

Lemma 4.4.1 (Kronecker). If $f(x) \in \mathbb{F}[x]$, $\deg f(x) > 0$, then there is a finite extension $\mathbb{E}/\mathbb{F}$ such that $f(x)$ has a root in $\mathbb{E}$.

Proof. Without loss of generality, we assume that $f(x)$ is irreducible. If there is an extension $\mathbb{E}/\mathbb{F}$ containing a root α of $f(x)$, then $\mathbb{F}[x]/\langle f(x) \rangle$ is isomorphic to the subfield $\mathbb{F}(\alpha)$ of $\mathbb{E}$ by the ring homomorphism (4.1). Hence we take $\mathbb{E} = \mathbb{F}[x]/\langle f(x) \rangle$ and $\alpha = x + \langle f(x) \rangle \in \mathbb{E}$. Then $\mathbb{E} = \mathbb{F}(\alpha)$ and $\mathrm{Irr}(\alpha, \mathbb{F}) = f(x)$. That is to say, α is a root of $f(x)$ in $\mathbb{E}$. Thus, we get the conclusion. $\square$

Furthermore, we should naturally consider the following problem.

Problem 4.4.2. Assume that $f(x)$ is a monic polynomial of degree $n > 0$. Is there an extension $\mathbb{E}/\mathbb{F}$ containing n roots of $f(x)$ such that $f(x)$ **splits** in $\mathbb{E}$, i.e., $f(x)$ can be factorized into the product of polynomials of degree 1:

$$f(x) = (x - \alpha_1)(x - \alpha_2) \cdots (x - \alpha_n), \quad \alpha_i \in \mathbb{E}, \quad i = 1, 2, \ldots, n?$$

Moreover, if such an extension exists, is it essentially unique?

If such an extension $\mathbb{E}$ exists, and $\mathbb{E} = \mathbb{F}(\alpha_1, \alpha_2, \ldots, \alpha_n)$, then $\mathbb{E}$ is called a **splitting field** of $f(x) \in \mathbb{F}[x]$. By the lemma above, we have the following.

Theorem 4.4.3. *Any polynomial $f(x) \in \mathbb{F}[x]$ of positive degree has a splitting field.*

Proof. We prove it by induction on $\deg f(x)$. If $\deg f(x) = 1$, then $\mathbb{F}$ itself is a splitting field of $f(x)$. We assume that the conclusion holds for every polynomial of degree $n - 1$. For $f(x)$ of degree n, there exists an extension $\mathbb{K} = \mathbb{F}(\alpha_1)$ containing a root α_1 of $f(x)$ by Lemma 4.4.1. Then we have the factorization

$$f(x) = (x - \alpha_1)f_1(x),$$

where $f_1(x) \in \mathbb{K}[x]$ and $\deg f_1(x) = n - 1$. By the induction hypothesis, $f_1(x)$ has a splitting field

$$\mathbb{E} = \mathbb{K}(\alpha_2, \ldots, \alpha_n) = \mathbb{F}(\alpha_1)(\alpha_2, \ldots, \alpha_n) = \mathbb{F}(\alpha_1, \alpha_2, \ldots, \alpha_n).$$

It is easily seen that $\mathbb{E}$ is a splitting field of $f(x)$. $\qquad\square$

The following corollaries follow from the theorem directly.

Corollary 4.4.4. *Let $\mathbb{E}$ be a splitting field of $f(x) \in \mathbb{F}[x]$, $\deg f(x) = n$. Then $[\mathbb{E} : \mathbb{F}] \leq n!$*

Corollary 4.4.5. *Let $\mathbb{E}$ be a splitting field of $f(x)$ over $\mathbb{F}$, $\mathbb{K}$ an intermediate field of $\mathbb{E}/\mathbb{F}$. Then $\mathbb{E}$ is also a splitting field of $f(x)$ over $\mathbb{K}$.*

Let's look at some examples.

Example 4.4.6. Let $f(x) = x^3 - 2 \in \mathbb{Q}[x]$, $\omega = \frac{-1+\sqrt{-3}}{2}$. Then the three complex roots of $f(x)$ are $\sqrt[3]{2}$, $\sqrt[3]{2}\omega$, and $\sqrt[3]{2}\omega^2$. Thus, $\mathbb{Q}(\sqrt[3]{2}, \sqrt[3]{2}\omega, \sqrt[3]{2}\omega^2) = \mathbb{Q}(\sqrt[3]{2}, \omega)$ is a splitting field of $f(x)$.

Example 4.4.7. Let $f(x) = x^p - 1 \in \mathbb{Q}[x]$, where p is a prime. The roots of $f(x)$ are $1, \omega, \ldots, \omega^{p-1}$, where $\omega = \cos\frac{2\pi}{p} + \sqrt{-1}\sin\frac{2\pi}{p}$ is a primitive pth root. Then $\mathbb{Q}(\omega)$ is a splitting field of $f(x)$. Noting that $x^{p-1} + x^{p-2} + \cdots + x + 1$ is irreducible over $\mathbb{Q}$ since p is prime, we conclude that $1, \omega, \ldots, \omega^{p-2}$ is a $\mathbb{Q}$-basis for $\mathbb{Q}(\omega)$. Hence, $[\mathbb{Q}(\alpha) : \mathbb{Q}] = p - 1$.

After solving the existence of splitting fields, we continue to deal with their uniqueness, that is, if there are two splitting fields $\mathbb{E}, \bar{\mathbb{E}}$ of $f(x) \in \mathbb{F}[x]$, is there a field isomorphism $\sigma : \mathbb{E} \to \bar{\mathbb{E}}$? The following process of constructing isomorphisms is crucial to our later-on research because this process can not only give the method to determine automorphism groups of fields but also help construct general field homomorphisms.

Let $\mathbb{E}, \bar{\mathbb{E}}$ be two splitting fields of $f(x) \in \mathbb{F}[x]$, where $\mathbb{E} = \mathbb{F}(\alpha_1, \alpha_2, \ldots, \alpha_n)$. To establish a field isomorphism between $\mathbb{E}$ and

$\overline{\mathbb{E}}$, it is natural to assume that $\sigma|_{\mathbb{F}} = \mathrm{id}_{\mathbb{F}}$, so that σ is a linear map between vector spaces. Thus, we need to define the images of the generators $\alpha_1, \alpha_2, \ldots, \alpha_n$, and then check whether σ is an isomorphism. Of course, it is not a good choice to consider the images of all these n elements simultaneously; the better way is to define the isomorphism step by step. We start with the image $\overline{\alpha}_1 = \sigma(\alpha_1)$. In other words, we consider the field homomorphism $\sigma_1 = \sigma|_{\mathbb{F}_1}$, where $\mathbb{F}_1 = \mathbb{F}(\alpha_1)$. Such σ_1, if exists, must be a field isomorphism between $\mathbb{F}_1$ and $\overline{\mathbb{F}}_1 = \mathbb{F}(\overline{\alpha}_1)$. Then we want to define the image $\overline{\alpha}_2$ of α_2, which is equivalent to define a homomorphism σ_2 from $\mathbb{F}_2 = \mathbb{F}_1(\alpha_2)$ to $\overline{\mathbb{E}}$ such that $\sigma_2|_{\mathbb{F}_1} = \sigma_1$. Obviously, σ_2 is an isomorphism between $\mathbb{F}_2$ and $\overline{\mathbb{F}}_2 = \overline{\mathbb{F}}_1(\overline{\alpha}_2)$. In this way, we will be able to define a homomorphism from $\mathbb{E}$ to $\overline{\mathbb{E}}$, as shown in the following diagram:

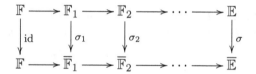

The following problem is crucial in the above procedure.

Problem 4.4.8. How to extend a homomorphism σ_i from $\mathbb{F}_i$ to $\overline{\mathbb{E}}$ (or $\overline{\mathbb{F}}_i$) to a homomorphism σ_{i+1} from $\mathbb{F}_{i+1} = \mathbb{F}_i(\alpha_{i+1})$ to $\overline{\mathbb{E}}$ (or $\overline{\mathbb{F}}_{i+1}$)? Here, $\mathbb{F}_0 = \mathbb{F}$, and $\sigma_{i+1}|_{\mathbb{F}_i} = \sigma_i$.

Before going further, we need some preparation first. In Section 2.9, we studied some properties of polynomials over rings, and now we will focus on polynomials over fields. Let $\varphi : \mathbb{K} \to \overline{\mathbb{K}}$ be a field isomorphism. For any $a \in \mathbb{K}$, set $\varphi(a) = \overline{a}$. Then for each $f(x) = \sum_{i=0}^{n} a_i x^i \in \mathbb{K}[x]$, define

$$\varphi(f(x)) = \overline{f}(x) = \sum_{i=0}^{n} \overline{a}_i x^i.$$

As we know, φ is a ring isomorphism. In addition, it is easy to verify the following lemma.

Lemma 4.4.9. *A polynomial $f(x) \in \mathbb{K}[x]$ is irreducible if and only if $\overline{f}(x) \in \overline{\mathbb{K}}[x]$ is irreducible.*

Furthermore, we have the following result.

Lemma 4.4.10. *Suppose $\varphi : \mathbb{K} \to \overline{\mathbb{K}}$ is a field isomorphism, $p(x)$ is a monic irreducible polynomial in $\mathbb{K}[x]$, and $\overline{p}(x) = \varphi(p(x)) \in \overline{\mathbb{K}}[x]$. Let α be a root of $p(x)$ in some extension $\mathbb{E}$ of $\mathbb{K}$, $\overline{\alpha}$ a root of $\overline{p}(x)$ in some extension of $\overline{\mathbb{K}}$.*

(1) *There exists a field isomorphism $\eta : \mathbb{K}[x]/\langle p(x) \rangle \to \overline{\mathbb{K}}[x]/\langle \overline{p}(x) \rangle$ such that $\eta|_{\mathbb{K}} = \varphi$.*
(2) *There exists a field isomorphism $\sigma : \mathbb{K}(\alpha) \to \overline{\mathbb{K}}(\overline{\alpha})$ such that $\sigma|_{\mathbb{K}} = \varphi$ and $\sigma(\alpha) = \overline{\alpha}$.*
(3) *If $\psi : \mathbb{K}(\alpha) \to \overline{\mathbb{K}}(\overline{\alpha})$ is a field homomorphism such that $\psi|_{\mathbb{K}} = \varphi$, then $\psi(\alpha)$ is a root of $\overline{p}(x)$.*

Proof. (1) Let $\overline{\pi} : \overline{\mathbb{K}}[x] \to \overline{\mathbb{K}}[x]/\langle \overline{p}(x) \rangle$ be the natural homomorphism. Then $\overline{\pi} \circ \varphi$ is a homomorphism from $\mathbb{K}[x]$ onto $\overline{\mathbb{K}}[x]/\langle \overline{p}(x) \rangle$ with $\langle p(x) \rangle$ being the kernel. By the Fundamental Theorem of Homomorphisms of rings, we get an isomorphism η with $\eta|_{\mathbb{K}} = \varphi$.

(2) The evaluation map $\nu_\alpha : \mathbb{K}[x] \to \mathbb{E}$, $f(x) \mapsto f(\alpha)$, induces an isomorphism $\varphi : \mathbb{K}[x]/\langle p(x) \rangle \to \mathbb{K}(\alpha)$ with $\varphi|_{\mathbb{K}} = \mathrm{id}_{\mathbb{K}}$. Similarly, there exists a canonical $\overline{\mathbb{K}}$-isomorphism from $\overline{\mathbb{K}}[x]/\langle \overline{p}(x) \rangle$ to $\overline{\mathbb{K}}(\overline{\alpha})$. Then the assertion follows from (1).

(3) Write $p(x) = x^k + a_{k-1}x^{k-1} + \cdots + a_0$. Since $p(\alpha) = 0$, we have

$$0 = \psi(p(\alpha)) = \psi(\alpha)^k + \psi(a_{k-1})\psi(\alpha)^{k-1} + \cdots + \psi(a_0) = \overline{p}(\psi(\alpha)),$$

which implies that $\psi(\alpha)$ is a root of $\overline{p}(x)$. $\square$

Using the above lemma, we get the following proposition quickly.

Proposition 4.4.11. *Let $\sigma : \mathbb{K} \to \overline{\mathbb{K}}$ be a field isomorphism, and let $\mathbb{K}(\alpha)/\mathbb{K}$ be a simple algebraic extension. Set $p(x) = \mathrm{Irr}(\alpha, \mathbb{K})$, $\overline{p}(x) = \sigma(p(x))$. If $\overline{\mathbb{E}}$ be an extension of $\overline{\mathbb{K}}$ such that $\overline{p}(x)$ splits in $\overline{\mathbb{E}}$, then there exists a field homomorphism $\varphi : \mathbb{K}(\alpha) \to \overline{\mathbb{E}}$ with $\varphi|_{\mathbb{K}} = \sigma$. The number of such isomorphisms is less than or equal to $[\mathbb{K}(\alpha) : \mathbb{K}]$, and the equality holds if and only if $p(x)$ has no multiple roots in $\mathbb{E}$.*

Proof. By Lemma 4.4.10, the field homomorphism φ is uniquely determined by $\varphi(\alpha)$, and $\varphi(\alpha)$ must be a root of $\overline{p}(x)$. By the above lemma, any root $\overline{\alpha}$ of $\overline{p}(x)$ determines a unique field homomorphism

φ such that $\varphi(\alpha) = \overline{\alpha}$. It follows that the number of isomorphisms is equal to that of different roots of $\overline{p}(x)$ (or $p(x)$). The assertion follows. □

Now, we are ready to discuss the uniqueness of splitting fields.

Theorem 4.4.12. *Suppose $f(x) \in \mathbb{F}[x]$ has no multiple irreducible factors and $\mathbb{E}$, $\overline{\mathbb{E}}$ are two splitting fields of $f(x)$. Then there exists a field isomorphism $\sigma : \mathbb{E} \to \overline{\mathbb{E}}$ such that $\sigma|_{\mathbb{F}} = \mathrm{id}_{\mathbb{F}}$. Furthermore, the number of such isomorphisms is $\leq [\mathbb{E} : \mathbb{F}]$, and the equality holds if and only if $f(x)$ has no multiple roots in $\mathbb{E}$.*

Proof.   Assume that $\mathbb{E} = \mathbb{F}(\alpha_1, \alpha_2, \ldots, \alpha_n)$, where $\alpha_1, \alpha_2, \ldots, \alpha_n$ are all roots of $f(x)$ in $\mathbb{E}$. Set $\mathbb{F}_0 = \mathbb{F}$, $\mathbb{F}_i = \mathbb{F}(\alpha_1, \alpha_2, \ldots, \alpha_i) = \mathbb{F}_{i-1}(\alpha_i)$, $i = 1, 2, \ldots, n$. Let $\sigma_0 : \mathbb{F} \to \overline{\mathbb{E}}$ be the natural homomorphism. Then, by the above lemma, we may find a sequence of field homomorphisms $\sigma_i : \mathbb{F}_i \to \overline{\mathbb{E}}$ such that $\sigma_i|_{\mathbb{F}_{i-1}} = \sigma_{i-1}$, $i = 1, 2, \ldots, n$. Thus, $\sigma = \sigma_n$ is a field homomorphism between $\mathbb{E}$ and $\overline{\mathbb{E}}$. Therefore, we have the factorization in $\overline{\mathbb{E}}$:

$$f(x) = (x - \sigma(\alpha_1))(x - \sigma(\alpha_2)) \cdots (x - \sigma(\alpha_n)).$$

Since $\overline{\mathbb{E}}$ is also a splitting field of $f(x)$, we have $\overline{\mathbb{E}} = \mathbb{F}(\sigma(\alpha_1), \ldots, \sigma(\alpha_n))$, which implies that σ is an epimorphism, hence an isomorphism.

Furthermore, for each σ_i, there are at most $[\mathbb{F}_i : \mathbb{F}_{i-1}]$ choices. Therefore, the number of the field isomorphisms σ is at most $[\mathbb{F}_1 : \mathbb{F}_0] \ldots [\mathbb{F}_n : \mathbb{F}_{n-1}] = [\mathbb{F}_n : \mathbb{F}_0] = [\mathbb{E} : \mathbb{F}]$. If each irreducible factor of $f(x)$ has no multiple roots in $\mathbb{E}$, then there are exactly $[\mathbb{F}_i : \mathbb{F}_{i-1}]$ choices for σ_i; otherwise, the choices of σ_i is less than $[\mathbb{F}_i : \mathbb{F}_{i-1}]$, which implies that the choice of σ is less than $[\mathbb{E} : \mathbb{F}]$. □

It is worth noting that the isomorphism σ constructed in the proof satisfies $\sigma_{\mathbb{F}} = \mathrm{id}_{\mathbb{F}}$. Therefore, σ is also a linear map between $\mathbb{F}$-vector spaces, which allows us to apply the theory of vector spaces and linear transformations. For convenience, we introduce the following definition.

Definition 4.4.13. Let $\mathbb{E}, \overline{\mathbb{E}}$ be extensions of $\mathbb{F}$. A field homomorphism (or isomorphism) $\varphi : \mathbb{E} \to \overline{\mathbb{E}}$ is called an $\mathbb{F}$-**homomorphism** (or $\mathbb{F}$-**isomorphism**) if $\varphi|_{\mathbb{F}} = \mathrm{id}_{\mathbb{F}}$. Denote by $\mathrm{Hom}_{\mathbb{F}}(\mathbb{E}, \overline{\mathbb{E}})$ the set of all $\mathbb{F}$-homomorphisms. If φ is an isomorphism, then we say that $\mathbb{E}, \overline{\mathbb{E}}$ are **equivalent extensions** over $\mathbb{F}$.

The isomorphism we constructed in the proof of Theorem 4.2.13 is an $\mathbb{F}$-isomorphism. $\mathbb{F}$-homomorphisms or isomorphisms are very common in field theory. When considering different extensions over $\mathbb{F}$, we need to find the relationship of those extensions. The natural way is to study homomorphisms between field extensions; such homomorphisms should be $\mathbb{F}$-homomorphisms.

In addition, we also estimate the number of such isomorphisms. We will soon find that this plays a crucial role in the study of splitting fields, determining whether a polynomial $f(x)$ is solvable by radicals. By Theorem 4.4.12, we should pay attention to whether the polynomial has multiple roots. Therefore, we introduce the following definition.

Definition 4.4.14. Let $\mathbb{F}$ be a field. If an irreducible polynomial $p(x) \in \mathbb{F}[x]$ has no multiple roots in any splitting field, then we say that $p(x)$ is a **separable polynomial**. If every irreducible factor of $f(x) \in \mathbb{F}[x]$ is separable, then $f(x)$ is said to be **separable**; otherwise, $f(x)$ is called **inseparable**.

Separable and inseparable polynomials will be discussed in Section 4.8.

Exercises

(1) Determine a splitting field $\mathbb{F}/\mathbb{Q}$ of the following polynomials, and determine the number of the automorphism of $\mathbb{F}/\mathbb{Q}$:

(i) $x^2 + 3$; (ii) $x^5 - 1$; (iii) $x^3 - 2$; (iv) $(x^2 - 2)(x^3 - 2)$; (v) $x^5 - 3$.

(2) Let p be a prime, and $\mathbb{F}_p(\alpha)$ a simple transcendental extension of $\mathbb{F}_p$. Determine a splitting field of $x^p - \alpha \in \mathbb{F}_p(\alpha)[x]$.

(3) Let $\mathbb{F}$ be a finite field. Show that there exists an algebraic extension $\mathbb{E}/\mathbb{F}$ such that $\mathbb{E} \neq \mathbb{F}$.

(4) Let $\mathbb{F}$ be a field of characteristic $\neq 2$. Show that every quadratic extension of $\mathbb{F}$ has the form $\mathbb{F}(\alpha)$, where $\alpha^2 \in \mathbb{F}$. What about the conclusion for Ch $\mathbb{F} = 2$?

(5) Let $\mathbb{F}$ be a field of characteristic $p \neq 0$, $c \in \mathbb{F}$. Show that $x^p - c$ is irreducible over $\mathbb{F}$ if and only if $x^p - c$ has no roots in $\mathbb{F}$.

Exercises (hard)

(6) Let $\mathbb{F}$ be of characteristic $p > 0$, $c \in \mathbb{F}$. Show that $x^p - x - c$ is irreducible over $\mathbb{F}$ if and only if $x^p - x - c$ has no roots in $\mathbb{F}$. Does the conclusion still hold for Ch $\mathbb{F} = 0$?

(7) Let $\mathbb{E}$ be a splitting field of $f(x) \in \mathbb{F}[x]$, $\deg f(x) = n$. Show that $[\mathbb{E} : \mathbb{F}] \mid n!$.

4.5   Galois groups

In the previous section, we proved that a splitting field of a polynomial $f(x) \in \mathbb{F}[x]$ is unique up to isomorphism. The structure of the splitting field will help us understand the roots of $f(x)$. The critical ingredient to studying splitting fields is the set of $\mathbb{F}$-homomorphisms. In general, we have the following definition.

Definition 4.5.1. Let $\mathbb{E}/\mathbb{F}$ be an extension. An $\mathbb{F}$-**endomorphism** or $\mathbb{F}$-**automorphism** of $\mathbb{E}/\mathbb{F}$ is an $\mathbb{F}$-homomorphism or $\mathbb{F}$-isomorphism from $\mathbb{E}$ to itself. The set of all $\mathbb{F}$-endomorphisms of $\mathbb{E}$ is a monoid, denoted by $\operatorname{Hom}_{\mathbb{F}}(\mathbb{E}, \mathbb{E})$. The set of all $\mathbb{F}$-automorphisms of $\mathbb{E}$ forms a group, which is called the **Galois group** of $\mathbb{E}/\mathbb{F}$, denoted by $\operatorname{Gal}(\mathbb{E}/\mathbb{F})$.

We know that an $\mathbb{F}$-endomorphism is always injective, but it is not necessarily an $\mathbb{F}$-automorphism. However, in many cases we are interested in, an $\mathbb{F}$-endomorphism is indeed an $\mathbb{F}$-automorphism. Let's first prove the following lemma.

Lemma 4.5.2. *Let $\mathbb{E}/\mathbb{F}$ be an extension, $\varphi \in \operatorname{Gal}(\mathbb{E}/\mathbb{F})$, and $f(x) \in \mathbb{F}[x]$. Let R be set of all roots of $f(x)$ in $\mathbb{E}$. Then φ is a permutation of R. In particular, if $\alpha \in \mathbb{E}$ is algebraic over $\mathbb{F}$, then so is $\varphi(\alpha)$.*

Proof.   Let $f(x) = x^n + a_{n-1}x^{n-1} + \cdots + a_0$, $\alpha \in R$. Then $f(\alpha) = 0$, and $\varphi(a_i) = a_i$. Thus,

$$\varphi(f(\alpha)) = (\varphi(\alpha))^n + a_{n-1}(\varphi(\alpha))^{n-1} + \cdots + a_0 = f(\varphi(\alpha)) = 0,$$

which implies that $\varphi(\alpha)$ is also a root of $f(x)$. Therefore, $\varphi(R) \subseteq R$. Since φ is injective and R is finite, $\varphi|_R$ is invertible, hence a permutation of R.

Since any algebraic element $\alpha \in \mathbb{E}$ over $\mathbb{F}$ is a root of its minimal polynomial $\mathrm{Irr}(\alpha, \mathbb{F})$, $\sigma(\alpha)$ is also a root of $\mathrm{Irr}(\alpha, \mathbb{F})$, hence algebraic.

$\square$

Definition 4.5.3. Let $\mathbb{E}/\mathbb{F}$ be a field extension. Two algebraic elements $\alpha, \beta \in \mathbb{E}/\mathbb{F}$ are called **conjugate**, if α, β are the roots of the same irreducible polynomial over $\mathbb{F}$, i.e., $\mathrm{Irr}(\alpha, \mathbb{F}) = \mathrm{Irr}(\beta, \mathbb{F})$.

Lemma 4.5.4.

(1) *If $\mathbb{E}/\mathbb{F}$ is a finite extension, then $\mathrm{Hom}_{\mathbb{F}}(\mathbb{E}, \mathbb{E}) = \mathrm{Gal}\,(\mathbb{E}/\mathbb{F})$.*
(2) *If $\mathbb{E}/\mathbb{F}$ is an algebraic extension, then $\mathrm{Hom}_{\mathbb{F}}(\mathbb{E}, \mathbb{E}) = \mathrm{Gal}\,(\mathbb{E}/\mathbb{F})$.*

Proof. (1) Let $\varphi \in \mathrm{Hom}_{\mathbb{F}}(\mathbb{E}, \mathbb{E})$. Then φ is an $\mathbb{F}$-linear transformation in $\mathbb{E}$. If $\mathbb{E}/\mathbb{F}$ is a finite extension, then $\mathbb{E}$ is a finite-dimensional $\mathbb{F}$-vector space. Since φ is injective, it must be a linear isomorphism. Thus, $\mathrm{Hom}_{\mathbb{F}}(\mathbb{E}, \mathbb{E}) = \mathrm{Gal}\,(\mathbb{E}/\mathbb{F})$.

(2) Let $\mathbb{E}/\mathbb{F}$ be an algebraic extension. For any $\alpha \in \mathbb{E}$, assume that $\alpha_1 = \alpha, \alpha_2, \ldots, \alpha_n$ are all roots in $\mathbb{E}$ of the minimal polynomial $f_\alpha(x)$ of α. Set $\mathbb{K} = \mathbb{F}(\alpha_1, \alpha_2, \ldots, \alpha_n)$. By Lemma 4.5.2, φ is a permutation of these α_i, and $\mathbb{K}$ is invariant under φ. Therefore, $\varphi|_{\mathbb{K}}$ is an automorphism, which implies that $\alpha \in \mathrm{Im}\,(\varphi)$, i.e., φ is surjective, hence an isomorphism. $\square$

By Corollary 4.1.10, if $\mathbb{F}$ is the prime field of $\mathbb{E}$, $\mathrm{Gal}\,(\mathbb{E}/\mathbb{F})$ is exactly $\mathrm{Aut}\,(\mathbb{E})$. Therefore, $\mathrm{Aut}\,(\mathbb{E})$ is a special Galois group. We will see that Galois groups play a crucial role. For any $\varphi \in \mathrm{Gal}\,(\mathbb{E}/\mathbb{F})$, define

$$\mathbb{K} = \mathbb{E}^\varphi = \{\alpha \in \mathbb{E} \mid \varphi(\alpha) = \alpha\}.$$

Obviously, $\mathbb{F} \subseteq \mathbb{K}$. For any $x, y \in \mathbb{K}$, $\varphi(x + y) = \varphi(x) + \varphi(y) = x + y$, which implies that $x + y \in \mathbb{K}$, and similarly we have $x - y, xy \in \mathbb{K}$ and if $y \neq 0$, $x/y \in \mathbb{K}$. Therefore, $\mathbb{K}$ is an intermediate field of $\mathbb{E}/\mathbb{F}$, called the **invariant subfield** of φ. Furthermore, $\mathbb{K}$ is not only φ-fixed, but is φ^m-fixed, for any $m \in \mathbb{Z}$. That is to say, $\mathbb{K}$ is the fixed subfield of the subgroup generated by φ. In general, let G be any subgroup of $\mathrm{Gal}\,(\mathbb{E}/\mathbb{F})$. Then

$$\mathbb{K} = \mathbb{E}^G = \{\alpha \in \mathbb{E} \mid \varphi(\alpha) = \alpha, \forall \varphi \in G\}$$

is an intermediate field of $\mathbb{E}/\mathbb{F}$. Hence, we get a relationship between subgroups of $\mathrm{Gal}\,(\mathbb{E}/\mathbb{F})$ and intermediate fields of $\mathbb{E}/\mathbb{F}$. Is this relationship a 1-1 correspondence? To answer this question, we need the following **Dedekind independence theorem.**

Theorem 4.5.5. *Let $\mathbb{E}/\mathbb{F}$ be an extension, and let $\sigma_1, \sigma_2, \ldots, \sigma_n \in$ $\mathrm{Gal}\,(\mathbb{E}/\mathbb{F})$ be different. Then $\sigma_1, \sigma_2, \ldots, \sigma_n$ are $\mathbb{E}$-linearly independent as $\mathbb{F}$-linear transformations, i.e., if $x_1, x_2, \ldots, x_n \in \mathbb{E}$ satisfy $x_1\sigma_1 + x_2\sigma_2 + \cdots + x_n\sigma_n = 0$, then $x_1 = x_2 = \cdots = x_n = 0$.*

Proof. We prove it by induction. The assertion is trivial for $n = 1$. Assume that the assertion holds for $n = k - 1$, i.e, any $k - 1$ different elements in $\mathrm{Gal}\,(\mathbb{E}/\mathbb{F})$ are $\mathbb{E}$-linearly independent. For $n = k$, if $\sigma_1, \sigma_2, \ldots, \sigma_k$ are $\mathbb{E}$-linearly dependent, then there exists $a_1, a_2, \ldots, a_k \in \mathbb{E}$ such that

$$a_1\sigma_1 + a_2\sigma_2 + \cdots + a_k\sigma_k = 0, \qquad (4.2)$$

where we must have $a_i \neq 0$ for any i, otherwise if $a_i = 0$, then $\sigma_1, \ldots, \sigma_{i-1}, \sigma_{i+1}, \ldots, \sigma_k$ are linearly independent. It contradicts the induction hypothesis. By multiplying a_k^{-1} in both sides of (4.2), we may assume that $a_k = 1$. Since $\sigma_1 \neq \sigma_k$, there exists $a \in \mathbb{E}$ such that $\sigma_1(a) \neq \sigma_k(a)$. Therefore, we have

$$a_1\sigma_1(x) + \cdots + a_{k-1}\sigma_{k-1}(x) + \sigma_k(x) = 0, \quad \forall x \in \mathbb{E}, \qquad (4.3)$$

and $\quad a_1\sigma_1(ax) + \cdots + \sigma_{k-1}(ax) + \sigma_k(ax) = 0$, i.e.,

$$a_1\sigma_1(a)\sigma_1(x) + \cdots + a_{k-1}\sigma_{k-1}(a)\sigma_{k-1}(x) + \sigma_k(a)\sigma_k(x) = 0. \quad (4.4)$$

Multiplying (4.3) by $\sigma_k(a)$ and subtracting (4.4), we get

$$a_1(\sigma_k(a) - \sigma_1(a))\sigma_1(x) + \cdots + a_{k-1}(\sigma_k(a) - \sigma_{k-1}(a))\sigma_{k-1}(x) = 0.$$

Since $a_1 \neq 0$ and $\sigma_k(a) - \sigma_1(a) \neq 0$, we have that $\sigma_1, \ldots, \sigma_{k-1}$ are $\mathbb{E}$-linearly dependent, which contradicts the induction hypothesis. The assertion follows. $\qquad \square$

By Dedekind independence theorem, we get the following.

Theorem 4.5.6. *Let* $\sigma_1, \sigma_2, \ldots, \sigma_n$ *be different elements in* $\mathrm{Gal}\,(\mathbb{E}/\mathbb{F})$. *Then* $[\mathbb{E} : \mathbb{F}] \geq n$.

Proof. Let $\alpha_1, \sigma_2, \ldots, \alpha_r$ be an $\mathbb{F}$-basis of $\mathbb{E}$, where $r = [\mathbb{E} : \mathbb{F}] < n$. Consider the following linear equations

$$\begin{cases} \sigma_1(\alpha_1)x_1 + \cdots + \sigma_n(\alpha_1)x_n = 0, \\ \ldots\ldots \\ \sigma_1(\alpha_r)x_1 + \cdots + \sigma_n(\alpha_r)x_n = 0. \end{cases} \tag{4.5}$$

Since $r < n$, the linear equations have a non-zero solution, say, $x_1, x_2, \ldots, x_n$. For any $\alpha \in \mathbb{E}$, there exist $a_1, a_2, \ldots, a_r \in \mathbb{F}$ such that $\alpha = a_1\alpha_1 + a_2\alpha_2 + \cdots + a_r\alpha_r$. Multiplying the ith equation in 4.5 by a_i, and then adding them together, we have

$$\sum_{i=1}^{r} a_i\sigma_1(\alpha_i)x_1 + \cdots + \sum_{i=1}^{r} a_i\sigma_n(\alpha_i)x_n = 0.$$

Thus,

$$x_1\sigma_1(\alpha) + \cdots + x_n\sigma_n(\alpha) = 0, \quad \forall \alpha \in \mathbb{E},$$

which implies that $\sigma_1, \sigma_2, \ldots, \sigma_n$ are linearly dependent, a contradiction. Hence, $[\mathbb{E} : \mathbb{F}] \geq n$. □

The theorem tells us that, for a finite extension $\mathbb{E}/\mathbb{F}$, the order of the Galois group $\mathrm{Gal}\,(\mathbb{E}/\mathbb{F})$ is not great than $[\mathbb{E} : \mathbb{F}]$. We have found some examples when the equality does not hold. In what conditions does the equality hold? Noticing that the fixed subfield $\mathbb{K}$ of $\mathrm{Gal}\,(\mathbb{E}/\mathbb{F})$ naturally contains $\mathbb{F}$, we have $\mathrm{Gal}\,(\mathbb{E}/\mathbb{F}) = \mathrm{Gal}\,(\mathbb{E}/\mathbb{K})$. Furthermore, we have the following.

Theorem 4.5.7. *Let* $\mathbb{E}$ *be a field, and let* G *be a finite subgroup of* $\mathrm{Aut}\,(\mathbb{E})$, $\mathbb{F} = \mathbb{E}^G$. *Then* $|G| = [\mathbb{E} : \mathbb{F}]$, *and* $G = \mathrm{Gal}\,(\mathbb{E}/\mathbb{F})$.

Proof. It is easy to see that G is a subgroup of $\mathrm{Gal}\,(\mathbb{E}/\mathbb{F})$. Thus, $|G| \leq |\mathrm{Gal}\,(\mathbb{E}/\mathbb{F})| \leq [\mathbb{E} : \mathbb{F}]$. We just need to prove $|G| \geq [\mathbb{E} : \mathbb{F}]$.

Assume that $G = \{\sigma_1 = \mathrm{id}, \sigma_2, \ldots, \sigma_n\}$. For any $\alpha_1, \alpha_2, \ldots, \alpha_{n+1} \in \mathbb{E}$, the linear equations

$$\begin{cases} \sigma_1(\alpha_1)x_1 + \cdots + \sigma_1(\alpha_{n+1})x_{n+1} = 0, \\ \cdots\cdots \\ \sigma_n(\alpha_1)x_1 + \cdots + \sigma_n(\alpha_{n+1})x_{n+1} = 0 \end{cases} \tag{4.6}$$

must have a non-zero solution. Assume that $(b_1, b_2, \ldots, b_{n+1})$ is one of the non-trivial solutions with as many zeros as possible. (*Note*: from the first equation we have $b_1\alpha_1 + b_2\alpha_2 + \cdots + b_{n+1}\alpha_{n+1} = 0$, which does not imply that $\alpha_1, \alpha_2, \ldots, \alpha_{n+1}$ are linearly dependent! Why?). Assume without loss of generality that $b_1 \neq 0$. Since $\left(1, \frac{b_2}{b_1}, \ldots, \frac{b_{n+1}}{b_1}\right)$ is also a solution to the linear equations, we may assume that $b_1 = 1$. Then for any $i = 1, 2, \ldots, n$,

$$\sigma_i(\alpha_1) + \sigma_i(\alpha_2)b_2 + \cdots + \sigma_i(\alpha_{n+1})b_{n+1} = 0. \tag{4.7}$$

If every b_i is in $\mathbb{F}$, we've done. Without loss of generality, assume that b_2 is not in $\mathbb{F}$. Then there exists $\sigma \in G$ such that $\sigma(b_2) \neq b_2$. Applying the action of σ on (4.7), we have

$$(\sigma\sigma_i)(\alpha_1) + (\sigma\sigma_i)(\alpha_2)\sigma(b_2) + \cdots + (\sigma\sigma_i)(\alpha_{n+1})\sigma(b_{n+1}) = 0.$$

Note that $\sigma\sigma_1, \sigma\sigma_2, \ldots, \sigma\sigma_n$ is a permutation of all elements in G. Thus, $(1, \sigma(b_2), \ldots, \sigma(b_{n+1}))$ is also a solution to (4.6). Therefore, $(0, b_2 - \sigma(b_2), \ldots, b_{n+1} - \sigma(b_{n+1}))$ is a non-trivial solution with more zeros, which contradicts the choice of $(b_1, b_2, \ldots, b_{n+1})$. Hence, $b_2, \ldots, b_{n+1} \in \mathbb{F}$. Since $\sigma_1 = \mathrm{id}$, we have $\alpha_1 + b_2\alpha_2 + \cdots + b_{n+1}\alpha_{n+1} = 0$, i.e., $\alpha_1, \alpha_2, \ldots, \alpha_{n+1}$ are linearly dependent. In other words, $[\mathbb{E} : \mathbb{F}] \leq n \leq |G|$. Combining with Theorem 4.5.6, we have $|G| = |\operatorname{Gal}(\mathbb{E}/\mathbb{F})| = [\mathbb{E} : \mathbb{F}]$, which implies that $G = \operatorname{Gal}(\mathbb{E}/\mathbb{F})$. $\qquad \square$

Combining all the above results, we have the following.

Theorem 4.5.8. *Let $\mathbb{E}/\mathbb{F}$ be a finite extension. Then the following conditions are equivalent:*

(1) $\mathbb{F}$ *is a fixed subfield of some finite subgroup of* $\operatorname{Aut}(\mathbb{E})$;
(2) $\mathbb{F}$ *is the fixed subfield of* $\operatorname{Gal}(\mathbb{E}/\mathbb{F})$;
(3) $[\mathbb{E} : \mathbb{F}] = |\operatorname{Gal}(\mathbb{E}/\mathbb{F})|$.

Definition 4.5.9. A finite extension $\mathbb{E}/\mathbb{F}$ satisfying one of the above equivalent conditions is called a (finite) **Galois extension**.

For a finite extension $\mathbb{E}/\mathbb{F}$, if the fixed subfield of its Galois group is $\mathbb{K}$, then $\mathbb{E}/\mathbb{K}$ is a Galois extension, and it is easy to see that $\text{Gal}(\mathbb{E}/\mathbb{K}) = \text{Gal}(\mathbb{E}/\mathbb{F})$. In general, how to determine whether an extension is a Galois extension? Why is a Galois extension so special? How to calculate the Galois group of a field? We will answer these questions in the rest of this chapter.

Exercises

(1) Determine $\text{Gal}(\mathbb{Q}(\sqrt{2} + \sqrt{3})/\mathbb{Q})$.
(2) Determine the Galois group of a splitting field of $x^3 - 2 \in \mathbb{Q}[x]$.
(3) Let $\mathbb{E}$ be a splitting field of $x^4 - 10x^2 + 1 \in \mathbb{Q}[x]$. Determine $\text{Gal}(\mathbb{E}/\mathbb{Q})$.
(4) Find a field extension $\mathbb{E}/\mathbb{F}$ such that $\text{Hom}_{\mathbb{F}}(\mathbb{E}, \mathbb{E}) \neq \text{Gal}(\mathbb{E}/\mathbb{F})$.

Exercises (hard)

(5) Let $\mathbb{E} = \mathbb{F}_q(t)$ be a simple transcendental extension.

    (i) For any $a \in \mathbb{F}_q$, define $\sigma_a(t) = t + a$. Show that $\sigma_a \in \text{Gal}(\mathbb{E}/\mathbb{F}_q)$;
    (ii) Show that $G = \{\sigma_a | a \in \mathbb{F}_q\}$ is a subgroup of $\text{Gal}(\mathbb{E}/\mathbb{F}_q)$;
    (iii) Let $\mathbb{K}$ be the fixed subfield of G. Show that $\mathbb{K} = \mathbb{F}(t^q - t)$;
    (iv) Determine $\text{Gal}(\mathbb{E}/\mathbb{F}_q)$.

4.6   Galois extensions and Galois correspondence

Galois extensions are valuable field extensions, and they will be crucial for the study of the solvability of equations by radicals. In this section, we will focus mainly on such extensions.

Let $\mathbb{E}/\mathbb{F}$ be a finite Galois extension, and $G = \text{Gal}(\mathbb{E}/\mathbb{F})$. For any $\alpha \in \mathbb{E}$, let $p(x) = \text{Irr}(\alpha, \mathbb{F})$ be the minimal polynomial of α. Then by Lemma 4.5.2, $\varphi(\alpha)$ is also a root of $p(x)$, for any $\varphi \in G$. Assume that the G-orbit of α is $O_\alpha = \{\varphi(\alpha) | \varphi \in G\} = \{\alpha_1, \alpha_2, \ldots, \alpha_n\}$.

Consider the polynomial

$$f(x) = \prod_{i=1}^{n}(x - \alpha_i) \in \mathbb{E}[x].$$

The $(n - k)$th coefficient of $f(x)$ is $(-1)^k \sigma_k(\alpha_1, \alpha_2, \ldots, \alpha_n)$, where $\sigma_k(\alpha_1, \alpha_2, \ldots, \alpha_n)$ is the kth elementary symmetric polynomials of α_i. Noting that $\varphi(\sigma_k(\alpha_1, \alpha_2, \ldots, \alpha_n)) = \sigma_k(\varphi(\alpha_1), \varphi(\alpha_2), \ldots, \varphi(\alpha_n))$, and $\varphi(\alpha_1), \varphi(\alpha_2), \ldots, \varphi(\alpha_n)$ is a permutation of $\alpha_1, \alpha_2, \ldots, \alpha_n$, we have $\sigma_k(\varphi(\alpha_1), \sigma(\alpha_2), \ldots, \varphi(\alpha_n)) = \sigma_k(\alpha_1, \alpha_2, \ldots, \alpha_n) \in \mathbb{E}^G = \mathbb{F}$. Hence, $\sigma_k(\alpha_1, \alpha_2, \ldots, \alpha_n) \in \mathbb{F}$. Thus, $f(x) \in \mathbb{F}[x]$. Furthermore, $f(x)$ and $p(x)$ are not coprime since they have a common factor $x - \alpha$ in $\mathbb{E}(x)$. Therefore, $f(x)$ and $p(x)$ are not coprime in $\mathbb{F}(x)$, which implies that $p(x) \mid f(x)$, since $p(x)$ is irreducible. But from the definition of $f(x)$, one has $f(x) \mid p(x)$, which implies that $p(x) = f(x)$. Therefore, for any Galois extension $\mathbb{E}/\mathbb{F}$, the minimal polynomial $\mathrm{Irr}(\alpha, \mathbb{F})$ splits in $\mathbb{E}$ for any $\alpha \in \mathbb{E}$. Now, we are ready to introduce two important notions in Galois theory.

Definition 4.6.1. Let $\mathbb{E}/\mathbb{F}$ be an algebraic extension, $\alpha \in \mathbb{E}$. If $\mathrm{Irr}(\alpha, \mathbb{F})$ is separable, then α is called **separable** over $\mathbb{F}$. If $\mathrm{Irr}(\alpha, \mathbb{F})$ is inseparable, then α is called **inseparable**. If every element in $\mathbb{E}$ is separable over $\mathbb{F}$, then $\mathbb{E}/\mathbb{F}$ is called a **separable extension**.

Definition 4.6.2. An algebraic extension $\mathbb{E}/\mathbb{F}$ is called **normal**, if the minimal polynomial of every element in $\mathbb{E}$ splits in $\mathbb{E}$, or equivalently, every irreducible polynomial in $\mathbb{F}[x]$ which has a root in $\mathbb{E}$ is a product of linear factors in $\mathbb{E}[x]$.

From the above discussion, we deduce the following theorem.

Theorem 4.6.3. *A finite Galois extension $\mathbb{E}/\mathbb{F}$ is separable and normal.*

Let $\mathbb{E} = \mathbb{F}(\alpha_1, \alpha_2, \ldots, \alpha_m)$ be a finite extension of $\mathbb{F}$. Set $p_i(x) = \mathrm{Irr}(\alpha_i, \mathbb{F})$, $f(x) = \prod_{i=1}^{m} p_i(x)$. If $\mathbb{E}/\mathbb{F}$ is a normal extension, then $p_i(x)$ splits in $\mathbb{E}[x]$. Since $x - \alpha_i$, $i = 1, 2, \ldots, m$ are factors of $f(x)$, we may assume $f(x) = \prod_{i=1}^{n}(x - \alpha_i)$ and $\mathbb{E} = \mathbb{F}(\alpha_1, \alpha_2, \ldots, \alpha_n)$. Thus, we have the following theorem.

Theorem 4.6.4. *Let $\mathbb{E}/\mathbb{F}$ be a finite normal extension. Then $\mathbb{E}$ is a splitting field of some polynomial in $\mathbb{F}[x]$.*

The converse statement is also true. We first prove the following lemma.

Lemma 4.6.5. *Let $\mathbb{E}$ be a splitting field of $f(x) \in \mathbb{F}[x]$ and let $\mathbb{E}$ be a subfield of $\mathbb{K}$. Then $\sigma(\mathbb{E}) = \mathbb{E}$, for any $\sigma \in \mathrm{Gal}\,(\mathbb{K}/\mathbb{F})$.*

Proof. Let $\alpha_1, \alpha_2, \ldots, \alpha_n \in \mathbb{E}$ be all roots of $f(x)$. Then $\mathbb{E} = \mathbb{F}(\alpha_1, \alpha_2, \ldots, \alpha_n)$. By Lemma 4.5.2, any σ in $\mathrm{Gal}\,(\mathbb{K}/\mathbb{F})$ permutes the roots of $f(x)$, which implies that

$$\sigma(\mathbb{E}) = \mathbb{F}(\sigma(\alpha_1), \sigma(\alpha_2), \ldots, \sigma(\alpha_n)) = \mathbb{E}. \qquad \square$$

Now, we are ready to prove the following theorem.

Theorem 4.6.6. *Let $\mathbb{E}$ be a splitting field of $f(x) \in \mathbb{F}[x]$. Then $\mathbb{E}/\mathbb{F}$ is a finite normal extension.*

Proof. Let $\alpha \in \mathbb{E}$ and $p(x) = \mathrm{Irr}(\alpha, \mathbb{F})$. Let $\mathbb{K}$ be a splitting field of $p(x)f(x) \in \mathbb{F}[x]$. For any root β of $p(x)$ in $\mathbb{K}$, there exists an $\mathbb{F}$-isomorphism $\sigma : \mathbb{F}(\alpha) \to \mathbb{F}(\beta)$ such that $\sigma(\alpha) = \beta$. By the proof of Theorem 4.4.12, σ can be extended to an $\mathbb{F}$-automorphism of $\mathbb{K}$. By the above lemma, we have $\sigma(\mathbb{E}) = \mathbb{E}$, which implies $\beta = \sigma(\alpha) \in \mathbb{E}$. Hence, $p(x)$ splits in $\mathbb{E}$, and $\mathbb{E}/\mathbb{F}$ is a finite normal extension. $\qquad \square$

A finite normal extension is exactly a splitting field of some polynomial. Generally, we have the following proposition, whose proof is left to the reader.

Proposition 4.6.7. *Let $\mathbb{E}/\mathbb{F}$ be a normal extension. Then $\mathbb{E}$ is a splitting field of a family of polynomials in $\mathbb{F}$, i.e., $\mathbb{E}$ is generated over $\mathbb{F}$ by the roots of a family of polynomials in $\mathbb{F}[x]$.*

For any finite extension $\mathbb{E} = \mathbb{F}(\alpha_1, \alpha_2, \ldots, \alpha_n)$ of $\mathbb{F}$, set $f(x) = \prod_{i=1}^{n} \mathrm{Irr}(\alpha_i, \mathbb{F})$. Then a splitting field of $f(x)$ is a normal extension of $\mathbb{F}$. Any normal extension $\mathbb{K}/\mathbb{F}$ containing $\mathbb{E}$ must contain a splitting field of $f(x)$. Since the splitting field of $f(x)$ is unique up to isomorphism, there exists a minimal normal extension of $\mathbb{F}$ containing $\mathbb{E}$, which is unique up to isomorphism. We call this minimal normal extension the **normal closure** of $\mathbb{E}$.

Now, we discuss whether a splitting field of $f(x) \in \mathbb{F}[x]$ is a Galois extension of $\mathbb{F}$. Thus, we need to determine $\mathrm{Gal}\,(\mathbb{E}/\mathbb{F})$. By letting $\mathbb{E} = \overline{\mathbb{E}}$ in Theorem 4.4.12, we have the following.

Corollary 4.6.8. *If $\mathbb{E}$ is a splitting field of $f(x) \in \mathbb{F}[x]$, then $|\mathrm{Gal}\,(\mathbb{E}/\mathbb{F})| \leq [\mathbb{E} : \mathbb{F}]$, and the equality holds if and only if $f(x)$ is separable.*

Therefore, we have the following result.

Theorem 4.6.9. *Let $f(x) \in \mathbb{F}[x]$ be separable, and let $\mathbb{E}$ be a splitting field of $f(x)$. Then $\mathbb{E}/\mathbb{F}$ is a Galois extension.*

Combining Theorem 4.6.3, 4.6.4, and 4.6.9, we get

Theorem 4.6.10. *Let $\mathbb{E}/\mathbb{F}$ be a finite extension. Then the following assertions are equivalent.*

(1) $\mathbb{E}/\mathbb{F}$ *is a finite Galois extension, i.e.,* $|\mathrm{Gal}\,(\mathbb{E}/\mathbb{F})| = [\mathbb{E} : \mathbb{F}]$.
(2) $\mathbb{F}$ *is the fixed subfield of* $\mathrm{Gal}\,(\mathbb{E}/\mathbb{F})$ *in* $\mathbb{E}$.
(3) *There exists a finite subgroup G of* $\mathrm{Aut}\,(\mathbb{E})$ *such that* $\mathbb{F} = \mathbb{E}^G$. *In this case,* $G = \mathrm{Gal}\,(\mathbb{E}/\mathbb{F})$.
(4) $\mathbb{E}/\mathbb{F}$ *is a finite separable normal extension.*
(5) $\mathbb{E}$ *is a splitting field of some separable polynomial in* $\mathbb{F}[x]$.

By the above theorem, we may give a more general definition of Galois extensions.

Definition 4.6.11. An algebraic extension $\mathbb{E}/\mathbb{F}$ is called a **Galois extension**, if $\mathbb{E}/\mathbb{F}$ is a separable normal extension.

For example, let $\overline{\mathbb{Q}}$ be the algebraic closure of $\mathbb{Q}$ in $\mathbb{C}$. Then $\overline{\mathbb{Q}}$ is a separable normal extension of $\mathbb{Q}$. Hence, $\overline{\mathbb{Q}}/\mathbb{Q}$ is a Galois extension. One of the most important problems in number theory is to understand the structure of $\mathrm{Gal}\,(\overline{\mathbb{Q}}/\mathbb{Q})$.

A difficult problem in field theory is to determine all the subextensions, or intermediate subfields, of a field extension $\mathbb{E}/\mathbb{F}$. For any $\alpha \in \mathbb{E}$, $\mathbb{F}(\alpha)$ is an intermediate subfield. For instance, for $\mathbb{F} = \mathbb{Q}$,

$\mathbb{E} = \mathbb{Q}(\sqrt[3]{2}, \omega)$, we can easily get the following six intermediate subfields

$$\mathbb{Q}, \quad \mathbb{Q}(\sqrt[3]{2}), \quad \mathbb{Q}(\sqrt[3]{2}\omega), \quad \mathbb{Q}(\sqrt[3]{2}\omega^2), \quad \mathbb{Q}(\omega), \quad \mathbb{Q}(\sqrt[3]{2}, \omega).$$

Are there any other intermediate subfields? Say, for any $k \in \mathbb{Q}, k \neq 0$, $\mathbb{Q}(\sqrt[3]{2} + k\omega)$ is also an intermediate subfield. Is $\mathbb{Q}(\sqrt[3]{2} + k\omega)$ different from the above six ones? In the following, we will see how Galois groups and Galois extensions help answer similar questions.

We begin with the relationship between the subgroups of Galois groups and intermediate subfields of Galois extensions. Let $\mathbb{E}/\mathbb{F}$ be a finite extension, $G = \mathrm{Gal}\,(\mathbb{E}/\mathbb{F})$. Denote by $\mathcal{S}$ the set of all subgroups of G, and $\mathcal{I}$ the set of intermediate subfields of $\mathbb{E}/\mathbb{F}$. Define two mappings

$$\mathrm{Inv} : \mathcal{S} \to \mathcal{I}, \quad \mathrm{Inv}\,(H) = \mathbb{E}^H,$$

$$\mathrm{Gal} : \mathcal{I} \to \mathcal{S}, \quad \mathrm{Gal}\,(\mathbb{K}) = \mathrm{Gal}\,(\mathbb{E}/\mathbb{K}).$$

Lemma 4.6.12.

(1) Let H_1, H_2 be subgroups of $\mathrm{Gal}\,(\mathbb{E}/\mathbb{F})$. Then $H_1 \subseteq H_2$ if and only if $\mathrm{Inv}\,(H_1) \supseteq \mathrm{Inv}\,(H_2)$.
(2) Let $\mathbb{K}_1, \mathbb{K}_2$ be intermediate subfields of $\mathbb{E}/\mathbb{F}$. Then $\mathbb{K}_1 \subseteq \mathbb{K}_2$ if and only if $\mathrm{Gal}\,(\mathbb{K}_1) \supseteq \mathrm{Gal}\,(\mathbb{K}_2)$.

For any $H < \mathrm{Gal}\,(\mathbb{E}/\mathbb{F})$, set $\mathbb{K} = \mathbb{E}^H$. Then by Theorem 4.5.7, $\mathbb{E}/\mathbb{K}$ is a Galois extension and $\mathrm{Gal}\,(\mathbb{E}/\mathbb{K}) = H$. Therefore, $\mathrm{Gal}\,(\mathrm{Inv}\,(H)) = H$, which proves the following lemma.

Lemma 4.6.13. $\mathrm{Gal} \circ \mathrm{Inv} = \mathrm{id}_{\mathcal{S}}$. *Consequently,* Inv *is injective, and* Gal *is surjective.*

In general, Gal is not necessarily injective, while Inv is not necessarily surjective. If Gal and Inv are 1-1 correspondences between $\mathcal{S}$ and $\mathcal{I}$, then the minimal intermediate field $\mathbb{F}$ corresponds to the largest subgroup $\mathrm{Gal}\,(\mathbb{E}/\mathbb{F})$, i.e., $\mathbb{F} = \mathrm{Inv}\,(\mathrm{Gal}\,(\mathbb{E}/\mathbb{F}))$, which implies that $\mathbb{F}$ is the fixed subfield of $\mathrm{Gal}\,(\mathbb{E}/\mathbb{F})$. Hence, $\mathbb{E}/\mathbb{F}$ is a Galois extension. We have the following **Fundamental Theorem of Galois theory**.

Theorem 4.6.14. *Let* $\mathbb{E}/\mathbb{F}$ *be a finite Galois extension,* $G = \mathrm{Gal}\,(\mathbb{E}/\mathbb{F})$.

(1) *Both* Inv *and* Gal *are bijections and* Gal $^{-1}$ = Inv. *Thus, there is a 1-1 correspondence, called **Galois correspondence**, between the set of subgroups of G and that of intermediate fields of $\mathbb{E}/\mathbb{F}$.*

(2) *For any $H < G$, $|H| = [\mathbb{E} : \text{Inv}\,(H)]$, $[G : H] = [\text{Inv}\,(H) : \mathbb{F}]$.*

(3) *A subgroup H of G is normal if and only if the extension $\text{Inv}\,(H)/\mathbb{F}$ is normal. In this case, $\text{Gal}\,(\text{Inv}\,(H)/\mathbb{F}) \simeq G/H$.*

Proof.   (1) Since $\mathbb{E}/\mathbb{F}$ is a finite Galois extension, $\mathbb{E}$ is a splitting field of some separable polynomial $f(x) \in \mathbb{F}[x]$. If $\mathbb{K}$ is an intermediate subfield, then $f(x)$ is a polynomial over $\mathbb{K}$, and $\mathbb{E}/\mathbb{K}$ is a finite Galois extension since $\mathbb{E}$ is a splitting field of $f(x) \in \mathbb{K}[x]$. Let $H = \text{Gal}\,(\mathbb{E}/\mathbb{K})$. Then $\mathbb{E}^H = \mathbb{K}$, i.e., $\text{Inv}\,(\text{Gal}\,(\mathbb{K})) = \mathbb{K}$. By Lemma 4.6.13, Inv and Gal are inverse to each other.

(2) We see from Theorem 4.5.7 that $\mathbb{E}/\text{Inv}\,(H)$ is a Galois extension and $\text{Gal}\,(\mathbb{E}/\text{Inv}\,(H)) = H$, which implies that $|H| = |\text{Gal}\,(\mathbb{E}/\text{Inv}\,(H))| = [\mathbb{E} : \text{Inv}\,(H)]$. Since $[\mathbb{E} : \mathbb{F}] = [\mathbb{E} : \text{Inv}\,(H)]$ $[\text{Inv}\,(H) : \mathbb{F}]$ and $|G| = |H||G/H|$, we have $[G : H] = [\text{Inv}\,(H) : \mathbb{F}]$.

(3) Set $\mathbb{K} = \text{Inv}\,(H)$. For any $\varphi \in G$, $\varphi(\mathbb{K})$ is also an intermediate subfield. It is easy to see $\text{Gal}\,(\varphi(\mathbb{K})) = \varphi H \varphi^{-1}$. Therefore, H is a normal subgroup of G if and only if $\varphi(\mathbb{K}) = \mathbb{K}$, for any $\varphi \in G$. In this case, $\varphi|_{\mathbb{K}}$ is an $\mathbb{F}$-automorphism of $\mathbb{K}$. Thus, $\varphi \mapsto \varphi|_{\mathbb{K}}$ is a group homomorphism from G to $\text{Gal}\,(\mathbb{K}/\mathbb{F})$ with the kernel H, i.e., G/H is isomorphic to a subgroup of $\text{Gal}\,(\mathbb{K}/\mathbb{F})$. Since $|G/H| = [\mathbb{K} : \mathbb{F}] \geq |\text{Gal}\,(\mathbb{K}/\mathbb{F})|$, $\text{Gal}\,(\mathbb{K}/\mathbb{F}) \simeq G/H$, and $\mathbb{K}/\mathbb{F}$ is a Galois extension, hence a normal extension.

Conversely, let $\mathbb{K}$ be an intermediate subfield. For any $\alpha \in \mathbb{K}$, $\varphi \in G$, $\varphi(\alpha)$ is also a root of $\text{Irr}(\alpha, \mathbb{F})$. If $\mathbb{K}/\mathbb{F}$ is normal, then $\varphi(\alpha) \in \mathbb{K}$, i.e., $\varphi(\mathbb{K}) \subseteq \mathbb{K}$, hence $\varphi(\mathbb{K}) = \mathbb{K}$. It follows that H is a normal subgroup of G. $\qquad\square$

Corollary 4.6.15. *Let $\mathbb{E}/\mathbb{F}$ be a finite Galois extension, $\mathbb{K}$ an intermediate subfield. Then $\mathbb{E}/\mathbb{K}$ is also a Galois extension.*

Exercises

(1) Let $\mathbb{K}$ be an intermediate subfield of a Galois extension $\mathbb{E}/\mathbb{F}$ with H the corresponding subgroup of $\text{Gal}\,(\mathbb{E}/\mathbb{F})$. For $\varphi \in \text{Gal}\,(\mathbb{E}/\mathbb{F})$, set $\varphi(\mathbb{K}) = \{\varphi(x) : x \in \mathbb{K}\}$. Show that the subgroup

corresponding to $\varphi(\mathbb{K})$ is $\varphi H \varphi^{-1}$. We call $\varphi(\mathbb{K})$ a **conjugate subfield** of $\mathbb{K}$.

(2) Let $\mathbb{E}$ be a splitting field of $x^3 - 3 \in \mathbb{Q}[x]$. Show that $\mathrm{Gal}\,(\mathbb{E}/\mathbb{Q}) \simeq S_3$. Furthermore, determine all subgroups of $\mathrm{Gal}\,(\mathbb{E}/\mathbb{Q})$ and the corresponding intermediate subfields.

(3) Let $\mathbb{E} = \mathbb{Q}(\sqrt{2}, \sqrt{3}, \sqrt{5})$. Determine all subgroups of $\mathrm{Gal}\,(\mathbb{E}/\mathbb{Q})$ and the corresponding intermediate subfields.

(4) Let α be a root of $x^3 + x^2 - 2x - 1 \in \mathbb{Q}[x]$. Show that $\alpha^2 - 2$ is also a root, and $\mathbb{Q}(\alpha)/\mathbb{Q}$ is normal. Determine $\mathrm{Gal}\,(\mathbb{Q}(\alpha)/\mathbb{Q})$.

(5) Let $\mathbb{E}/\mathbb{F}$ be a finite Galois extension, and let $\mathbb{K}_1, \mathbb{K}_2$ be two intermediate subfields of $\mathbb{E}/\mathbb{F}$. Denote by $\mathbb{K}_1 \vee \mathbb{K}_2$ the minimal subfield containing $\mathbb{K}_1$ and $\mathbb{K}_2$, called the **sum field** of $\mathbb{K}_1, \mathbb{K}_2$.

   (i) Show that $\mathbb{K}_1 \vee \mathbb{K}_2 = \mathbb{E}$ if and only if $\mathrm{Gal}\,(\mathbb{E}/\mathbb{K}_1) \cap \mathrm{Gal}\,(\mathbb{E}/\mathbb{K}_2) = \{\mathrm{id}\}$.

   (ii) If $\mathbb{K}_1/\mathbb{F}$ is normal, show that $\mathrm{Gal}\,(\mathbb{K}_1 \vee \mathbb{K}_2/\mathbb{K}_2)$ is isomorphic to a subgroup of $\mathrm{Gal}\,(\mathbb{K}_1/\mathbb{F})$.

(6) Let $\mathbb{E}/\mathbb{F}$ be a finite Galois extension, G_1, G_2 two subgroups of $\mathrm{Gal}\,(\mathbb{E}/\mathbb{F})$. Show that

   (i) $\mathrm{Inv}\,(G_1 \cap G_2) = \mathrm{Inv}\,(G_1) \vee \mathrm{Inv}\,(G_2)$;

   (ii) $\mathrm{Inv}\,(\langle G_1, G_2 \rangle) = \mathrm{Inv}\,(G_1) \cap \mathrm{Inv}\,(G_2)$.

(7) Let $\mathbb{F}$ be a field of characteristic $p \neq 0$, $a \in \mathbb{F}$. Assume that α is a root of $x^p - x - a$. Show that $\mathbb{F}(\alpha)/\mathbb{F}$ is a Galois extension, and determine $\mathrm{Gal}\,(\mathbb{F}(\alpha)/\mathbb{F})$.

(8) Let $\mathbb{F}, \mathbb{K}$ be two subfields of $\mathbb{E}$. Assume that $\mathbb{F}/(\mathbb{F} \cap \mathbb{K})$ is a finite Galois extension. Show that $\mathbb{F} \vee \mathbb{K}/\mathbb{K}$ is also a finite Galois extension, and $\mathrm{Gal}\,(\mathbb{F} \vee \mathbb{K}/\mathbb{K}) \simeq \mathrm{Gal}\,(\mathbb{F}/(\mathbb{F} \cap \mathbb{K}))$.

(9) Show that quadratic extensions are normal. What about extensions of degree 3?

(10) Let $\mathbb{K}$ be an intermediate subfield of $\mathbb{E}/\mathbb{F}$ with $\mathbb{K}/\mathbb{F}$ normal, $\sigma \in \mathrm{Gal}\,(\mathbb{E}/\mathbb{F})$. Show that $\sigma(\mathbb{K}) = \mathbb{K}$.

(11) Let $\mathbb{E}/\mathbb{F}$ be a finite normal extension. Show that $\alpha, \beta \in \mathbb{E}/\mathbb{F}$ are conjugate if and only if $\sigma(\alpha) = \beta$, for some $\sigma \in \mathrm{Gal}\,(\mathbb{E}/\mathbb{F})$.

(12) (i) If $\mathbb{E}/\mathbb{K}, \mathbb{K}/\mathbb{F}$ are normal, is $\mathbb{E}/\mathbb{F}$ normal?

   (ii) If $\mathbb{E}/\mathbb{F}$ is normal, $\mathbb{K}$ an intermediate subfield of $\mathbb{E}/\mathbb{F}$, are $\mathbb{E}/\mathbb{K}, \mathbb{K}/\mathbb{F}$ normal?

Exercises (hard)

(13) Let $\mathbb{E}/\mathbb{F}$ be a finite extension. Show that $\mathbb{E}/\mathbb{F}$ is normal if and only if for any irreducible polynomial $f(x) \in \mathbb{F}[x]$, all irreducible factors of $f(x)$ in $\mathbb{E}[x]$ have the same degree.

(14) Let $\mathbb{E}/\mathbb{F}$ be a finite normal extension, $G = \mathrm{Gal}\,(\mathbb{E}/\mathbb{F})$, and $\mathbb{K}$ an intermediate subfield of $\mathbb{E}/\mathbb{F}$. Show that $\mathbb{K}/\mathbb{F}$ is normal if and only if $\sigma(\mathbb{K}) = \mathbb{K}$, for any $\sigma \in G$.

(15) Let $\overline{\mathbb{Q}}$ be the algebraic closure of $\mathbb{Q}$ in $\mathbb{C}$, $G = \mathrm{Gal}\,(\overline{\mathbb{Q}}/\mathbb{Q})$, and $\sigma \in G$.

   (i) Show that $\overline{\mathbb{Q}}/\mathbb{Q}$ is normal and $[\overline{\mathbb{Q}} : \mathbb{Q}] = +\infty$.

   (ii) For any Galois extension $\mathbb{E}/\mathbb{Q}$, $\sigma \mapsto \sigma|_{\mathbb{E}}$ is a group homomorphism from G to $\mathrm{Gal}\,(\mathbb{E}/\mathbb{Q})$.

   (iii) The group homomorphism in (2) is surjective.

4.7 Finite fields

Before further exploring Galois theory, let us first look at the structure of finite fields, where we may apply the Fundamental Theorem of Galois theory. It was Galois who first introduced finite fields, so they are also called Galois fields. If $\mathbb{E}$ is a finite field with characteristic p, then $\mathbb{E}$ can be regarded as an extension of $\mathbb{F}_p$. We will soon find that this extension is a Galois extension. First, we give a general conclusion.

Lemma 4.7.1. *Let $\mathbb{E}$ be a field, and let G be a finite subgroup of $\mathbb{E}^*$. Then G is a cyclic group.*

Proof. Assume that $|G| = n$, and m is the maximum of orders of elements in G. By Exercise 7 in Section 1.2, we have $b^m = 1$, for any $b \in G$. Therefore, the polynomial $x^m - 1$ has n roots in $\mathbb{E}$, which implies $n \leq m$ and, consequently, $n = m$. Thus, G has an element of order n. $\qquad\square$

In particular, for a finite field $\mathbb{E}$, $\mathbb{E}^*$ is cyclic with a generator α. Assume that the prime field of $\mathbb{E}$ is $\mathbb{F}_p$. Then $\mathbb{E} = \mathbb{F}_p(\alpha)$ is a simple algebraic extension. Let $[\mathbb{E} : \mathbb{F}_p] = n$. Assume that $\alpha_1, \alpha_2, \ldots, \alpha_n$ is a basis of $\mathbb{E}/\mathbb{F}_p$. Thus, any element $\alpha \in \mathbb{E}$ can be written uniquely as

$$\alpha = a_1\alpha_1 + a_2\alpha_2 + \cdots + a_n\alpha_n,$$

where $a_1, a_2 \ldots, a_n \in \mathbb{F}_p$. Noticing that $\mathbb{F}_p$ has p elements, we conclude that $|\mathbb{E}| = p^n$ and that $\mathbb{E}^*$ is a cyclic group of order $p^n - 1$. Hence, the elements in $\mathbb{E}$ are exactly the roots of the polynomial $x^{p^n} - x \in \mathbb{F}_p[x]$. In other words, $\mathbb{E}$ is a splitting field of the separable polynomial $x^{p^n} - x$. Since the splitting field of a polynomial is unique up to isomorphism, we have the following result.

Theorem 4.7.2. *Let $\mathbb{E}$ be a finite field of characteristic p with the prime field $\mathbb{F}_p$. Set $n = [\mathbb{E} : \mathbb{F}_p]$. Then*

(1) $\mathbb{E}$ *has p^n element;*
(2) $\mathbb{E}^*$ *is a cyclic group and $\mathbb{E}/\mathbb{F}_p$ is a simple extension;*
(3) $\mathbb{E}$ *is a splitting field of the polynomial $f(x) = x^{p^n} - x$ over $\mathbb{F}_p$;*
(4) *up to isomorphism, there is a unique finite field of order p^n, usually denoted by $\mathbb{F}_{p^n}$.*

Since $(x^{p^n} - x)' = -1$, $x^{p^n} - x$ is separable. We have the following.

Corollary 4.7.3. *Let $\mathbb{E}$ be a finite field of characteristic p. Then $\mathbb{E}/\mathbb{F}_p$ is a Galois extension, and its Galois group is cyclic with Frobenius isomorphism Fr as a generator. Here, $\mathrm{Fr}(x) = x^p$, for any $x \in \mathbb{E}$.*

Proof. Since $\mathbb{E}$ is a splitting field of the separable polynomial $x^{p^n} - x$, $\mathbb{E}/\mathbb{F}_p$ is a Galois extension. It follows that Fr is an isomorphism since $\mathbb{E}$ is finite. Denote by G the cyclic group generated by Fr. Then the fixed field $\mathbb{K}$ of G in $\mathbb{E}$ is exactly the fixed field of Fr. Thus,

$$\mathbb{K} = \{x \in \mathbb{E} | \mathrm{Fr}(x) = x\} = \{x \in \mathbb{E} | x^p - x = 0\}.$$

Hence, $\mathbb{K}$ consists of all the solutions of $x^p - x$, and it has at most p elements. Therefore, $\mathbb{K} = \mathbb{F}_p$ since $\mathbb{K}$ contains $\mathbb{F}_p$, which implies $\mathrm{Gal}(\mathbb{E}/\mathbb{F}_p) = G$. $\qquad\square$

Since $\mathbb{E}/\mathbb{F}_p$ is a Galois extension, $\mathbb{E}/\mathbb{F}$ is also a Galois extension for any intermediate field $\mathbb{F}$. Furthermore, we have the following.

Corollary 4.7.4. *Let $\mathbb{E}$ be a finite field with p^n elements. If $\mathbb{F}$ is a subfield of $\mathbb{E}$, then $|\mathbb{F}| = p^m$ with $m \mid n$. Conversely, for any $m \mid n$, there exists a unique subfield $\mathbb{F}$ of $\mathbb{E}$ such that $|\mathbb{F}| = p^m$. In this case, $\mathbb{E}/\mathbb{F}$ is a Galois extension, and $\mathrm{Gal}(\mathbb{E}/\mathbb{F})$ is cyclic with generator Fr^m.*

Proof. If $\mathbb{F}$ is a subfield of $\mathbb{E}$, then $\mathbb{E}$ is an $\mathbb{F}$-vector space with dimensional k. Thus, $|\mathbb{E}| = |\mathbb{F}|^k = p^{mk}$, which implies $mk = n$, i.e., $m \mid n$.

Conversely, let $\mathbb{F}_p$ be the prime field of $\mathbb{E}$. Then $\mathbb{E}/\mathbb{F}_p$ is a Galois extension, and the Galois group is the cyclic group $\langle \mathrm{Fr} \rangle$ of order n. For $m \mid n$, $\langle \mathrm{Fr} \rangle$ has a unique subgroup $H = \langle \mathrm{Fr}^m \rangle$ of order $\frac{n}{m}$. It is easy to see that $\mathbb{F} = \mathbb{E}^H$ is an intermediate field with p^m elements. Therefore, $\mathbb{E}/\mathbb{F}$ is a Galois extension, and $\mathrm{Gal}\,(\mathbb{E}/\mathbb{F}) = H = \langle \mathrm{Fr}^m \rangle$. $\qquad\square$

We are now ready to study the irreducible polynomials over finite fields.

Lemma 4.7.5. *Let $\mathbb{F}_p$ be a finite group of order p. Then for any $n \in \mathbb{N}^*$, there exists a monic irreducible polynomial $f(x) \in \mathbb{F}_p[x]$ of degree n.*

Proof. Let $\mathbb{E}$ be a finite field with p^n elements. Then the prime field of $\mathbb{E}$ is isomorphic to $\mathbb{F}_p$. Hence, $\mathbb{E}/\mathbb{F}_p$ is a simple extension of degree n. Set $\mathbb{E} = \mathbb{F}_p(\alpha)$. Thus, $\deg \mathrm{Irr}(\alpha, \mathbb{F}_p) = [\mathbb{E} : \mathbb{F}_p] = n$, i.e., $\mathrm{Irr}(\alpha, \mathbb{F}_p)$ is a monic irreducible polynomial over $\mathbb{F}_p$ of degree n. $\quad\square$

Lemma 4.7.6. *Let $f(x) \in \mathbb{F}_p[x]$ be a monic irreducible polynomial of degree n. Then $\mathbb{E} = \mathbb{F}_p[x]/\langle f(x) \rangle$ is a finite field with p^n elements, and $f(x)$ splits in $\mathbb{E}$. Furthermore, $f(x) \mid (x^{p^n} - x)$.*

Proof. Since $\mathbb{E} = \mathbb{F}_p[x]/\langle f(x) \rangle$ is an extension of $\mathbb{F}_p$ of degree n, it must have p^n elements. As a Galois extension, $\mathbb{E}/\mathbb{F}_p$ is also normal, which implies that $f(x)$ splits in $\mathbb{E}$ since $f(x)$ has a root $x + \langle f(x) \rangle$ in $\mathbb{E}$. Furthermore, every element in $\mathbb{E}$ is a root of $x^{p^n} - x$, hence $f(x)$ and $x^{p^n} - x$ are not coprime over $\mathbb{E}$, and they are not coprime over $\mathbb{F}_p$. Thus, $f(x) \mid (x^{p^n} - x)$ since $f(x)$ is irreducible over $\mathbb{F}_p$. $\qquad\square$

The above lemma tells us how to decompose $x^{p^n} - x$.

Corollary 4.7.7. *$x^{p^n} - x$ is the product of all monic irreducible polynomials in $\mathbb{F}_p[x]$ of degree m with $m \mid n$.*

Proof. Let $\mathbb{E}$ be a finite field with p^n elements. All elements in $\mathbb{E}$ are exactly the roots of $x^{p^n} - x$. If $g(x)$ is a monic irreducible factor of $x^{p^n} - x$ of degree m, and if $\alpha \in \mathbb{E}$ is a root of $g(x)$, then $\mathbb{F}_p(\alpha)/\mathbb{F}_p$ is a field extension of degree m. By Corollary 4.7.4, we have $m \mid n$. Conversely, if $g(x)$ is a monic irreducible polynomial over $\mathbb{F}_p$

of degree m such that $m \mid n$, then $\mathbb{K} = \mathbb{F}_p[x]/\langle g(x) \rangle$ is an extension of $\mathbb{F}_p$ of degree m. By Corollary 4.7.4, there exists an $\mathbb{F}_p$-homomorphism $\varphi : \mathbb{K} \to \mathbb{E}$. Thus, $\mathbb{E}$ contains some roots of $g(x)$, which implies that $g(x)$ and $x^{p^n} - x$ are not coprime. Therefore, $g(x) \mid (x^{p^n} - x)$ since $g(x)$ is irreducible.

Furthermore, $x^{p^n} - x$ has no multiple roots. Hence, every irreducible factor of $x^{p^n} - x$ must be of multiplicity one. The conclusion follows. $\qquad\square$

Example 4.7.8. Noticing that $-1 = 1$ in $\mathbb{F}_2$, one is easy to determine irreducible polynomials over $\mathbb{F}_2$ with degree ≤ 3:

$$x, \quad x + 1, \quad x^2 + x + 1, \quad x^3 + x^2 + 1, \quad x^3 + x + 1.$$

Therefore,

$$x^8 + x = x(x + 1)(x^3 + x^2 + 1)(x^3 + x + 1).$$

In the classification of finite simple groups, there is a class of simple groups called finite groups of Lie type, which are closely related to finite fields. We give one example.

Example 4.7.9. The group $G = \mathrm{SL}(3, \mathbb{F}_2)$ is a simple group of order 168.

Proof. Firstly, we determine the order of G. Since $\mathrm{SL}(3, \mathbb{F}_2) = \mathrm{GL}(3, \mathbb{F}_2)$, we need to determine the order of $\mathrm{GL}(3, \mathbb{F}_2)$.

We consider the general case. For each $A \in \mathrm{GL}(n, \mathbb{F}_q)$, the column vectors $A_1, A_2, \ldots, A_n \in \mathbb{F}_q^n$ of A are linearly independent. There are q^n elements in $\mathbb{F}_q^n$. Let A_1 be any non-zero element. Then we have $q^n - 1$ choices for A_1. Since A_2 must be linearly independent with A_1, i.e., $A_2 \neq kA_1$, for any $k \in \mathbb{F}_q$, we have $q^n - q$ choices for A_2. Similarly, A_{i+1} can not be a linear combination of $A_1, A_2, \ldots, A_i$, i.e., $A_{i+1} \neq k_1 A_1 + k_2 A_2 + \cdots + k_i A_i$, for any $k_i \in \mathbb{F}_q$, which implies that we have $q^n - q^i$ choices for A_{i+1}. Therefore, there are

$$(q^n - 1)(q^n - q) \cdots (q^n - q^{n-1})$$

elements in $\mathrm{GL}(n, \mathbb{F}_q)$. Hence, the order of $G = \mathrm{SL}(3, \mathbb{F}_2)$ is $(2^3 - 1)(2^3 - 2)(2^3 - 2^2) = 168 = 2^3 \times 3 \times 7$.

Secondly, to prove that G is simple, we need to show that any union of conjugacy classes of G is not a non-trivial subgroup of G. Thus, in the following, we will determine conjugacy classes of G.

For any $A \in G$, the characteristic polynomial $f(x) = |xI_3 - A|$ is a monic polynomial of degree 3 with a non-zero constant term and must be one of the following:

(1) $x^3 + x^2 + x + 1 = (x+1)^3$; (2) $x^3 + x + 1$;
(3) $x^3 + x^2 + 1$; (4) $x^3 + 1 = (x+1)(x^2+x+1)$.

We deal with it case by case. Denote by $m(x)$ the minimal polynomial of A.

(1) $f(x) = x^3 + x^2 + x + 1 = (x+1)^3$.

(a) If $m(x) = x + 1$, then $A = I_3$.

(b) If $m(x) = (x+1)^2 = x^2 + 1$, then A is of order 2; conversely, for every A of order 2 we have $m(x) = x^2 + 1$. It follows that every element of order 2 is conjugate with

$$T_2 = \begin{pmatrix} 1 & 0 & 1 \\ 0 & 1 & 0 \\ 0 & 0 & 1 \end{pmatrix}.$$

Hence, all elements of order 2 are conjugate. It is easy to see that

$$C_G(T_2) = \left\{ \begin{pmatrix} 1 & a & c \\ 0 & 1 & b \\ 0 & 0 & 1 \end{pmatrix} \middle| a, b, c \in \mathbb{F}_2 \right\},$$

which is a subgroup of order 8, hence a Sylow 2-subgroup of G. Therefore, there are $168/8 = 21$ elements of order 2.

(c) If $m(x) = (x+1)^3$, then $(A+I_3)^3 = 0$, and $A^4 + I_3 = (A+I_3)^4 = 0$, i.e., A is of order 4. Conversely, if A is of order 4, i.e., $A^4 = I_3$, then $x^4 + 1 = (x+1)^4$ is an annihilating polynomial of A, which implies that $f(x) = (x+1)^3$. Therefore, any element of order 4 must be conjugate with

$$T_4 = \begin{pmatrix} 1 & 1 & 0 \\ 0 & 1 & 1 \\ 0 & 0 & 1 \end{pmatrix}.$$

Furthermore, it is easy to see that

$$C_G(T_4) = \left\{ \begin{pmatrix} 1 & a & b \\ 0 & 1 & a \\ 0 & 0 & 1 \end{pmatrix} \middle| a, b \in \mathbb{F}_2 \right\},$$

which implies that there are $168/4 = 42$ elements of order 4.

(2) If $f(x) = x^3 + x + 1$ or $x^3 + x^2 + 1$, then $f(x) \mid (x^7 + 1)$, which implies that A is of order 7. Conversely, if A is of order 7, then $f(X)$ must be one of $x^3 + x + 1$ or $x^3 + x^2 + 1$. Since $f(x)$ is irreducible, A must be conjugate with $T_7 = \begin{pmatrix} 0 & 0 & 1 \\ 1 & 0 & 1 \\ 0 & 1 & 0 \end{pmatrix}$ or $S_7 = \begin{pmatrix} 0 & 0 & 1 \\ 1 & 0 & 0 \\ 0 & 1 & 1 \end{pmatrix}$. Noticing that every element in $C_G(T_7)$ must be a polynomial of T_7 (why?), we conclude that $C_G(T_7)$ has at most 7 element. Hence, $C_G(T_7) = \langle T_7 \rangle$ since it contains $\langle T_7 \rangle$. Similarly, $C_G(S_7) = \langle S_7 \rangle$. Thus, elements of order 7 run into two conjugacy classes, each having $168/7 = 21$ elements. It is easy to see that an element A of order 7 is not conjugate with its inverse A^{-1} since their characteristic polynomials are different.

(3) If $f(x) = x^3 + 1 = (x + 1)(x^2 + x + 1)$, then A is of order 3, and it is conjugate with

$$T_3 = \begin{pmatrix} 1 & 0 & 0 \\ 0 & 0 & 1 \\ 0 & 1 & 1 \end{pmatrix}.$$

Thus, all elements of order 3 are conjugate. It is easy to see that $C_G(T_3) = \langle T_3 \rangle$, which implies that there are $168/3 = 56$ elements of order 3.

We are now ready to prove that G is simple. Note that a normal subgroup of G must be a union of conjugacy classes. If a normal subgroup H of G contains elements of order 2, then it contains $I_3 + E_{ij}$, $i \neq j$. Here, E_{ij} denotes the matrix with 1 in the (i, j) entry and 0 elsewhere. But $I_3 + E_{ij}$ generate G. Hence, $H = G$. Therefore, a proper normal subgroup does not contain elements of order 2 and 4. Noticing that the order of a subgroup of G must be a divisor of $|G|$, one can quickly see that G is simple. □

Problem 4.7.10. Show that a simple group of order 168 must be isomorphic to $SL(3, \mathbb{F}_2)$.

Exercises

(1) Find all irreducible polynomials in $\mathbb{F}_3[x]$ of degree at most 2.
(2) Generalize all the results in this section to finite fields $\mathbb{F}_q$ with $q = p^n$.

(3) Let p be a prime, $q = p^k$, $k \in \mathbb{N}^*$, and let V be an n-dimensional vector space over $\mathbb{F}_q$.

   (i) Determine the number of one-dimensional subspaces of V.

   (ii) Show that $\mathrm{GL}(n, \mathbb{F}_q)$ has a cyclic subgroup of order $q^n - 1$.

(4) Find an isomorphism between $\mathbb{F}_3(u)$ and $\mathbb{F}_3(v)$, where u, v are roots of $x^2 + 1, x^2 + x + 2 \in \mathbb{F}_3[x]$, respectively.

(5) (i) In a field $\mathbb{E} = \mathbb{F}_p(u)$ of p^n elements, is u necessarily a generator of the cyclic group $\mathbb{E}^*$?

   (ii) Let $2^n - 1$ be a prime, and let $\mathbb{E} = \mathbb{F}_2(u)$ be a field of order 2^n. Is u a generator of the cyclic group $\mathbb{E}^*$?

(6) A monic irreducible polynomial $f(x) \in \mathbb{F}_q[x]$ of degree n is called **primitive** if some root u of $f(x)$ is a generator of the cyclic group $\mathbb{F}_q(u)^*$.

   (i) Show that $x^4 + x + 1 \in \mathbb{F}_2[x]$ is primitive.

   (ii) Show that $x^4 + x^3 + x^2 + x + 1 \in \mathbb{F}_2[x]$ is irreducible but not primitive.

   (iii) Let u be a root of $x^4 + x^3 + x^2 + x + 1 \in \mathbb{F}_2[x]$. Determine all the generators of the cyclic group $\mathbb{F}_2(u)^*$.

(7) Show that the set $\left\{ \begin{pmatrix} 1 & a & b \\ 0 & 1 & c \\ 0 & 0 & 1 \end{pmatrix} \,\middle|\, a, b, c \in \mathbb{F}_q \right\}$ is a non-abelian group of order q^3 under matrix multiplication.

(8) Determine the number of Sylow p-subgroups of $G = \mathrm{GL}_n(\mathbb{F}_p)$.

(9) Let $f(x)$ be a monic polynomial with integer coefficients. Regard $f(x)$ as a polynomial over $\mathbb{F}_p$, denoted by $f_p(x)$.

   (i) If $f_p(x)$ is irreducible for some prime p, then $f(x)$ is irreducible.

   (ii) If $f_p(x)$ is reducible for any prime p, is $f(x)$ reducible?

Exercises (hard)

(10) Let $\mathbb{F}$ be a field. Show that $\mathbb{F}$ is finite if and only if $\mathbb{F}^*$ is cyclic.

(11) Let R be a finite ring without zero divisors.

   (i) Show that R is a division ring.

   (ii) Show that there exists a prime p and a positive integer n such that $|R| = p^n$.

   (iii) Prove **Wedderburn's theorem**: R is a field.

4.8 Separable polynomials and perfect fields

To study separable extensions, we need to know whether a polynomial is separable or whether an irreducible polynomial has multiple roots. In linear algebra, when we study multiple factors of polynomials, we introduce the notion of formal derivative, which can be generalized as follows.

Definition 4.8.1. Let $f(x) = a_n x^n + a_{n-1} x^{n-1} + \cdots + a_0 \in \mathbb{F}[x]$. The polynomial

$$f'(x) = na_n x^{n-1} + (n-1)a_{n-1} x^{n-2} + \cdots + a_1$$

is call the **formal derivative** of $f(x)$.

The following results follow quickly.

Proposition 4.8.2. Let $f(x), g(x) \in \mathbb{F}[x]$, $a, b \in \mathbb{F}$.

(1) If $\deg f(x) \leq 0$, then $f'(x) = 0$. If $\mathrm{Ch}\ \mathbb{F} = 0$ and $f'(x) = 0$, then $\deg f(x) \leq 0$.
(2) $x' = 1$.
(3) $(af(x) + bg(x))' = af'(x) + bg'(x)$.
(4) $(f(x)g(x))' = f'(x)g(x) + f(x)g'(x)$.

The above properties completely determine the formal derivation (see Exercise 1). It is worth noting that if $\mathrm{Ch}\ \mathbb{F} = p > 0$, the second assertion in (1) is false. For example, $(x^p)' = px^{p-1} = 0$.

Theorem 4.8.3. *Let $\mathbb{E}$ be a splitting field of $f(x) \in \mathbb{F}[x]$, and let $\alpha \in \mathbb{E}$ be a root of $f(x)$ with multiplicity k. If $\mathrm{Ch}\ \mathbb{F} \nmid k$, then α is a root of $f'(x)$ with multiplicity $k - 1$; if $\mathrm{Ch}\ \mathbb{F} \mid k$, then α is a root of $f'(x)$ with multiplicity at least k.*

Proof. Let $f(x) = (x - \alpha)^k g(x)$, where $g(x) \in \mathbb{E}[x]$ and $g(\alpha) \neq 0$. Then

$$f'(x) = k(x - \alpha)^{k-1} g(x) + (x - \alpha)^k g'(x)$$
$$= (x - \alpha)^{k-1}(kg(x) + (x - \alpha)g'(x)).$$

If $\mathrm{Ch}\ \mathbb{F} \nmid k$, then $kg(\alpha) + (\alpha - \alpha)g'(\alpha) = kg(\alpha) \neq 0$, which implies that α is a root of $f'(x)$ with multiplicity $k - 1$. If $\mathrm{Ch}\ \mathbb{F} \mid k$, then

$kg(x) = 0$, which implies $f'(x) = (x - \alpha)^k g'(x)$. Thus, α is a root of $f'(x)$ with multiplicity at least k. $\qquad\square$

Let $f(x), g(x) \in \mathbb{F}[x]$. Denote by $(f(x), g(x))$ the (monic) greatest common divisor of $f(x), g(x)$. Since $\mathbb{F}[x]$ is a Euclidean domain, $(f(x), g(x))$ can be obtained via the Euclidean algorithm. It follows that $(f(x), g(x))$ is independent of the base field, i.e., if $\mathbb{E}$ is an extension of $\mathbb{F}$, then the greatest common divisor of $f(x), g(x)$ in $\mathbb{E}[x]$ is the same as that in $\mathbb{F}[x]$.

Now, we are ready to give the following criteria.

Theorem 4.8.4. *Let $\mathbb{E}$ be a splitting field of $f(x) \in \mathbb{F}$. Then $f(x)$ has no multiple roots in $\mathbb{E}$ if and only if $(f(x), f'(x)) = 1$.*

Proof. If α is a root of $f(x)$ with multiplicity $k > 1$. By Theorem 4.8.3, α is a root of $f'(x)$ with multiplicity at least $k - 1$. Thus, in $\mathbb{E}[x]$, we have $(x - \alpha)^{k-1} \mid (f(x), f'(x))$. Hence, $(f(x), f'(x)) \neq 1$.

Conversely, if $f(x)$ has no multiple roots in $\mathbb{E}$, then

$$f(x) = c(x - \alpha_1)(x - \alpha_2) \cdots (x - \alpha_n),$$

where $c \in \mathbb{K}$, and $\alpha_i \neq \alpha_j$ for $i \neq j$. By Theorem 4.8.3, α_i ($i = 1, 2, \ldots, n$) are not roots of $f'(x)$. Hence, $(f(x), f'(x)) = 1$. $\qquad\square$

Corollary 4.8.5. *Let $p(x) \in \mathbb{F}[x]$ be irreducible. Then $p(x)$ has multiple roots in a splitting field if and only if $p'(x) = 0$.*

Proof. We may assume that $p(x)$ is monic. Then the monic factors of $p(x)$ are itself and 1. Thus, $(p(x), p'(x)) = p(x)$ or 1. Hence, $p(x)$ has multiple roots if and only if $p(x) \mid p'(x)$ if and only if $p'(x) = 0$, since $\deg p'(x) < \deg p(x)$. $\qquad\square$

If $\mathrm{Ch}\,\mathbb{F} = 0$, then $\deg p'(x) \geq 0$ for any irreducible polynomial $p(x)$. Thus, $p'(x) \neq 0$. Hence, we have the following.

Corollary 4.8.6. *If $\mathrm{Ch}\,\mathbb{F} = 0$, any irreducible polynomial in $\mathbb{F}[x]$ is separable. Consequently, any polynomial over $\mathbb{F}$ is separable.*

For convenience, we introduce the following definition.

Definition 4.8.7. If any polynomial over $\mathbb{F}$ is separable, then $\mathbb{F}$ is called a **perfect field**.

Therefore, fields with characteristic 0 are complete. If $\text{Ch } \mathbb{F} = p > 0$, and if $f(x) = a_n x^n + a_{n-1} x^{n-1} + \cdots + a_0 \in \mathbb{F}[x]$ is an inseparable irreducible polynomial, then by Corollary 4.8.5 we have $f'(x) = n a_n x^{n-1} + \cdots + a_1 = 0$. Hence, $l a_l = 0$, $l = 1, 2, \ldots, n$, i.e., $a_l = 0$ if $l \nmid p$, i.e.,

$$f(x) = a_{mp} x^{mp} + a_{(m-1)p} x^{(m-1)p} + \cdots + a_p x^p + a_0.$$

Let $f_1(x) = a_{mp} x^m + a_{(m-1)p} x^{m-1} + a_p x + a_0$, which must be irreducible. Then $f(x) = f_1(x^p)$. If $f_1(x)$ is inseparable, we may find $f_2(x)$ such that $f_1(x) = f_2(x^p)$ and $f(x) = f_2(x^{p^2})$. After finite steps, we obtain in $\mathbb{F}[x]$ a separable irreducible polynomial

$$h(x) = a_{mp^k} x^m + a_{(m-1)p^k} x^{m-1} + \cdots + a_{p^k} x + a_0 \qquad (4.8)$$

such that

$$f(x) = h(x^{p^k}),$$

and in a splitting field of $h(x)$, we have

$$h(x) = c(x - \beta_1)(x - \beta_2) \cdots (x - \beta_r),$$

where $\beta_i \neq \beta_j$ for $i \neq j$. Therefore,

$$f(x) = c(x^{p^k} - \beta_1)(x^{p^k} - \beta_2) \cdots (x^{p^k} - \beta_r).$$

Assume that α_i is a root of $x^{p^k} - \beta_i$. Then $\beta_i = \alpha_i^{p^k}$. Hence, we have the following.

Theorem 4.8.8. *Let $\text{Ch } \mathbb{F} = p > 0$, and let $f(x) \in \mathbb{F}[x]$ be an inseparable irreducible polynomial. Then in a splitting field of $f(x)$, we have the decomposition*

$$f(x) = c(x - \alpha_1)^{p^k}(x - \alpha_2)^{p^k} \cdots (x - \alpha_r)^{p^k},$$

where $k \in \mathbb{N}^*$, $\alpha_i \neq \alpha_j$, for $i \neq j$, and

$$h(x) = c(x - \alpha_1^{p^k})(x - \alpha_2^{p^k}) \cdots (x - \alpha_r^{p^k})$$

is a separable irreducible polynomial in $\mathbb{F}[x]$.

Now, we focus on polynomials over a field $\mathbb{F}$ of characteristic p. Noticing that $\mathrm{Fr}(a) = a^p$ is the Frobenius endomorphism on $\mathbb{F}$, we have

$$(a + b)^p = a^p + b^p, \quad \forall a, b \in \mathbb{F}.$$

If Fr is an automorphism, i.e., $\mathbb{F} = \mathbb{F}^p = \{b^p | b \in \mathbb{F}\}$, then in (4.8), we have $b_i \in \mathbb{F}$ such that $a_{ip^k} = b_i^p$, $i = 0, 1, \ldots, m$. Thus,

$$f(x) = b_m^p x^{mp^k} + \cdots + b_1^p x^{p^k} + b_0^p = (b_m x^{mp^{k-1}} + \cdots + b_1 x^{p^{k-1}} + b_0)^p,$$

which leads to a contradiction since $f(x)$ is irreducible. Therefore, any polynomial over $\mathbb{F}$ is separable, i.e., $\mathbb{F}$ is perfect.

Conversely, if $\mathbb{F} \neq \mathbb{F}^p$, then there exists $a \in \mathbb{F}$ such that $a \neq b^p$ for any $b \in \mathbb{F}$. Set $f(x) = x^p - a$. Then $f'(x) = 0$. Let $\mathbb{E}$ be a splitting field of $f(x)$, $\alpha \in \mathbb{E}$ a root of $f(x)$, i.e., $a = \alpha^p$. Then in $\mathbb{E}[x]$, we have

$$f(x) = (x - \alpha)^p.$$

If $f(x)$ is reducible in $\mathbb{F}[x]$, say, $f(x) = g(x)h(x)$, where $g(x), h(x)$ are monic polynomials in $\mathbb{F}[x]$ of positive degree, then we must have

$$g(x) = (x - \alpha)^r, \quad h(x) = (x - \alpha)^{p-r},$$

where $0 < r < p$. Hence, $\alpha^r \in \mathbb{F}$. Since p is a prime number, there exist $u, v \in \mathbb{Z}$ such that $up + vr = 1$. Then

$$\alpha = \alpha^{(up+vr)} = (\alpha^p)^u (\alpha^r)^v = a^u (\alpha^r)^v \in \mathbb{F},$$

which implies that $a = \alpha^p \in \mathbb{F}^p$, a contradiction. Thus, $f(x)$ is irreducible and inseparable over $\mathbb{F}$.

In conclusion, we have the following criteria.

Theorem 4.8.9. *A field $\mathbb{F}$ of characteristic $p > 0$ is perfect if ana only if $\mathbb{F} = \mathbb{F}^p$, i.e., the Frobenius endomorphism of $\mathbb{F}$ is an automorphism.*

The Frobenius endomorphism of a finite field $\mathbb{F}$ must be an automorphism. Hence, we have the following.

Corollary 4.8.10. *A finite field is perfect.*

Furthermore, we have the following.

Corollary 4.8.11. *Algebraic extensions of perfect fields are also perfect.*

Proof. Let $\mathbb{F}$ be a perfect field, $\mathbb{E}/\mathbb{F}$ an algebraic extension. If Ch $\mathbb{F} = 0$, then Ch $\mathbb{E} = 0$ and $\mathbb{E}$ is perfect. Now, we consider the case when Ch $\mathbb{F} = p > 0$. Since $\mathbb{F}$ is perfect, we have $\mathrm{Fr}(\mathbb{F}) = \mathbb{F}$. For any $\alpha \in \mathbb{E}$, $\mathbb{K} = \mathbb{F}(\alpha)$ is a finite extension of $\mathbb{F}$. Thus, $\mathrm{Fr}(\mathbb{K}) = \mathbb{F}(\alpha^p)$, i.e., $\mathbb{F}(\alpha^p)$ is an intermediate field of $\mathbb{K}/\mathbb{F}$. By

$$[\mathrm{Fr}(\mathbb{K}) : \mathbb{F}] = [\mathrm{Fr}(\mathbb{K}) : \mathrm{Fr}(\mathbb{F})] = [\mathbb{K} : \mathbb{F}],$$

we get that, as $\mathbb{F}$-vector spaces, the subspace $\mathrm{Fr}(\mathbb{K})$ has the same dimension as $\mathbb{K}$, which implies $\mathrm{Fr}(\mathbb{K}) = \mathbb{K}$, i.e., Fr is an automorphism. Hence, $\mathbb{E}$ is perfect. □

To find a non-perfect field, we must consider transcendental extensions of a field of characteristic p. The simplest case is the following.

Example 4.8.12. The simple transcendental extension $\mathbb{E} = \mathbb{F}_p(t)$ of a finite field $\mathbb{F}_p$ is not perfect.

If there exists $f(t)/g(t) \in \mathbb{E}$, where $f(t), g(t) \in \mathbb{F}_p[t]$, such that $(f(t)/g(t))^p = t$, then $f(t)^p - tg(t)^p = 0$, which contradicts the fact that t is transcendental over $\mathbb{F}_p$. Hence, $\mathbb{E}^p \neq \mathbb{E}$, i.e., $\mathbb{F}_p(t)$ is not perfect. Similar to the proof of Theorem 4.8.9, we can prove that $x^p - t$ is an inseparable irreducible polynomial.

Exercises

(1) Let $\mathbb{F}$ be a field. A linear transformation D on $\mathbb{F}[x]$ is called a **derivative**, if

$$D(f(x)g(x)) = D(f(x))g(x) + f(x)D(g(x)).$$

   (i) Determine the set Der $\mathbb{F}[x]$ of all derivatives of $\mathbb{F}[x]$.
   (ii) Show that Der $\mathbb{F}[x]$ is a free left $\mathbb{F}[x]$-module.
   (iii) If $D_1, D_2 \in$ Der $\mathbb{F}[x]$, show that $[D_1, D_2] = D_1 D_2 - D_2 D_1 \in$ Der $\mathbb{F}[x]$. Do we have $D_1 D_2 \in$ Der $\mathbb{F}[x]$?

(2) Let Ch $\mathbb{F} = p$, $\alpha \in \mathbb{F}$, $\alpha \notin \mathbb{F}^p$. Show that, for any $n \in \mathbb{N}$, $x^{p^n} - \alpha$ is irreducible in $\mathbb{F}[x]$.

(3) Let $\mathrm{Ch}\,\mathbb{F} = p$, $\mathbb{F}(\alpha_1, \alpha_2, \ldots, \alpha_m)$ an extension of $\mathbb{F}$, where $\alpha_1, \alpha_2, \ldots, \alpha_m$ are algebraically independent over $\mathbb{F}$. Show that

  (i) $[\mathbb{F}(\alpha_1, \alpha_2, \ldots, \alpha_m) : \mathbb{F}(\alpha_1^p, \alpha_2^p, \ldots, \alpha_m^p)] = m$;
  (ii) $\mathrm{Gal}\,(\mathbb{F}(\alpha_1, \alpha_2, \ldots, \alpha_m)/\mathbb{F}(\alpha_1^p, \alpha_2^p, \ldots, \alpha_m^p))$ is a trivial group.

Exercises (hard)

(4) Let $\mathbb{E}/\mathbb{F}$ be a finite extension. Show that $\mathbb{E}$ is perfect if and only if so is $\mathbb{F}$. Does the conclusion still hold for any algebraic extension $\mathbb{E}/\mathbb{F}$?

4.9   Separable extensions

Since every polynomial over a perfect field is separable, any algebraic extension over a perfect field is separable. In particular, algebraic extensions over finite fields or fields of characteristic 0 are separable. The following result is helpful for any finite extension $\mathbb{E}/\mathbb{F}$.

Proposition 4.9.1. *Let* $\mathbb{E} = \mathbb{F}(\alpha_1, \alpha_2, \ldots, \alpha_n)$ *be a finite extension of* $\mathbb{F}$, *and let* $\alpha_1, \alpha_2, \ldots, \alpha_n$ *be separable over* $\mathbb{F}$. *Then* $\mathbb{E}/\mathbb{F}$ *is a separable extension.*

Proof.   Since α_i, $i = 1, 2, \ldots, n$, are separable, all of $\mathrm{Irr}(\alpha_i, \mathbb{F})$ are irreducible separable polynomials. Then $f(x) = \prod_{i=1}^{n} \mathrm{Irr}(\alpha_i, \mathbb{F})$ is separable. Let $\mathbb{K}$ be a splitting field of $f(x)$ over $\mathbb{E}$. Then $\mathbb{K}$ is a splitting field of $f(x)$ over $\mathbb{F}$. By Theorem 4.6.10, $\mathbb{K}/\mathbb{F}$ is separable. Hence, every element in $\mathbb{E}$ is separable over $\mathbb{F}$, which implies that $\mathbb{E}/\mathbb{F}$ is separable.   □

We may deduce the following result quickly from the above proposition.

Corollary 4.9.2. *Let* $\mathbb{E}/\mathbb{F}$ *be an algebraic extension,* $\alpha, \beta \in \mathbb{E}$. *If* α, β *are separable over* $\mathbb{F}$, *then* $\alpha \pm \beta$, $\alpha\beta$, $\alpha\beta^{-1}(\beta \neq 0)$ *are all separable over* $\mathbb{F}$.

Let $\mathbb{E}/\mathbb{F}$ be an algebraic extension, and let $\mathbb{E}_0$ be the set of elements in $\mathbb{E}$ which are separable over $\mathbb{F}$. By the above corollary, $\mathbb{E}_0$ is a subfield of $\mathbb{E}$, called the **separable closure** of $\mathbb{F}$ in $\mathbb{E}$. If $\mathrm{Ch}\,\mathbb{F} = 0$, then $\mathbb{E}_0 = \mathbb{E}$. Therefore, we only need to pay attention to fields of positive characteristic. The following theorem reveals a special property of finite separable extensions.

Theorem 4.9.3. *Let* $\mathbb{E}/\mathbb{F}$ *be a finite separable extension. Then* $\mathbb{E}/\mathbb{F}$ *is a simple algebraic extension, i.e., there exists* $\alpha \in \mathbb{E}$ *such that* $\mathbb{E} = \mathbb{F}(\alpha)$. *We call* α *a **primitive element** of* $\mathbb{E}/\mathbb{F}$.

Proof. If $\mathbb{F}$ is finite, then so is $\mathbb{E}$ and $\mathbb{E}^*$ is cyclic with one generator. Hence, $\mathbb{E}/\mathbb{F}$ is a simple extension. In the following, we assume that $\mathbb{F}$ is infinite.

Method 1. Let $\mathbb{E} = \mathbb{F}(\alpha_1, \alpha_2, \ldots, \alpha_n)$ be a finite separable extension over $\mathbb{F}$. Set $f(x) = \prod_{i=1}^{n} \mathrm{Irr}(\alpha_i, \mathbb{F})$. Then $f(x)$ is separable over $\mathbb{F}$. Let $\mathbb{K}/\mathbb{F}$ be the splitting field of $f(x)$. Then $\mathbb{K}/\mathbb{F}$ is a Galois extension and has only finitely many intermediate fields. Therefore, $\mathbb{E}/\mathbb{F}$ has only finitely many intermediate fields. Since $\mathbb{E}$ is a vector space over $\mathbb{F}$, the union of finitely many proper subspaces of $\mathbb{E}$ is not the whole space $\mathbb{E}$ (Why?). Thus, there exists $\alpha \in \mathbb{E}$, which is not contained in any proper intermediate field of $\mathbb{E}/\mathbb{F}$. Hence, $\mathbb{F}(\alpha) = \mathbb{E}$.

Method 2. By induction, we need to prove that $\mathbb{E} = \mathbb{F}(\beta_1, \gamma_1)$ is a simple extension over $\mathbb{F}$. Suppose $\mathbb{E} \neq \mathbb{F}(\gamma_1)$. If $\mathbb{E} = \mathbb{F}(\alpha)$, then α must be of the form $\beta_1 + c\gamma_1$ for some $c \in \mathbb{F}$. We want to find $c \in \mathbb{F}$ such that $\gamma_1 \in \mathbb{F}(\alpha)$, i.e., $\mathrm{Irr}(\gamma_1, \mathbb{F}(\alpha))$ is of degree 1.

Set $f(x) = \mathrm{Irr}(\beta_1, \mathbb{F})$, $g(x) = \mathrm{Irr}(\gamma_1, \mathbb{F})$. Then $\mathrm{Irr}(\gamma_1, \mathbb{F}(\alpha)) \mid g(x)$. Considering the polynomial $h(x) = f(\alpha - cx)$ over $\mathbb{F}(\alpha)$, one has $h(\gamma_1) = f(\alpha - c\gamma_1) = f(\beta_1) = 0$. Hence, γ_1 is a common root of $h(x)$ and $g(x)$. Let $\mathbb{K}$ be a splitting field of $f(x)g(x)$ over $\mathbb{F}$. Then $(x - \gamma_1) \mid (g(x), h(x))$ in $\mathbb{K}[x]$. If we can choose c such that γ_1 is the only common root of $g(x), h(x)$, then $(g(x), h(x)) = x - \gamma_1$. Since the greatest common divisor is independent of the choice of base fields, it follows that $x - \gamma_1 \in \mathbb{F}(\alpha)[x]$, i.e., $\gamma_1 \in \mathbb{F}(\alpha)$.

Therefore, we need to choose suitable c such that γ_1 is the only common root of $h(x)$ and $g(x)$. Consider the following factorization in $\mathbb{K}[x]$

$$f(x) = \prod_{i=1}^{s}(x - \beta_i), \quad g(x) = \prod_{j=1}^{r}(x - \gamma_j).$$

Since γ_1 is separable over $\mathbb{F}$, $g(x)$ has no multiple roots. Furthermore,

$$h(x) = f(\alpha - cx) = \prod_{i=1}^{n}(\alpha - cx - \beta_i) = \prod_{i=1}^{n}((\beta_1 - \beta_i) + c(\gamma_1 - x)),$$

and all roots of $h(x)$ are $\gamma_1, \gamma_1 + \frac{\beta_1-\beta_2}{c}, \ldots, \gamma_1 + \frac{\beta_1-\beta_s}{c}$. Consider the finite subset of $\mathbb{F}$:

$$\left\{ \frac{\beta_1 - \beta_i}{\gamma_j - \gamma_1} \,\middle|\, 1 \le i \le s, 2 \le j \le r \right\}.$$

There must be a non-zero element $c \in \mathbb{F}$ such that

$$c \ne \frac{\beta_1 - \beta_i}{\gamma_j - \gamma_1}, \quad 1 \le i \le s, \quad 2 \le j \le r.$$

Hence, $g(x), h(x)$ has only one common root γ_1 in $\mathbb{K}$, which finishes the proof of the theorem. $\qquad\square$

Now, we shall focus on simple extensions.

Lemma 4.9.4. *Let* $\text{Ch } \mathbb{F} = p > 0$. *Then a simple algebraic extension* $\mathbb{F}(\beta)/\mathbb{F}$ *is separable if and only if* $\mathbb{F}(\beta) = \mathbb{F}(\beta^p)$.

Proof. Let β be separable over $\mathbb{F}$. Then β is also separable over $\mathbb{F}(\beta^p)$. Set $f(x) = x^p - \beta^p \in \mathbb{F}(\beta^p)[x]$. Hence, $f(\beta) = 0$, and $f(x) = (x - \beta)^p$ over any splitting field of $f(x)$. Thus, $\text{Irr}(\beta, \mathbb{F}(\beta^p)) = x - \beta$, i.e., $\beta \in \mathbb{F}(\beta^p)$. Consequently, $\mathbb{F}(\beta) = \mathbb{F}(\beta^p)$.

If β is inseparable over $\mathbb{F}$, then there exists a separable irreducible polynomial $g(x) \in \mathbb{F}[x]$ such that $\text{Irr}(\beta, \mathbb{F}) = g(x^{p^e})$. Hence, β^p is a root of $g(x^{p^{e-1}})$, and $\deg(\beta, \mathbb{F}) > \deg(\beta^p, \mathbb{F})$. $\qquad\square$

By the above lemma, we can get the following result.

Lemma 4.9.5. *Let* $\mathbb{F}(\alpha, \beta)/\mathbb{F}$ *be an algebraic extension with* α *separable over* $\mathbb{F}$ *and* β *separable over* $\mathbb{F}(\alpha)$. *Then* $\mathbb{F}(\alpha, \beta)/\mathbb{F}$ *is a separable extension.*

Proof. We may assume that $\text{Ch } \mathbb{F} = p$, and only need to prove that β is separable over $\mathbb{F}$, i.e., $\mathbb{F}(\beta) = \mathbb{F}(\beta^p)$. Since α is separable over $\mathbb{F}$, α is also separable over $\mathbb{F}(\beta)$ and $\mathbb{F}(\beta^p)$. Set $g(x) = \text{Irr}(\alpha, \mathbb{F}(\beta))$,

$h(x) = \mathrm{Irr}(\alpha, \mathbb{F}(\beta^p))$. Then $g(x) \mid h(x)$. Noticing that $g(x)^p \in \mathbb{F}(\beta^p)[x] \subseteq \mathbb{F}(\beta)[x]$, we have $h(x) \mid g(x)^p$, which implies that $g(x) = h(x)$ since both $g(x)$ and $h(x)$ are separable. Therefore,

$$[\mathbb{F}(\alpha, \beta^p) : \mathbb{F}(\beta^p)] = [\mathbb{F}(\alpha, \beta) : \mathbb{F}(\beta)].$$

Since β is separable over $\mathbb{F}(\alpha)$, we have

$$\mathbb{F}(\alpha, \beta^p) = \mathbb{F}(\alpha)(\beta^p) = \mathbb{F}(\alpha)(\beta) = \mathbb{F}(\alpha, \beta).$$

Thus, $\mathbb{F}(\beta^p) = \mathbb{F}(\beta)$. $\qquad\square$

Now, we are ready to prove the transitivity of separable extensions.

Theorem 4.9.6. *If* $\mathbb{F} \subseteq \mathbb{K} \subseteq \mathbb{E}$*, then* $\mathbb{E}/\mathbb{K}, \mathbb{K}/\mathbb{F}$ *are separable if and only if* $\mathbb{E}/\mathbb{F}$ *is separable.*

Proof. It is easy to see that if $\mathbb{E}/\mathbb{F}$ is separable, then so are $\mathbb{E}/\mathbb{K}, \mathbb{K}/\mathbb{F}$.

Conversely, let $\beta \in \mathbb{E}$. Since β is separable over $\mathbb{K}$, the polynomial

$$\mathrm{Irr}(\beta, \mathbb{K}) = x^n + a_1 x^{n-1} + \cdots + a_n$$

is separable over $\mathbb{K}$. Noticing that $a_1, \ldots, a_n \in \mathbb{K}$ are separable over $\mathbb{F}$, we have $\mathbb{F}(a_1, \ldots, a_n) = \mathbb{F}(\alpha)$ for some $\alpha \in \mathbb{K}$. Hence, $\mathrm{Irr}(\beta, \mathbb{K}) \in \mathbb{F}(\alpha)[x]$, and β is separable over $\mathbb{F}(\alpha)$. Therefore, $\mathbb{F}(\alpha, \beta)$ is separable over $\mathbb{F}$ by the above lemma, and consequently, β is separable over $\mathbb{F}$, which implies that $\mathbb{E}/\mathbb{F}$ is separable. $\qquad\square$

Exercises

(1) Find a primitive element of $\mathbb{Q}(\sqrt{2}, \sqrt{3})/\mathbb{Q}$.

(2) Let $\mathbb{E}$ be a splitting field of $x^5 - 2 \in \mathbb{Q}[x]$. Find θ such that $\mathbb{E} = \mathbb{Q}(\theta)$.

(3) Let $\mathrm{Ch}\,\mathbb{F} = p > 0$, $\mathbb{F}(\alpha, \beta)$ an algebraic extension over $\mathbb{F}$ with α separable and β inseparable. Suppose $\deg(\alpha, \mathbb{F}) = n$ and $\deg(\beta, \mathbb{F}) = p$. Determine $[\mathbb{F}(\alpha, \beta) : \mathbb{F}]$.

(4) Let $\mathrm{Ch}\,\mathbb{F} = p \neq 0$. Suppose $\mathbb{K}/\mathbb{F}$ is an algebraic extension, and $\mathbb{K}_0$ is the separable closure of $\mathbb{F}$ in $\mathbb{K}$. Show that $\mathbb{K}/\mathbb{K}_0$ is an inseparable extension, and for any $\alpha \in \mathbb{K} \backslash \mathbb{K}_0$, there exists $s \in \mathbb{N}$ such that $\alpha^{p^s} \in \mathbb{K}_0$.

(5) Let $\mathbb{E}/\mathbb{F}$ be an algebraic extension, Ch $\mathbb{F} = p \neq 0$. We call $\alpha \in \mathbb{E}$ a **purely inseparable element** over $\mathbb{F}$, if $\alpha^{p^s} \in \mathbb{F}$ for some $s \in \mathbb{N}$. If every element in $\mathbb{E}$ is purely inseparable over $\mathbb{F}$, then $\mathbb{E}/\mathbb{F}$ is called a **purely inseparable extension**.

   (i) Show that $\alpha \in \mathbb{E}$ is purely inseparable over $\mathbb{F}$ if and only if $\mathrm{Irr}(\alpha, \mathbb{F})$ is of the form $x^{p^s} - a$, for some $a \in \mathbb{F}$;

   (ii) If $\alpha \in \mathbb{K}$ is both separable and purely inseparable over $\mathbb{F}$, then $\alpha \in \mathbb{F}$;

   (iii) Let $\mathbb{E}_0$ be the separable closure of $\mathbb{E}$ over $\mathbb{F}$. Show that $\mathbb{E}/\mathbb{E}_0$ is a purely inseparable extension.

(6) Let Ch $\mathbb{F} = p > 0$, $\mathbb{E} = \mathbb{F}(\alpha)$, and $\alpha^{p^n} \in \mathbb{F}$ for some $n \in \mathbb{N}$. Show that $\mathbb{E}/\mathbb{F}$ is a purely inseparable extension.

(7) Let Ch $\mathbb{F} = p > 0$, and let $\mathbb{K}$ be an intermediate field of $\mathbb{E}/\mathbb{F}$.

   (i) Show that $\mathbb{E}/\mathbb{F}$ is a purely inseparable extension if and only if so are $\mathbb{K}/\mathbb{F}$ and $\mathbb{E}/\mathbb{K}$.

   (ii) If $\mathbb{E}/\mathbb{F}$ is a finite normal extension, $\mathbb{K} = \mathrm{Inv}\,(\mathrm{Gal}\,(\mathbb{E}/\mathbb{F}))$, show that $\mathbb{K}/\mathbb{F}$ is a purely inseparable extension, and $\mathbb{E}/\mathbb{K}$ is a separable extension.

(8) Let $\mathbb{E}$ be an algebraic extension over a field $\mathbb{F}$ of prime characteristic, and $\mathbb{K}$ the set of purely inseparable elements in $\mathbb{E}$ over $\mathbb{F}$. Show that $\mathbb{K}$ is an intermediate field of $\mathbb{E}/\mathbb{F}$.

(9) Show that the Galois group of a finite purely inseparable extension is trivial.

(10) Let $\mathbb{K}/\mathbb{F}$ be a normal extension, and let $\mathbb{K}_0$ be the normal closure of $\mathbb{F}$ in $\mathbb{K}$. Show that $\mathbb{K}_0/\mathbb{F}$ is normal.

Exercises (hard)

(11) Let $\mathbb{E}/\mathbb{F}$ be a finite extension. Show that $\mathbb{E}/\mathbb{F}$ has only finitely many intermediate fields if and only if $\mathbb{E}/\mathbb{F}$ is a simple extension.

(12) Construct a finite extension which is not a simple extension.

(13) Let $\mathbb{E}/\mathbb{F}$ be a separable extension of degree n, $\overline{\mathbb{F}}$ an algebraic closure of $\mathbb{F}$. For any homomorphism $\sigma : \mathbb{F} \to \overline{\mathbb{F}}$, show that σ can be extended to exactly n homomorphisms from $\mathbb{E}$ into $\overline{\mathbb{F}}$. In particular, let $\sigma = \mathrm{id}_{\mathbb{F}}$. Then there exist exact n $\mathbb{F}$-homomorphisms from $\mathbb{E}$ to $\overline{\mathbb{F}}$.

(14) **(Fundamental Theorem of Algebra).** Let $\mathbb{K}$ be a finite extension of $\mathbb{C}$.

   (i) Show that there exists an extension $\mathbb{E}$ of $\mathbb{K}$ such that $\mathbb{E}/\mathbb{R}$ is a Galois extension.

  (ii) Let H be a Sylow 2-subgroup of $\text{Gal}\,(\mathbb{E}/\mathbb{R})$, and $\mathbb{F} = \mathbb{E}^H$. Show that $[\mathbb{F} : \mathbb{R}]$ is odd.

 (iii) Show that $\mathbb{F} = \mathbb{R}$. Consequently, $\text{Gal}\,(\mathbb{E}/\mathbb{R}) = H$.

 (iv) If $[\mathbb{K} : \mathbb{C}] > 1$, then there exists a subfield $\mathbb{K}_1$ of $\mathbb{K}$ such that $[\mathbb{K}_1 : \mathbb{C}] = 2$.

  (v) Show that $\mathbb{C}$ has no extension of degree 2. Hence, $\mathbb{C}$ is algebraically closed.

4.10 The inverse Galois problem

We know that a finite Galois extension or a splitting field of a separable polynomial corresponds to a finite group, the Galois group. A natural inverse problem is: is any finite group the Galois group of some finite Galois extension? Since a finite Galois extension $\mathbb{E}/\mathbb{F}$ is always a splitting field of a separable polynomial $f(x) \in \mathbb{F}[x]$, we also call $\text{Gal}\,(\mathbb{E}/\mathbb{F})$ the **Galois group** of $f(x)$, denoted by $G(f, \mathbb{F})$ or G_f.

We may assume that $f(x)$ has no multiple roots, otherwise, we replace $f(x)$ by $f(x)/(f(x), f'(x))$. Let $X = \{\alpha_1, \alpha_2, \ldots, \alpha_n\}$ be the set of all roots of $f(x)$ in a splitting field $\mathbb{E}$. By Lemma 4.5.2, G_f acts on X naturally, and the action is effective since $\mathbb{E} = \mathbb{F}(\alpha_1, \alpha_2, \ldots, \alpha_n)$. Thus, we have the following.

Lemma 4.10.1. *Let the notation be as above.*

(1) $G_f < S_X \simeq S_n$.

(2) *There is a 1-1 correspondence between irreducible factors of $f(x)$ and G_f-orbits of X. In particular, $f(x) \in \mathbb{F}[x]$ is irreducible if and only if G_f acts on X transitively.*

Proof. We only need to prove (2). Let $p(x)$ be a monic irreducible factor of $f(x)$ with the set of roots $X_1 = \{\alpha_1, \alpha_2, \ldots, \alpha_l\}$. It is easy to see that $\sigma(\alpha_1) \in X_1$, for any $\sigma \in G_f$. Assume that the orbit α_1 is

$O_1 = \{\alpha_1, \alpha_2, \ldots, \alpha_k\}$. Consider the polynomial

$$g(x) = \prod_{i=1}^{k} (x - \alpha_i).$$

Each coefficient of $g(x)$ is a symmetric polynomial in $\alpha_1, \alpha_2, \ldots, \alpha_k$, hence $g(x)$ is G_f-invariant. It follows that $g(x) \in \mathbb{F}[x]$. Thus, $g(x) = p(x)$ since $g(x) \mid p(x)$ and $p(x)$ is a monic irreducible polynomial. Therefore, there is a 1-1 correspondence between G_f-orbits in X and irreducible factors of $f(x)$. Consequently, the action of G_f on X is transitive if and only if $f(x)$ is irreducible. $\qquad\square$

Definition 4.10.2. A subgroup G of S_n is called **transitive**, if the action of G on $\{1, 2, \ldots, n\}$ is transitive.

Every finite group G is isomorphic to a subgroup of some symmetric group S_n. By the Galois correspondence, if there exists a Galois extension $\mathbb{E}/\mathbb{F}$ such that $\mathrm{Gal}\,(\mathbb{E}/\mathbb{F}) = S_n$, then $G = \mathrm{Gal}\,(\mathbb{E}/\mathbb{K})$, where $\mathbb{K} = \mathrm{Inv}\,(G)$. To find a Galois extension $\mathbb{E}/\mathbb{F}$ with the Galois group S_n, we need to find a field $\mathbb{E}$ such that S_n is a subgroup of $\mathrm{Aut}\,(\mathbb{E})$.

We consider the polynomial ring $R = \mathbb{F}[x_1, x_2, \ldots, x_n]$, where $\mathbb{F}$ is a field. There is a natural action of S_n on R defined by $\sigma(x_i) = x_{\sigma(i)}$ for any $\sigma \in S_n$. The fixed points of S_n are the set of all symmetric polynomials. Consider the fractional field $\mathbb{E} = \mathbb{F}(x_1, x_2, \ldots, x_n)$ of R. Then σ can be naturally extended to an automorphism of $\mathbb{E}$. Let

$$p_1 = x_1 + x_2 + \cdots + x_n,$$

$$p_2 = \sum_{1 \leq i < j \leq n} x_i x_j,$$

$$\ldots\ldots$$

$$p_n = \prod_{i=1}^{n} x_i$$

be the elementary symmetric polynomials. Then the intermediate field $\mathbb{K} = \mathbb{F}(p_1, p_2, \ldots, p_n)$ generated by $p_1, p_2, \ldots, p_n$ is contained in

Inv (S_n). Consider the polynomial

$$g(x) = (x - x_1)(x - x_2) \cdots (x - x_n)$$
$$= x^n - p_1 x^{n-1} + p_2 x^{n-2} - \cdots + (-1)^n p_n.$$

Then $g(x) \in \mathbb{K}[x]$. It follows that $\mathbb{E}$ is a splitting field of $g(x)$. Hence, $\mathbb{E}/\mathbb{K}$ is a Galois extension and $S_n < \mathrm{Gal}\,(\mathbb{E}/\mathbb{K})$. Since $\deg g(x) = n$, we have $[\mathbb{E} : \mathbb{K}] \le n!$ Thus,

$$n! = |S_n| \le |\,\mathrm{Gal}\,(\mathbb{E}/\mathbb{K})| = [\mathbb{E} : \mathbb{K}] \le n!,$$

which implies $\mathrm{Gal}\,(\mathbb{E}/\mathbb{K}) = S_n$. In summary, we have proved the following result.

Theorem 4.10.3. *Let the notation be as above. Then $\mathbb{E}/\mathbb{K}$ is a Galois extension with the Galois group S_n.*

In number theory, we pay more attention to finite extensions over $\mathbb{Q}$. The natural question is whether there is a Galois extension over $\mathbb{Q}$ such that the Galois group is S_n. The answer is affirmative. We will only consider the case when $n = p$ is a prime. The following lemma is crucial.

Lemma 4.10.4. *Let p be a prime. Then the symmetric group S_p is generated by any transposition and any p-cycle.*

Proof. By rearranging the order of $1, 2, \ldots, p$, we may assume that the transposition is $\tau = (12)$. For any p-cyclic $\gamma = (1i_2 \ldots i_p)$. Assume $i_k = 2$. Since p is a prime, γ^{k-1} is also a p-cycle and is of the form $(12j_3 \cdots j_p)$. Rearranging the order of $3, 4, \ldots, p$ again, we may assume that $\gamma^{k-1} = (123 \cdots p)$. Hence, we just need to prove that S_p can be generated by $\tau = (12)$ and $\sigma = (123 \cdots p)$. Noticing that

$$\sigma^i (12) \sigma^{-i} = (\sigma^i(1)\sigma^i(2)) = (i+1, i+2),$$

and $(12), (23), \ldots, (p-1, p)$ are generators of S_p, we conclude that (12) and $(12 \cdots p)$ are generators of S_p. $\qquad\square$

Proposition 4.10.5. *Let p be a prime, and let $f(x) \in \mathbb{Q}[x]$ be an irreducible polynomial of degree p. If $f(x)$ has exactly two non-real complex roots, then $G_f \simeq S_p$.*

Proof. Let $\mathbb{E}$ be a splitting field of $f(x)$, and let $\alpha \in \mathbb{E}$ be a root of $f(x)$. Then $[\mathbb{Q}(\alpha) : \mathbb{Q}] = p$, which implies that $p \mid [\mathbb{E} : \mathbb{Q}]$. By Cauchy's theorem, G_f has an element of order p. Furthermore, the complex conjugate $\tau(\alpha) = \overline{\alpha}$ preserves roots of $f(x)$, and consequently, $\tau(\mathbb{E}) = \mathbb{E}$, i.e., $\tau|_{\mathbb{E}} \in G_f$. Since $f(x)$ has only two non-real complex roots, τ is a transposition on the set of roots of $f(x)$. Thus, $G_f \simeq S_p$ by the above lemma. $\qquad\square$

With the above preparation, we can find many irreducible polynomials $f(x) \in \mathbb{Q}[x]$ with $G_f \simeq S_p$.

Example 4.10.6. Let $p \geq 5$ be a prime, m a positive number, and let $n_1 < n_2 < \cdots < n_{p-2}$ be integers. Then the polynomial

$$g(x) = (x^2 + m)(x - n_1) \cdots (x - n_{p-2}) \in \mathbb{Q}[x]$$

has exactly two non-real complex roots. However, $g(x)$ is not what we need as it is reducible. Since $n_1, n_2, \ldots, n_{p-2}$ are simple roots of $g(x)$, they are not extreme points of $g(x)$, and 0 is not an extreme value of $g(x)$. Set $a = \min_{g'(x)=0} |g(x)|$. Then $a > 0$. Choose a positive odd number n such that $\frac{1}{2n} < a$, and set

$$f(x) = g(x) + \frac{1}{2n}.$$

Then $f(x) \in \mathbb{Q}[x]$ still has $p - 2$ real roots and 2 non-real complex roots. By Eisenstein's criteria, the polynomial $2nf(x)$ is irreducible, so is $f(x)$. Hence, $G_f \simeq S_p$.

Some polynomials may not have exactly two non-real complex roots but still have S_p as their Galois groups. For example, $x^5 - 5x^3 + 4x + 1$ has five real roots, but its Galois group is S_5. In general, it is not easy to determine the Galois groups for such polynomials.

The inverse Galois problem concerns whether or not every finite group appears as the Galois group of some polynomial $f(x) \in \mathbb{Q}[x]$ or the Galois group of a Galois extension over $\mathbb{Q}$. Hilbert and his student Noether first proposed this problem, which is still open. Hilbert proved that S_n and A_n are the Galois groups of some rational polynomials, while Shafarevich showed that every finite solvable group is the Galois group of some Galois extension $\mathbb{E}/\mathbb{Q}$.

Exercises

(1) Assume that $f(x) = a_n x^n + a_{n-1} x^{n-1} + \cdots + a_0 \in \mathbb{F}[x]$ $(a_n \neq 0)$ is decomposed in a splitting field $\mathbb{E}$ as $f(x) = a_n \prod_{i=1}^{n} (x - \alpha_i)$. Set

$$\Delta(f) = \prod_{1 \leq i < j \leq n} (\alpha_j - \alpha_i), \quad D(f) = a_n^{2n-2} \Delta(f)^2.$$

We call $D(f)$ the **discriminant** of f. It is easy to see that $D(f) \neq 0$ if and only if f has no multiple roots.

  (i) Show that the discriminant of $ax^2 + bx + c$ is $b^2 - 4ac$.

 (ii) Show that the discriminant of $ax^3 + bx^2 + cx + d$ is $b^2 c^2 - 4ac^3 - 4b^3 d - 27a^2 d^2 + 18abcd$; in particular, if $a = 1$, $b = 0$, then the discriminant is $-4c^3 - 27d^2$.

(iii) Let $\sigma \in G(f, \mathbb{F}) < S_n$. Show that $\sigma \Delta(f) = \mathrm{sgn}(\sigma) \Delta(f)$ and $\sigma D(f) = D(f)$.

(iv) If $f(x) \in \mathbb{F}[x]$ has no multiple roots, show that $D(f) \in \mathbb{F}$, and $G(f, \mathbb{F}) < A_n$ if and only if $\Delta(f) \in \mathbb{F}$ if and only if $D(f)$ is a square in $\mathbb{F}$.

(2) Assume that $f(x) \in \mathbb{F}[x]$ has no multiple roots, $\mathbb{K}$ is an extension over $\mathbb{F}$, and $\mathbb{E}, \mathbb{E}'$ are the splitting fields of $f(x)$ over $\mathbb{F}$ and $\mathbb{K}$, respectively. Show that $\mathrm{Gal}\,(\mathbb{E}'/\mathbb{K})$ is isomorphic to a subgroup of $\mathrm{Gal}\,(\mathbb{E}/\mathbb{F})$.

(3) Assume that $f(x) \in \mathbb{F}[x]$ has no multiple roots, $\deg f(x) = n$, and $\mathbb{K}$ is a splitting field of $f(x)$. Let $u_1, u_2, \ldots, u_n$ be indeterminants, $\overline{\mathbb{F}} = \mathbb{F}(u_1, u_2, \ldots, u_n)$, and $\overline{\mathbb{K}}$ a splitting field of $f(x)$ over $\overline{\mathbb{F}}$. Show that $\mathrm{Gal}\,(\overline{\mathbb{K}}/\overline{\mathbb{F}}) \simeq \mathrm{Gal}\,(\mathbb{K}/\mathbb{F})$.

(4) Determine the Galois groups of the following $f(x) \in \mathbb{Q}[x]$:

   (i) $f(x) = x^5 - x - 1$; (ii) $f(x) = x^5 - 15x^2 + 9$.

(5) Let $\mathrm{Ch}\,\mathbb{F} = 2$. Determine the Galois groups of the following $f(x) \in \mathbb{F}[x]$:

   (i) $f(x) = x^3 + x + 1$; (ii) $f(x) = x^3 + x^2 + 1$.

Exercises (hard)

(6) (Dedekind's theorem) Let p be a prime, $f(x) \in \mathbb{Z}[x]$. Regard $f(x)$ as a polynomial over $\mathbb{F}_p = \mathbb{Z}/p\mathbb{Z}$, denoted by $f_p(x)$. If $f_p(x)$ has no multiple roots, show that $\mathrm{Gal}\,(f_p, \mathbb{F}_p)$ is isomorphic to a subgroup of $\mathrm{Gal}\,(f, \mathbb{Q})$. In particular, if $f_p(x) = g_1(x)g_2(x)\cdots g_r(x)$, where $g_i(x) \in \mathbb{F}_p[x]$ is an irreducible polynomial of degree n_i, then $G(f, \mathbb{Q})$ contains a permutation which is the product of r cycles of length $n_1, n_2, \ldots, n_r$, respectively.

(7) Let $g(x) \in \mathbb{F}_2[x], h(x) \in \mathbb{F}_3[x], k(x) \in \mathbb{F}_5[x]$ be monic irreducible polynomials such that $\deg g(x) = n \geq 5$, $\deg h(x) = n - 1$ and $\deg k(x) = 2$. Regard $g(x), h(x), k(x)$ as polynomials with integer coefficients, and set $f(x) = -15g(x) + 10xh(x) + 6x(x+1)\cdots(x+ n-3)k(x)$. Show that

   (i) $f(x)$ is irreducible;

   (ii) $G(f, \mathbb{Q})$ contains an n-cyclic, an $(n-1)$-cycle and a transposition;

   (iii) $G(f, \mathbb{Q}) \simeq S_n$.

4.11 Abelian extensions

Let $\mathbb{E}/\mathbb{F}$ be a Galois extension. We say that $\mathbb{E}/\mathbb{F}$ is a **solvable** (**abelian**, or **cyclic**) **extension** if $\mathrm{Gal}\,(\mathbb{E}/\mathbb{F})$ is a solvable (abelian, or cyclic) group. In this section, we will mainly deal with abelian extensions, including cyclotomic, Kummer, and cyclic extensions, which will be used to study the solvability of algebraic equations in the next section. Abel found that some equations of degree ≥ 5 are solvable by radicals (see Section 4.12). The key point is that their Galois groups are commutative, and this is the reason why commutative groups are called abelian groups.

We first discuss cyclotomic extensions, which are a special class of abelian extensions and play a fundamental role in number theory.

Definition 4.11.1. Let $\mathbb{E}$ be a splitting field of $x^n - 1 \in \mathbb{F}[x]$. A root of $x^n - 1$ is called an nth **root of unity**. All the nth roots of unity in $\mathbb{E}$ form a cyclic group. If this group is of order n, then any element θ_n of order n is a generator, called a **primitive nth root of unity**.

If $\mathbb{E}$ contains a primitive nth root of unity, then $\mathbb{E}/\mathbb{F}$ is called an nth **cyclotomic extension**, or we call $\mathbb{E}$ an nth **cyclotomic field**.

Although $\mathbb{E}$ is a splitting field of $x^n - 1$, the primitive nth roots of unity may not exist. If $\mathrm{Ch}\ \mathbb{F} = p$, $n = p^k m$, $(m, p) = 1$, then

$$x^n - 1 = (x^m)^{p^k} - 1 = (x^m - 1)^{p^k}.$$

Hence, $x^n - 1$ has multiple factors. Thus, we have the following.

Lemma 4.11.2. *If a field of characteristic $p > 0$ has primitive nth root of unity, then $(n, p) = 1$.*

If $(n, p) = 1$, then $(x^n - 1)' = nx^{n-1}$, and $(x^n - 1, nx^{n-1}) = 1$. Thus, $x^n - 1$ has no multiple roots. Consequently, any splitting field $\mathbb{E}$ of $x^n - 1$ has n different nth roots of unity, which form a cyclic group of order n. Let $\theta_n \in \mathbb{E}$ be a primitive nth root of unity. Then the set of roots of $x^n - 1$ is $1, \theta_n, \theta_n^2, \ldots, \theta_n^{n-1}$. If θ_n^k is also a primitive nth root of unity, then there exists $l \in \mathbb{N}$ such that $\theta_n = (\theta_n^k)^l = \theta_n^{kl}$, i.e., $kl \equiv 1 \pmod{n}$, or equivalently, $(k, n) = 1$. Conversely, if $(k, n) = 1$, then there exist $l, m \in \mathbb{Z}$ such that $kl + nm = 1$. Then $\theta_n^{kl} = \theta_n$. Thus, we have the following result.

Lemma 4.11.3. *θ_n^k is a primitive nth root of unity if and only if $(k, n) = 1$.*

The number of integers k in $\{1, 2, \ldots, n\}$ such that $(k, n) = 1$ is denoted by $\varphi(n)$, called the **Euler's φ function**. Assume that

$$n = p_1^{e_1} p_2^{e_2} \cdots p_r^{e_r},$$

where $p_1, p_2, \ldots, p_r$ are different primes, $e_1, e_2, \ldots, e_r$ are positive integers. Then

$$\varphi(n) = n \left(1 - \frac{1}{p_1}\right) \left(1 - \frac{1}{p_2}\right) \cdots \left(1 - \frac{1}{p_r}\right).$$

It is easy to see that integers k in $\{1, 2, \ldots, n\}$ with $(k, n) = 1$ are just the invertible elements (under multiplication) in $\mathbb{Z}/n\mathbb{Z}$, and they form a group, denoted by U_n. Thus, $\varphi(n) = |U_n|$. We will see that U_n is closely related to the Galois groups of cyclotomic extensions.

Proposition 4.11.4. *Let $\mathbb{E}/\mathbb{F}$ be an nth cyclotomic extension. Then* $\mathrm{Gal}\,(\mathbb{E}/\mathbb{F})$ *is isomorphic to a subgroup of* U_n; *consequently, cyclotomic extensions are abelian extensions.*

Proof. Since $\mathbb{E} = \mathbb{F}(\theta_n)$, any element $\sigma \in \mathrm{Gal}\,(\mathbb{E}/\mathbb{F})$ is completely determined by $\sigma(\theta_n)$, which is a primitive nth root of unity. Thus, there exists a unique $k \in \{1, 2, \ldots, n\}$ with $(k, n) = 1$ such that $\sigma(\theta_n) = \theta_n^k$. Then $\sigma \mapsto k$ is an injective map from $\mathrm{Gal}\,(\mathbb{E}/\mathbb{F})$ to U_n; it is easy to check that this map is a group homomorphism. The assertion follows. □

In general, $\mathrm{Gal}\,(\mathbb{E}/\mathbb{F}) \not\cong U_n$. For example, if $\mathbb{F}$ contains a primitive nth root of unity, then $\mathbb{E} = \mathbb{F}$, and $\mathrm{Gal}\,(\mathbb{E}/\mathbb{F})$ is trivial. In which cases do we have $\mathrm{Gal}\,(\mathbb{E}/\mathbb{F}) \simeq U_n$? To answer this question, consider the polynomial

$$\Psi_n(x) = \prod_{k \in U_n} (x - \theta_n^k).$$

Since the coefficients of $\Psi_n(x)$ are symmetric polynomials of the primitive nth roots of unity, which are preserved by each element of $\mathrm{Gal}\,(\mathbb{E}/\mathbb{F})$, one has $\Psi_n(x) \in \mathbb{F}[x]$, called the nth **cyclotomic polynomial**, whose degree is $\varphi(n)$.

Lemma 4.11.5. *Let $\mathbb{E} = \mathbb{F}(\theta_n)$ be an nth cyclotomic field. Then the followings are equivalent*:

(1) $[\mathbb{E} : \mathbb{F}] = \varphi(n)$;
(2) $\mathrm{Gal}\,(\mathbb{E}/\mathbb{F}) \simeq U_n$;
(3) $\Psi_n(x)$ *is irreducible over* $\mathbb{F}$.

Proof. $(3) \Rightarrow (1)$, $(1) \Rightarrow (2)$ follow from the definitions.
$(2) \Rightarrow (3)$ Since $\mathrm{Gal}\,(\mathbb{E}/\mathbb{F}) \simeq U_n$, for any $k \in U_n$, there exists $\sigma \in \mathrm{Gal}\,(\mathbb{E}/\mathbb{F})$ such that $\sigma(\theta_n) = \theta_n^k$. In other words, the set of the primitive nth roots of unity form an orbit under the action of $\mathrm{Gal}\,(\mathbb{E}/\mathbb{F})$. By Lemma 4.10.1, $\Psi_n(x)$ is irreducible. □

Therefore, the irreducibility of $\Psi_n(x)$ is crucial. Since $\Psi_n(x) \mid x^n - 1$, all roots of $x^n - 1$ form a cyclic group of order n, and the order of each root is a divisor of n. Conversely, for any $d \mid n$, a

dth root of unity must be an nth root of unity. Hence, $\Psi_d(x) \mid x^n - 1$ if and only if $d \mid n$. Thus, we have the following.

Proposition 4.11.6. *The polynomial $x^n - 1$ can be factorized as*

$$x^n - 1 = \prod_{d \mid n} \Psi_d(x).$$

In particular, if p is a prime, then

$$\Psi_p(x) = \frac{x^p - 1}{x - 1} = x^{p-1} + x^{p-2} + \cdots + 1.$$

This proposition gives us the recursion formula for $\Psi_n(x)$:

$$\Psi_n(x) = \frac{x^n - 1}{\prod_{d \mid n, d < n} \Psi_d(x)}.$$

For example,

$$\Psi_1(x) = x - 1,$$

$$\Psi_2(x) = \frac{x^2 - 1}{\Psi_1(x)} = x + 1,$$

$$\Psi_3(x) = x^2 + x + 1,$$

$$\Psi_4(x) = \frac{x^4 - 1}{(x - 1)(x + 1)} = x^2 + 1,$$

$$\Psi_5(x) = x^4 + x^3 + x^2 + x + 1,$$

$$\Psi_6(x) = \frac{x^6 - 1}{(x - 1)(x + 1)(x^2 + x + 1)} = x^2 - x + 1.$$

Are the cyclotomic polynomials irreducible over $\mathbb{Q}$? A simple case is the following.

Example 4.11.7. If p is a prime, then $\Psi_p(x)$ is irreducible over $\mathbb{Q}$ by Eisenstein's criterion. Hence, $\mathrm{Gal}\,(\mathbb{Q}(\theta_p)/\mathbb{Q})$ is isomorphic to U_p.

More generally, we have the following.

Proposition 4.11.8. *For any $n \in \mathbb{N}^*$, $\Psi_n(x) \in \mathbb{Z}[x]$ and $\Psi_n(x)$ is irreducible over $\mathbb{Q}$. Consequently, $\mathrm{Gal}\,(\mathbb{Q}(\theta_n)/\mathbb{Q}) \simeq U_n$.*

Proof. By Gauss's lemma, the product of primitive polynomials is also a primitive polynomial. Then we conclude that $\Psi_d(x) \in \mathbb{Z}[x]$ since $x^n - 1 = \prod_{d|n} \Psi_d(x)$, $x^n - 1 \in \mathbb{Z}[x]$ is a primitive polynomial, and $\Psi_d(x) \in \mathbb{Q}[x]$ is monic.

Assume that $\Psi_n(x)$ is reducible, say, $\Psi_n(x) = g(x)h(x)$, where $g(x), h(x) \in \mathbb{Z}[x]$ are monic polynomials with $1 \leq \deg g(x) < \varphi(n)$. Since every root of $g(x)$ is a primitive nth root of unity, there exists a root ξ of $g(x)$ and $p \in U_n$ such that ξ^p is a root of $h(x)$. We may choose p to be a prime (why?). Then ξ is a root of $h(x^p)$. Hence, $h(x^p)$ and $g(x)$ have a non-trivial common divisor. Suppose $k(x) = (h(x^p), g(x)) \neq 1$, where $k(x)$ is a monic polynomial with integer coefficients. Extend the natural ring homomorphism $\mathbb{Z} \to \mathbb{Z}/p\mathbb{Z} = \mathbb{F}_p$ to $\mathbb{Z}[x] \to \mathbb{F}_p[x]$, and denote the image of any $f(x) \in \mathbb{Z}[x]$ by $\overline{f}(x)$. Then $\overline{k}(x) \mid (\overline{h}(x^p), \overline{g}(x))$. For any $a \in \mathbb{F}_p$, one has $a^p = a$, which implies that $\overline{h}(x^p) = (\overline{h}(x))^p$. Hence, $(\overline{h}(x), \overline{g}(x)) \neq 1$. It follows that $\overline{\Psi}_n(x) = \overline{g}(x)\overline{h}(x)$ has multiple factors. Therefore, $x^n - 1 \in \mathbb{F}_p[x]$ has multiple roots. A contradiction. $\square$

Remark 4.11.9. The idea to prove the irreducibility of $\Psi_n(x)$ was proposed by Dedekind in 1857. It is very important since it reveals the connection among algebraic number theory, finite fields, and p-adic number fields.

In the following, we will study the structure of U_n. Since U_n is the set of (multiplicative) invertible elements in $\mathbb{Z}/n\mathbb{Z}$, we first consider $\mathbb{Z}/n\mathbb{Z}$.

Theorem 4.11.10. *Let $n = n_1 n_2 \cdots n_t$ with $(n_i, n_j) = 1$ for $i \neq j$. Then*

$$\mathbb{Z}/n\mathbb{Z} \simeq \mathbb{Z}/n_1\mathbb{Z} \oplus \mathbb{Z}/n_2\mathbb{Z} \oplus \cdots \oplus \mathbb{Z}/n_t\mathbb{Z}.$$

Proof. Let $\pi_i : \mathbb{Z} \to \mathbb{Z}/n_i\mathbb{Z}$ be the canonical ring homomorphism. Define a map

$$\pi : \mathbb{Z} \to \mathbb{Z}/n_1\mathbb{Z} \oplus \mathbb{Z}/n_2\mathbb{Z} \oplus \cdots \oplus \mathbb{Z}/n_t\mathbb{Z}$$

by $\pi(k) = (\pi_1(k), \ldots, \pi_t(k))$. It is quick to check that π is a ring homomorphism. Since n_i are coprime to each other, the kernel $\operatorname{Ker} \pi = n\mathbb{Z}$. By the Chinese Remainder Theorem, for any $k_1, k_2, \ldots, k_t \in \mathbb{Z}$, there exists $n \in \mathbb{Z}$ such that $n \equiv k_i (\bmod n_i)$.

Thus, π is surjective. The assertion follows from the Fundamental Theorem of Homomorphisms of rings. □

Considering the invertible elements in $\mathbb{Z}/n\mathbb{Z}$ and $\mathbb{Z}/n_1\mathbb{Z} \oplus \mathbb{Z}/n_2\mathbb{Z} \oplus \cdots \oplus \mathbb{Z}/n_t\mathbb{Z}$, we have the following.

Corollary 4.11.11. $U_n \simeq U_{n_1} \times U_{n_2} \times \cdots \times U_{n_t}$. *In particular, if* $n = p_1^{a_1} p_2^{a_2} \cdots p_t^{a_t}$ *with* $p_1, p_2, \ldots, p_t$ *prime, then* $U_n \simeq U_{p_1^{a_1}} \times U_{p_2^{a_2}} \times \cdots \times U_{p_t^{a_t}}$.

If p is a prime, then $U_p = \mathbb{F}_p^*$ is a cyclic group. For the structure of U_{p^k}, see Exercise 7 in this section.

Based on cyclotomic extensions, we are ready to discuss a class of abelian extensions that will play a vital role in solving algebraic equations.

Definition 4.11.12. Suppose $\mathbb{F}$ contains a primitive nth root of unity. By a **Kummer field** we mean a splitting field $\mathbb{E}$ of a polynomial $(x^n - a_1)(x^n - a_2) \cdots (x^n - a_r)$, for some $a_1, a_2, \ldots, a_r \in \mathbb{F}$. We call $\mathbb{E}/\mathbb{F}$ a **Kummer extension**.

If Ch $\mathbb{F} = p$, then $(p, n) = 1$ since $\mathbb{F}$ contains a primitive nth root of unity. Hence, $x^n - a_i$ are separable polynomials, which implies that $\mathbb{E}/\mathbb{F}$ is a Galois extension. If Ch $\mathbb{F} = 0$, $\mathbb{E}/\mathbb{F}$ is also a Galois extension.

Let us look at some examples.

Example 4.11.13. If $n = 2$, then Ch $\mathbb{F} \neq 2$; and any field of characteristic $\neq 2$ contains the primitive 2nd root of unity -1. In this case, $\mathbb{E} = \mathbb{F}(\sqrt{a_1}, \sqrt{a_2}, \ldots, \sqrt{a_r})$. If $r = 1$, then $\mathbb{E}/\mathbb{F}$ is a quadratic extension. Conversely, any quadratic extension over $\mathbb{F}$ is a Kummer field of the form $\mathbb{F}(\sqrt{a})$, where $a \in \mathbb{F}$ is not a square.

Example 4.11.14. Consider a splitting field $\mathbb{E}$ of $x^3 - a$ over $\mathbb{Q}$. If $a = b^3$ for some $b \in \mathbb{Q}$, then $\mathbb{E}/\mathbb{Q}$ is a quadratic extension, hence a Kummer extension. If $a \neq b^3$ for any $b \in \mathbb{Q}$, then choosing different roots α, β of $x^3 - a$, we have $(\alpha/\beta)^3 = 1$. Hence, $\mathbb{E}$ contains a primitive 3rd root of unity θ_3. Therefore, $\mathbb{E}$ is a Kummer extension over $\mathbb{Q}(\theta_3)$.

Theorem 4.11.15. *Suppose $\mathbb{F}$ contains a primitive nth root of unity, and $\mathbb{E}/\mathbb{F}$ is a Kummer extension. Then $\mathbb{E}/\mathbb{F}$ is an abelian extension, and the order of every element in* Gal $(\mathbb{E}/\mathbb{F})$ *is a divisor of n.*

Proof.   Assume that $\mathbb{E}$ is a splitting field of $(x^n - a_1)(x^n - a_2)$ $\cdots (x^n - a_r) \in \mathbb{F}[x]$, and that $\alpha_i \in \mathbb{E}$ $(i = 1, 2, \ldots, r)$ is a root of $x^n - a_i$. Then $\alpha_i, \alpha_i\theta_n, \ldots, \alpha_i\theta_n^{n-1}$ are all the roots of $x^n - a_i$. Since $\theta_n \in \mathbb{F}$, we have $\mathbb{E} = \mathbb{F}(\alpha_1, \alpha_2, \ldots, \alpha_r)$. For any $\sigma, \tau \in \mathrm{Gal}\,(\mathbb{E}/\mathbb{F})$, $\sigma(\alpha_i)$, $\tau(\alpha_i)$ are both roots of $x^n - a_i$, say, $\sigma(\alpha_i) = \alpha_i\theta_n^{j_i}$, $\tau(\alpha_i) = \alpha_i\theta_n^{k_i}$. Hence, $\sigma(\tau(\alpha_i)) = \alpha_i\theta_n^{k_i+j_i} = \tau(\sigma(\alpha_i))$. Since $\alpha_1, \alpha_2, \ldots, \alpha_r$ are generators of $\mathbb{E}/\mathbb{F}$, we have $\sigma\tau = \tau\sigma$, which implies that $\mathrm{Gal}\,(\mathbb{E}/\mathbb{F})$ is abelian. Furthermore, $\sigma^n(\alpha_i) = \alpha_i\theta_n^{nj_i} = \alpha_i$. It follows that $\sigma^n = \mathrm{id}$. $\qquad\square$

Corollary 4.11.16. *Assume that p is a prime, $\mathbb{F}$ is a field contains a primitive pth root of unity, $\mathbb{E}$ is a splitting field of $x^p - a \in \mathbb{F}[x]$, and $\mathbb{E} \neq \mathbb{F}$. Then $x^p - a$ is irreducible, and $\mathrm{Gal}\,(\mathbb{E}/\mathbb{F})$ is the cyclic group of order p.*

Proof.   It is easy to see that $|\mathrm{Gal}\,(\mathbb{E}/\mathbb{F})| = [\mathbb{E} : \mathbb{F}] \leq p$. By the above theorem, every element in $\mathrm{Gal}\,(\mathbb{E}/\mathbb{F})$ is of order p. It follows that $\mathrm{Gal}\,(\mathbb{E}/\mathbb{F})$ is a cyclic group of order p, which implies that $x^p - a$ is irreducible. $\qquad\square$

The converse statement of the above corollary is also true.

Lemma 4.11.17. *Assume that p is a prime, $\mathbb{F}$ contains a primitive pth root of unity, and $\mathbb{E}/\mathbb{F}$ is a cyclic extension of degree p. Then $\mathbb{E} = \mathbb{F}(d)$, for some $d \in \mathbb{E}$ with $d^p \in \mathbb{F}$.*

Proof.   Fix any $\alpha \in \mathbb{E} \setminus \mathbb{F}$. Then $\mathbb{E} = \mathbb{F}(\alpha)$. Hence, α is a root of an irreducible polynomial $f(x) \in \mathbb{F}[x]$ of degree p. Let φ be a generator of $\mathrm{Gal}\,(\mathbb{E}/\mathbb{F})$. It follows that all roots of $f(x)$ are

$$\alpha_0 = \alpha, \alpha_1 = \varphi(\alpha), \ldots, \alpha_{p-1} = \varphi^{p-1}(\alpha),$$

which form a basis of $\mathbb{E}$ over $\mathbb{F}$. The element d we are looking for must be of the form

$$d = c_0\alpha_0 + c_1\alpha_1 + \cdots + c_{p-1}\alpha_{p-1}, \quad c_i \in \mathbb{F}.$$

Thus, $d^p \in \mathbb{F}$ if and only if $d^p = \varphi(d^p) = \varphi(d)^p$. Therefore, $\theta = \frac{\varphi(d)}{d}$ is a (primitive) pth root of unity. It follows that $\varphi(d) = \theta d$, and $c_i = \theta c_{i-1}$, for $i = 0, 1, \ldots, p$, where $c_p = c_0$. Therefore, up to a constant, we have p choices of the element d:

$$d_i = \alpha_0 + \theta^i \alpha_1 + \cdots + \theta^{i(p-1)} \alpha_{p-1}, \quad i = 0, 1, \ldots, p-1.$$

By Vandermonde determinant, one can see that $d_0, d_1, \ldots, d_{p-1}$ form a basis of $\mathbb{E}$ over $\mathbb{F}$. Hence, there must be some $d_i \notin \mathbb{F}$, and we have $\mathbb{E} = \mathbb{F}(d_i)$. $\qquad\square$

Exercises

(1) Show that $\varphi(n) = \sum_{d|n} \mu(n/d)d$, where μ is the Möbius function.
(2) Show that $\Psi_n(x) = \prod_{d|n}(x^d - 1)^{\mu(n/d)}$.
(3) Let $(m, n) = 1$. Show that the splitting field of $x^{mn} - 1 \in \mathbb{Q}[x]$ is isomorphic to that of $(x^m - 1)(x^n - 1) \in \mathbb{Q}[x]$, and $\mathrm{Gal}\,(x^{mn} - 1, \mathbb{Q}) \simeq \mathrm{Gal}\,(x^m - 1, \mathbb{Q}) \times \mathrm{Gal}\,(x^n - 1, \mathbb{Q})$.
(4) Let $x^p - a \in \mathbb{Q}[x]$ be irreducible. Show that $\mathrm{Gal}\,(x^p - a, \mathbb{Q})$ is isomorphic to the transformation group $\{\sigma_{kl} | k \neq 0\}$ of $\mathbb{F}_p$, where $\sigma_{kl}(y) = ky + l$, $y \in \mathbb{F}_p$.
(5) Show that an extension $\mathbb{E}/\mathbb{F}$ is a Kummer extension if and only if $\mathbb{E}/\mathbb{F}$ is an abelian extension, and $\mathbb{F}$ contains a primitive rth root of unity, where r is the maximum of the orders of elements in $\mathrm{Gal}\,(\mathbb{E}/\mathbb{F})$.

Exercises (hard)

(6) Let $\mathbb{E}$ be a number field, and $\mathbb{E}/\mathbb{Q}$ a finite Galois extension.
   (i) Show that every element in $\mathbb{E}$ is constructible if and only if $[\mathbb{E} : \mathbb{Q}] = 2^m$ for some $m \in \mathbb{N}$;
   (ii) Let $\theta \in \mathbb{C}$ be a primitive nth root of unity. Show that the followings are equivalent:
      (a) The regular n-gon is constructible;
      (b) $[\mathbb{Q}(\theta) : \mathbb{Q}] = 2^m$ for some $m \in \mathbb{N}$;
      (c) $n = 2^k p_1 p_2 \cdots p_l$, where $k \in \mathbb{N}$, $p_1, p_2, \ldots, p_l$ are different **Fermat primes**, i.e., primes of the form $2^{2^m} + 1$.

(7) Show that if $k \geq 3$, U_{2^k} is the direct product of a group of order 2 and a cyclic group of order 2^{k-2}. Moreover, if p is an odd prime, show that U_{p^k} is a cyclic group.

(8) (i) Let G be a finite abelian group. Show that there exists a subgroup H of some U_n such that $G \simeq U_n/H$;

    (ii) Any finite abelian group is the Galois group of a Galois extension $\mathbb{E}/\mathbb{Q}$.

4.12 Solvability by radicals

In the early 1900s, archaeologists found many Babylonian clay tablets dating back to about 3800 years ago in southern Iraq. Some clay tablets showed that ancient Babylonians knew how to solve quadratic equations. Until the Renaissance, several Italian mathematicians made progress in solving cubic and quartic equations. Ferro and Tartaglia discovered the cubic formula, later named after Cardano. Then Ferrari, a student of Cardano, found the method for solving quartic equations. In 1770, Lagrange dealt with these equations in a unified way; his method embodies the idea of root permutations. However, Lagrange's method could not be applied to solve equations of higher degrees. The exploration of Lagrange and his contemporaries suggested that there may not exist a formula for solving equations of degree ≥ 5. In 1799, Ruffini proved that general equations of degree ≥ 5 have no solutions in radicals, but there were flaws in the proof. In 1824, Abel filled the gap, and the result was now called the Abel-Ruffini theorem.

However, there are many equations of higher degrees that are solvable. Galois gives a perfect answer to this problem with the idea of groups. Now, we are ready to appreciate Galois's ideal. First, we give a rigorous definition of the solvability of algebraic equations. In this section, we only consider fields of characteristic 0. Hence, every extension in this section is separable, and normal extensions are Galois extensions.

Definition 4.12.1. We call $\mathbb{E}/\mathbb{F}$ a **simple radical extension** if $\mathbb{E} = \mathbb{F}(\alpha)$, and α is a root of $x^n - \beta \in \mathbb{F}[x]$. Call $\mathbb{E}/\mathbb{F}$ a **radical extension** if there exist intermediate fields $\mathbb{F}_0 = \mathbb{F}, \mathbb{F}_1, \ldots, \mathbb{F}_r = \mathbb{E}$ such that $\mathbb{F}_{i+1}/\mathbb{F}_i$, $i = 0, 1, \ldots, r-1$, are all simple radical extensions, and call $\mathbb{F}_0, \mathbb{F}_1, \ldots, \mathbb{F}_r$ a **tower of radical extensions**. If a splitting field of $f(x) \in \mathbb{F}[x]$ is a subfield of some radical extension of $\mathbb{F}$, then we say that $f(x)$ is **solvable by radicals**.

By the definition, one may get the following.

Lemma 4.12.2. *If* $\mathbb{K}/\mathbb{F}$, $\mathbb{E}/\mathbb{K}$ *are radical extensions, then so is* $\mathbb{E}/\mathbb{F}$.

Example 4.12.3.

(1) Cyclotomic extensions are simple radical extensions. In fact, if $\mathbb{E}/\mathbb{F}$ is an nth cyclotomic extension, and θ_n a primitive nth root of unity, then $\mathbb{E} = \mathbb{F}(\theta_n)$ is a simple radical extension.

(2) Kummer extensions are radical ones. Let $\mathbb{E}/\mathbb{F}$ be a Kummer extension, say, $\mathbb{E}$ is the splitting field of $\prod_{i=1}^{r}(x^n - a_i) \in \mathbb{F}[x]$. Suppose $\alpha_i \in \mathbb{E}$ is a root of $x^n - a_i$. Then $\mathbb{K} = \mathbb{F}(\alpha_1, \alpha_2, \ldots, \alpha_n)$ since $\mathbb{F}$ contains a primitive nth root of unity. Thus, $\mathbb{F} \subseteq \mathbb{F}(\alpha_1) \subseteq \mathbb{F}(\alpha_1, \alpha_2) \subseteq \cdots \subseteq \mathbb{F}(\alpha_1, \ldots, \alpha_n) = \mathbb{K}$ is a tower of radical extensions. Hence, $\mathbb{K}/\mathbb{F}$ is a radical extension.

We need to consider Galois extensions to apply Galois theory to study the solvability by radicals of equations. However, radical extensions may not be Galois extensions, which should be separable and normal. Since every extension over a field of characteristic 0 is separable, we need to find normal radical extensions.

Let $\mathbb{F}_0 = \mathbb{F} \subseteq \mathbb{F}_1 \subseteq \cdots \subseteq \mathbb{F}_r = \mathbb{E}$ be a tower of radical extensions, where $\mathbb{F}_{i+1} = \mathbb{F}_i(\alpha_i)$, and α_i is a root of $x^{n_i} - a_i \in \mathbb{F}_i[x]$, for $i = 0, 1, \ldots, r - 1$. Regarding α_i as a root of $x^n - a_i^{n/n_i} \in \mathbb{F}_i[x]$, we may replace n_i by $n = n_0 n_1 \cdots n_{r-1}$. Without loss of generality, we assume that $n_0 = n_1 = \cdots = n_{r-1} = n$, and α_i is a root of $x^n - a_i$. Let $\mathbb{K}_0$ be a splitting field of $x^n - 1 \in \mathbb{F}_0[x]$. Then $\mathbb{K}_1 = \mathbb{K}_0(\alpha_0) \supseteq \mathbb{F}_0(\alpha_0) = \mathbb{F}_1$. Since $\mathbb{K}_0$ contains all nth roots of unity, $\mathbb{K}_0(\alpha_0)$ is a splitting field of $x^n - a_0$, which implies that $\mathbb{K}_1/\mathbb{F}$ is a normal radical extension. Similarly, $\mathbb{K}_1(\alpha_1)$ is a splitting field of $x^n - a_1$. Noting that $x^n - a_1 \in \mathbb{F}_1[x]$, one gets that $\mathbb{K}_1(\alpha_1)/\mathbb{K}_1$ is a normal radical extension; but $x^n - a_1$ may not lie in $\mathbb{F}[x]$, thus $\mathbb{K}_1(\alpha_1)/\mathbb{F}$ may not be normal. Here, we need to be more careful. Since $a_1 \in \mathbb{F}_1 \subseteq \mathbb{K}_1$ and $\mathbb{K}_1/\mathbb{F}$ is a normal extension, $f_{a_1}(x) = \mathrm{Irr}(a_1, \mathbb{F})$ splits in $\mathbb{K}_1$, i.e., there exist $b_1, b_2, \ldots, b_k \in \mathbb{K}_1$ such that $f_{a_1}(x) = (x-b_1)(x-b_2)\cdots(x-b_k)$. Since $(x^n-b_1)\cdots(x^n-b_k) = f_{a_1}(x^n) \in \mathbb{F}[x]$, the splitting field $\mathbb{K}_2$ of $f_{a_1}(x^n)$ is normal over $\mathbb{F}$. It follows that $\mathbb{K}_2/\mathbb{K}_1$ is a Kummer extension and, consequently, a radical extension. Repeating the procedure, we have the following.

Theorem 4.12.4. *Let* $\mathbb{F}_0 = \mathbb{F} \subseteq \mathbb{F}_1 \subseteq \cdots \subseteq \mathbb{F}_r = \mathbb{E}$ *be a tower of radical extensions. Then there exists a sequence of field extensions*

$$\mathbb{F} \subseteq \mathbb{K}_0 \subseteq \mathbb{K}_1 \subseteq \cdots \subseteq \mathbb{K}_r = \mathbb{K},$$

such that $\mathbb{F}_i \subseteq \mathbb{K}_i$, $\mathbb{K}_i/\mathbb{F}$ *is a normal extension, and* $\mathbb{K}_i/\mathbb{K}_{i-1}$ *is a Kummer extension. In particular, a radical extension* $\mathbb{E}/\mathbb{F}$ *is always a subextension of a normal radical extension* $\mathbb{K}/\mathbb{F}$.

Since $\mathbb{K}/\mathbb{F}$ is a normal extension, it is a Galois extension. We now consider its Galois group $G = \mathrm{Gal}\,(\mathbb{K}/\mathbb{F})$ and the sequence of subgroups corresponding to the tower of radical extensions. Set $G_{-1} = G$, $G_i = \mathrm{Gal}\,(\mathbb{K}_i) = \mathrm{Gal}\,(\mathbb{K}/\mathbb{K}_i)$. By the Fundamental Theorem of Galois theory, one gets a sequence of subgroups

$$G_{-1} \supseteq G_0 \supseteq G_1 \supseteq \cdots \supseteq G_r = \{0\}.$$

Since $\mathbb{K}_i/\mathbb{F}$ is a normal extension, $G_i \lhd G$, and the above sequence is a normal series of G. Furthermore, since $G_i = \mathrm{Gal}\,(\mathbb{K}/\mathbb{K}_i)$ and $G_i \lhd G_{i-1}$, we have $G_{i-1}/G_i \simeq \mathrm{Gal}\,(\mathbb{K}_i/\mathbb{K}_{i-1})$. Noticing that $\mathbb{K}_i/\mathbb{K}_{i-1}$ is a Kummer extension, we get that $\mathrm{Gal}\,(\mathbb{K}_i/\mathbb{K}_{i-1})$ is an abelian group. It follows that $G_{i-1}^{(1)} < G_i$, i.e., G is a solvable group.

Therefore, we have the following result.

Theorem 4.12.5. *Let* $f(x) \in \mathbb{F}[x]$. *Then* $f(x) = 0$ *is solvable by radicals if and only if the Galois group* G_f *is solvable.*

Proof. Let $\mathbb{E}$ be a splitting field of $f(x)$. If $f(x) = 0$ is solvable by radicals, then $\mathbb{E}$ is contained in a radical extension $\mathbb{K}$ over $\mathbb{F}$. We may assume that $\mathbb{K}/\mathbb{F}$ is normal. Thus, $G_f = \mathrm{Gal}\,(\mathbb{E}/\mathbb{F}) \simeq \mathrm{Gal}\,(\mathbb{K}/\mathbb{F})/\mathrm{Gal}\,(\mathbb{K}/\mathbb{E})$, which implies that G_f is solvable since $\mathrm{Gal}\,(\mathbb{K}/\mathbb{F})$ is solvable.

Conversely, suppose $n = |G_f|$ and $\mathbb{K}/\mathbb{E}$ is a splitting field of $x^n - 1$. Then $\mathbb{K} = \mathbb{E}(\theta)$ for some primitive nth root of unity $\theta \in \mathbb{K}$. Set $\mathbb{F}_1 = \mathbb{F}(\theta)$, which is also a cyclotomic field. Since $\mathbb{E}/\mathbb{F}$ is a normal extension, one has $\sigma(\mathbb{E}) = \mathbb{E}$ for any $\sigma \in G = \mathrm{Gal}\,(\mathbb{K}/\mathbb{F}_1)$. It follows that the map $\pi : G \to G_f$, $\pi(\sigma) = \sigma|_{\mathbb{E}}$, is a group homomorphism. If $\sigma|_{\mathbb{E}} = \mathrm{id}_{\mathbb{E}}$, then $\sigma = \mathrm{id}_{\mathbb{K}}$ since $\sigma|_{\mathbb{F}_1} = \mathrm{id}$. Hence, π is injective, and G is isomorphic to a subgroup of G_f. Since G_f is solvable, so is G, and there exists a subnormal series of G: $G = G_1 \supset G_2 \supset \cdots \supset G_r = \{e\}$

such that G_i/G_{i+1} is a group of order p_i, where p_i is a prime, $i = 1, 2, \ldots, r-1$. Consequently, we have a tower of fields

$$\mathbb{F}_1 \subset \mathbb{F}_2 \subset \cdots \subset \mathbb{F}_r = \mathbb{K},$$

such that $\mathbb{F}_i = \mathbb{K}^{G_i}$. Since $\mathrm{Gal}\,(\mathbb{K}/\mathbb{F}_{i+1}) = G_{i+1} \triangleleft G_i = \mathrm{Gal}\,(\mathbb{K}/\mathbb{F}_i)$, $\mathbb{F}_{i+1}/\mathbb{F}_i$ is a normal extension and $\mathrm{Gal}\,(\mathbb{F}_{i+1}/\mathbb{F}_i) \simeq G_i/G_{i+1}$ is a cyclic group of prime order. Furthermore, since $\mathbb{F}_i$ contains an nth root of unity and $p \mid n$, $\mathbb{F}_{i+1}/\mathbb{F}_i$ is a simple radical extension by Lemma 4.11.17. It follows that $\mathbb{K}/\mathbb{F}_1$ is a radical extension, and so is $\mathbb{K}/\mathbb{F}$. Since $\mathbb{E}$ is contained in a radical extension $\mathbb{K}/\mathbb{F}$, $f(x) = 0$ is solvable by radicals. □

Remark 4.12.6. If $n \leq 4$, S_n is solvable. Hence, polynomials of degree ≤ 4 are solvable by radicals. If $n \geq 5$, A_n is simple. Thus, S_n is not solvable. For any prime $p \geq 5$, we have constructed polynomials of degree p such that their Galois groups are S_p. Therefore, there exist equations of degree ≥ 5 which are not solvable by radicals.

Exercises

(1) From the identity $(p+q)^3 = 3pq(p+q) + (p^3+q^3)$ one gets that $p+q$ is a root of $x^3 - 3pqx - (p^3+q^3) = 0$. Use this fact to find the solutions to cubic equations.

(2) Rewrite a quartic equation $x^4 + bx^2 + cx + d = 0$ as follows:

$$(x^2 - t)^2 = (-b - 2t)x^2 - cx + (t^2 - d).$$

Show that the right-hand side of the above equation is a perfect square if and only if $c^2 + 4(b + 2t)(t^2 - d) = 0$, and try to find the solutions to quartic equations.

(3) Let x_1, x_2 be two solutions of a quadratic equation $ax^2 + bx + c = 0$. Use Vieta's formulas to show that $x_1 - x_2 = \pm\frac{\sqrt{b^2-4ac}}{a}$ and deduce the quadratic formula.

(4) Let x_1, x_2, x_3 be three roots of the cubic equation $x^3 + bx + c = 0$, ω a primitive 3rd root of unity, $u = x_1 + \omega x_2 + \omega^2 x_3$, $v = x_1 + \omega^2 x_2 + \omega^4 x_3$.

   (i) Show that $u^3 + v^3$ and uv are symmetric polynomials in x_1, x_2, x_3; find their expressions in b, c.
   (ii) Use u, v and $x_1 + x_2 + x_3 = 0$ to deduce the cubic formula.

Exercises (hard)

(5) Let x_1, x_2, x_3, x_4 be the roots of $x^4 + bx^2 + cx + d = 0$.

(i) Set $\begin{cases} u = x_1 + \omega x_2 + \omega^2 x_3 + \omega^3 x_4, \\ v = x_1 + \omega^2 x_2 + \omega^4 x_3 + \omega^6 x_4, \\ w = x_1 + \omega^3 x_2 + \omega^6 x_3 + \omega^9 x_4, \end{cases}$ where ω is a primitive 4th root of unity. Try to find the formula for roots of quartic equations.

(ii) Let $u = x_1 + x_2 - x_3 - x_4$, $v = x_1 - x_2 + x_3 - x_4$, $w = x_1 - x_2 - x_3 + x_4$. Try to find the formula for roots of quartic equations.

4.13   Summary of the chapter

Galois theory developed when mathematicians explored whether an algebraic equation $f(x) = 0$ is solvable by radicals. All the roots of the equation naturally generate a splitting field $\mathbb{E}$ of $f(x) \in \mathbb{F}[x]$. The solvability of $f(x) = 0$ by radicals means that the splitting field $\mathbb{E}$ is contained in some radical extension $\mathbb{K}$ of $\mathbb{F}$. We may assume that $\mathbb{K}/\mathbb{F}$ is a Galois extension, which implies that both $\mathrm{Gal}\,(\mathbb{K}/\mathbb{F})$ and $\mathrm{Gal}\,(\mathbb{E}/\mathbb{F})$ are solvable groups. Therefore, the solvability of $f(x) = 0$ by radicals is equivalent to the solvability of the Galois group $\mathrm{Gal}\,(\mathbb{E}/\mathbb{F})$. Studying splitting fields of polynomials and their Galois groups is crucial. For separable polynomials, their splitting fields and Galois groups are perfectly combined, which leads to the study of finite Galois extensions. In this process, the notions of finite extensions, algebraic extensions, normal extensions, separable extensions, and complete fields have been introduced one after another in a very natural way, along with the research on transcendental extensions and purely inseparable extensions. Interestingly, the problems of straightedge and compass construction are closely related to field theory, for example, the solution to the three construction problems in ancient Greece. The further application of Galois theory can solve the problem of constructing regular n-gons.

Bibliography

M. Artin (1991). *Algebra*, Prentice-Hall, New Jersey.

M.F. Atiyah and I.G. Macdonald (1969). *Introduction to Commutative Algebra*, Addison-Wesley Publishing Co., Reading, Massachusetts.

J. Baez (2001). *The Octonions*, Bulletins of the American Mathematical Society, Providence, Rhode Island, Vol. 39, pp. 145–205.

D.A. Cox (2012). *Galois Theory*. John Wiley & Sons, Inc., Hoboken, New Jersey.

H. Edwards (1997). *Galois Theory*, 3rd printing, Springer, New York.

P. Gu and S. Deng (2003). *Succinct Abstract Algebra*, High Education Press, Beijing (in Chinese).

L.K. Hua (1949). On the Automorphisms of a Sfield, *Proc. Nat. Sci., USA*, Vol. 35, pp. 386–389.

I.M. Isaacs (2009). Algebra: A Graduate Course, *Amer. Math. Soc.*, Providence, Rhode Island.

N. Jacobson (1985). *Basic Algebra I*, Dover Publications, New York.

I. Kleiner (2007). *A History of Abstract Algebra*. Birkhäuser, Boston.

N. Koblitz (1984). *p-adic Numbers, p-adic Analysis, and Zeta-functions*, 2nd edition, Springer, New York.

S. Lang (2002). *Algebra*, Revised 3rd edition, Springer, New York.

G. Malle, The proof of Ore's conjecture (after Ellers-Gordeev and Liebeck-O'Brien-Shalev-Tiep), *Astérisque*, No. 361 (2014), Exp. No. 1069, ix, 325–348.

D. Meng, L. Chen, Y. Shi, and R. Bai (2010). *Abstract Algebra I-Foundations of Algebra*, Science Press, Beijing (in Chinese).

J.S. Milne (2012). *Fields and Galois Theory*, Kea Books, Ann Arbor, MI.

L. Nie and S. Ding (1988). *Introduction to Algebra*, High Education Press, Beijing (in Chinese).

A.M. Robert (2000). *A Course in p-adic Analysis*, Springer, New York.

J. Stillwell (2010). *Mathematics and Its History*. 3rd edition, Springer, New York.

J. Tignol (2001). *Galois' theory of Algebraic Equations*. World Scientific Publishing Co., Inc., Singapore, River Edge, New Jersey.

O. Zariski and P. Samuel (1958). *Commutative Algebra*, Vols. I and II, Springer, New York.

Index